# Fire and Polymers V

ACS SYMPOSIUM SERIES **1013**

# Fire and Polymers V

# Materials and Concepts for Fire Retardancy

**Charles A Wilkie,** Editor
*Marquette University*

**Alexander B. Morgan,** Editor
*University of Dayton Research Institute*

**Gordon L. Nelson,** Editor
*Florida Institute of Technology*



**Sponsored by the**
**ACS Division of Polymeric Materials: Science and Engineering, Inc.**

American Chemical Society, Washington, DC

ISBN 978–0–8412–6988–0

The paper used in this publication meets the minimum requirements of American National Standard for Information Sciences—Permanence of Paper for Printed Library Materials, ANSI Z39.48–1984.



Distributed by Oxford University Press



PRINTED IN THE UNITED STATES OF AMERICA

# Foreword

The ACS Symposium Series was first published in 1974 to provide a mechanism for publishing symposia quickly in book form. The purpose of the series is to publish timely, comprehensive books developed from ACS sponsored symposia based on current scientific research. Occasionally, books are developed from symposia sponsored by other organizations when the topic is of keen interest to the chemistry audience.

Before agreeing to publish a book, the proposed table of contents is reviewed for appropriate and comprehensive coverage and for interest to the audience. Some papers may be excluded to better focus the book; others may be added to provide comprehensiveness. When appropriate, overview or introductory chapters are added. Drafts of chapters are peer-reviewed prior to final acceptance or rejection, and manuscripts are prepared in camera-ready format.

As a rule, only original research papers and original review papers are included in the volumes. Verbatim reproductions of previously published papers are not accepted.

**ACS Books Department**

# Contents

## Conventional Fire Retardants

## Metal-Containing Systems



## Toxicity



## Thermal Degradation Studies



## Indexes

# Preface

The possibility always exists worldwide that a modern material will catch fire, which can lead to loss of property or life. Although fire risk is not a new problem to society, as new polymeric materials are introduced into more and more everyday activities, the potential for fire risk sometimes increases or changes in such a way that one must remain vigilant in addressing this risk. The huge benefits of polymeric materials in modern life cannot be stressed enough, but as these materials are put into new applications, the fire risk must be addressed. The typical method of addressing this risk is through the use of fire-retardant additives, which interact with polymers through physical and chemical mechanisms to either prevent the polymer from burning or to slow the rate of polymer burning so that it can be easily extinguished. As a class of industrial additives, flame retardants are the largest group of materials used and sold worldwide and growth continues in this area. Due to this ever-increasing use as well as new regulations on the use and sale of these additives, a clear need exists for a peer-reviewed book showing the latest advances in flame retardant additives, chemistry, and approaches to the study of fire and polymers.

Because fire and polymers are important social issues for safety in modern civilization and due to the complex scientific issues and multiple disciplines of science involved, a symposium on the subject of fire and polymers was organized and held in New Orleans, Louisiana in April of 2008. The symposium was the 5$^{th}$ in this series, building from previous symposia held in 1989, 1994, 2000, and 2004. A total of 44 papers from researchers from around the world were presented, representing academia, government, and industry research in this field. From those presentations, 22 chapters were selected for incorporation into this book.

## Acknowledgments

We gratefully acknowledge support from the ACS Division of Polymeric Materials: Science and Engineering, Inc. for providing the venue for the symposium as well as for providing financial support for the symposium activities and for support in getting foreign speakers to present at the symposium. We also thank the Petroleum Research Foundation for their grant to invite and support travel costs for foreign researchers to speak at this symposium. Finally, we also gratefully acknowledge support from our industrial sponsors who helped with many of the other expenses we had and helped make our sym-posium a great success: The Govmark Organization, Minelco, Ashland Inc., Rio Tinto Minerals, Akzo Nobel, Albemarle Corporation, Nabeltec AG, Elkem AS, ICL Industrial Products, Nanocor, Clariant GmbH, Ram Technologies, and Southern Clay products.

**Charles A. Wilkie**
Department of Chemistry
Marquette University
P.O. Box 1881
Milwaukee, WI 53201–1881

**Gordon L. Nelson**
College of Science
Florida Institute of Technology
150 West University Boulevard
Melbourne, FL 32901–6975

**Alexander B. Morgan**
Multiscale Composites and Polymers Division
University of Dayton Research Institute
300 College Park
Dayton, OH 45469–0160

# Chapter 1

# Fire Retardancy in 2009

**Gordon L. Nelson[1], Alexander B. Morgan[2], and Charles A. Wilkie[3]**

**[1]Florida Institute of Technology, 150 West University Boulevard, Melbourne, FL 32901-6975**
**[2]University of Dayton Research Institute, 300 College Park, Dayton, OH 45469-0160**
**[3]Department of Chemistry, Marquette University, PO Box 1881, Milwaukee, WI 53201-1881**

Fire continues to be a worldwide problem. It claims thousands of lives and causes tens of billions of dollars in loss of property every year. In this 5th volume of Fire and Polymers, current fire-related problems are discussed and solutions delineated. This peer-reviewed volume is designed to be representative of the state-of-the-art and places current work in perspective.

## Introduction

In the United States every 20 seconds a fire department responds to a fire somewhere in the country. A fire occurs in a structure at the rate of one every 59 seconds. A residential fire occurs every 76 seconds. A fire occurs in a vehicle every 122 seconds. There is a fire in an outside property every 41 seconds. The result is 1.6 million fires per year (2007) attended by public fire departments. In 2007 those fires accounted for $14.6 billion in property damage and 3430 civilian fire deaths (one every 153 minutes) and 17,675 injuries (one every 30 minutes). Some 105 fire fighters died in the line of duty in 2003. Fires have declined over the period 1977 to 2007, most notably structural fires, from 1,098,000 to 530,500. Civilian fire deaths in the home (80% of all fire

declined from 6,015 in 1978 to 2,865 in 2007. While those declines are progress, recent years have been static, and the United States still maintains one of the highest rates of fire in the world. Just as the U.S. has a high fire rate, the fire death rate in the US varies by state, from 30/million population in Arkansas to 3/million in Utah. Fire deaths vary by size of community, from 10/million in communities above 25,000 people to 26/million for communities under 2500 people. Importantly, 70% of fire deaths in the U.S. occur in homes without working smoke alarms, this despite years of effort to achieve a high penetration (*1-3*).

The higher rate of fire in the United States versus most industrialized countries is probably a product of five factors: (1) the U.S. commits fewer resources to fire prevention activities; (2) there is a greater tolerance in the U.S. for "accidental" fires (no one is at fault); (3) Americans practice riskier and more careless behavior than people in other countries (example, the use of space heaters); (4) homes in the U.S. are not built with the same degree of fire resistance and compartmentation as in some countries; and (5) most importantly, people in the U.S. have more contents or "stuff" than those in other countries (i.e., higher fire load) as well as a higher number of ignition sources (higher use of energy).

Polymers form a major part of the built environment. Fire safety depends upon those materials. Polymers are "enabling technology" for almost all of modern society as we know it today. Advances in numerous technologies depend on appropriate advances in polymers for success. While polymers are both natural and synthetic, this book focuses entirely on the fire safety aspects of synthetic polymers. Production of synthetic plastic resins totaled over 169 million metric tons worldwide in 2003 (*4*). The U.S. constitutes about one quarter of worldwide plastic consumption, the European Union only slightly less and Japan about 9%. In Table I, one finds 2007 plastic production figures by resin for North America (*5*). In Table II, one finds 2007 plastics use data by resin for North America (*5*).

All organic polymers are combustible. They decompose when exposed to heat, their decomposition products burn, smoke is generated, and the products of combustion are highly toxic. The prime toxic product is CO in concert with $CO_2$. While at times the heat from a fire can kill far sooner than toxic combustion gases, the preponderance of fire victims die in post-flashover fires in a room outside the room of origin, from toxic gases in smoke. That toxicity is made more complex by the pervasive presence of alcohol on the part of fire victims, 52% of young adult fire victims and 74% of middle-aged fire victims have significant blood alcohol. Alcohol got them into the incident. Fire kills young children, the old and infirm, and the drunk, not healthy unimpared adults *(6)*.

Fire is not a single material property and is regulated by many material parameters. Fire behavior from a material combines ease of thermal

**Table I. North American Plastics Production[a] – 1999 and 2007 (millions of pounds, dry weight basis)**

| *Resin* | *1999 Production* | *2007 Production* |
|---|---|---|
| Epoxy | 657 | 642 |
| Urea and melamine[c] | 2,985 | 3,471 |
| Phenolics (gross wt)[c] | 4,388 | 4,838 |
| **Total Thermosets** | **8,030** | **8,951** |
| LDPE[c] | 7,700 | 7,927 |
| LLDPE[c] | 8,107 | 13,584 |
| HDPE[c] | 13,864 | 18,222 |
| PP[c] | 15,493 | 19,445 |
| ABS[c,m] | 1,455 | 1,270 |
| Other Styrenics[c,m] | 1,767 | 1,726 |
| PS | 6,471 | 6,015 |
| Nylon[c,m] | 1,349 | 1,295 |
| PVC[c] | 14,912 | 14,606 |
| Thermoplastic Polyester | 4,846 | 8,745 |
| **Total Thermoplastics** | **75,964** | **92,835** |
| All Other Resins | 13,467 | 14,007 |
| **Grand Total** | **97,461** | **115,793** |

NOTES: [a] US, Canada and Mexico as noted, [c] Canada Included, [m] Mexico Included

**Table II. Resins Sales By Major Markets (millions of pounds)**

| **Major Market** | **2007[1]** | **%** |
|---|---|---|
| Transportation | 3,312 | 4.0 |
| Packaging | 26,527 | 32.2 |
| Building & Construction | 14,289 | 17.4 |
| Electrical/Electronic | 1,980 | 2.4 |
| Furniture & Furnishings | 3,091 | 3.8 |
| Consumer & Institutional | 17,193 | 20.9 |
| Industrial/Machinery | 943 | 1.1 |
| Adhesives/Inks/Coatings | 1,069 | 1.3 |
| All Others | 1,604 | 1.9 |
| Exports | 12,346 | 15.0 |
| **Total Selected Plastics[a]** | **82,354** | **100.0** |

NOTE: [1]Data for ABS, SAN, Other Styrene-Based Polymers, and Engineering Resins are not included.

decomposition, ease of ignition, flame spread, heat release, ease of extinction, smoke generation, toxic potency and other properties.

Since fire risk has been known for a very long time, the field of fire safety engineering looks at specific scenarios where fires can occur, and works with government through codes and standards to minimize fire events and subsequent losses should the fire blaze out of control. These codes and standards use specific tests covering the properties of material flammability described above. The results of these tests are combined with engineering assessments for materials and systems as deemed appropriate for a particular application. Thus, for example, it is appropriate in small appliances to only worry about ignitability by a Bunsen burner flame or a needle flame, since in the application, from an internal point of view, that is the size of a fire source (from an electrical short) possible in real appliance failures.

While there are some polymers that are very difficult to ignite or consume with flame, most other polymers require flame retardant additives to pass the regulatory tests required for safe use and sale of polymeric material containing goods. The total market for flame retardants in the United States, Europe, and Asia in 2004 amounted to about 1.5 million metric tons and was valued at $2.8-2.9 billion. The market is expected to grow at an average 3% annual rate on a volume basis (*7,8*). Figure 1 shows the consumption of flame retardants by region (2004). Different flame retardants are prominent in different regions. Figure 2 shows relative consumption of different flame retardants by region (*7*).

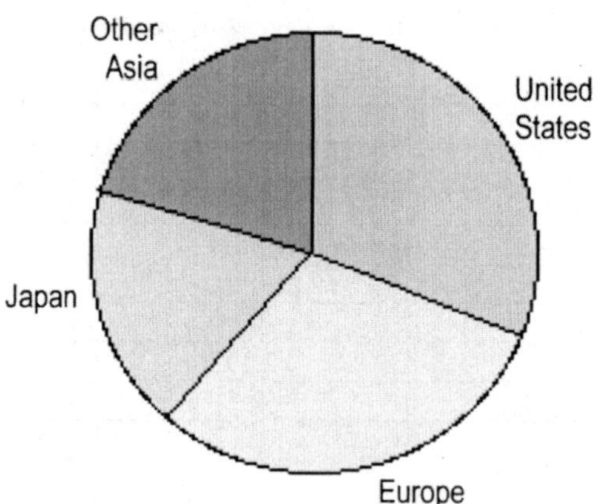


*Figure 1. Consumption of Flame Retardants by Region - 2004*

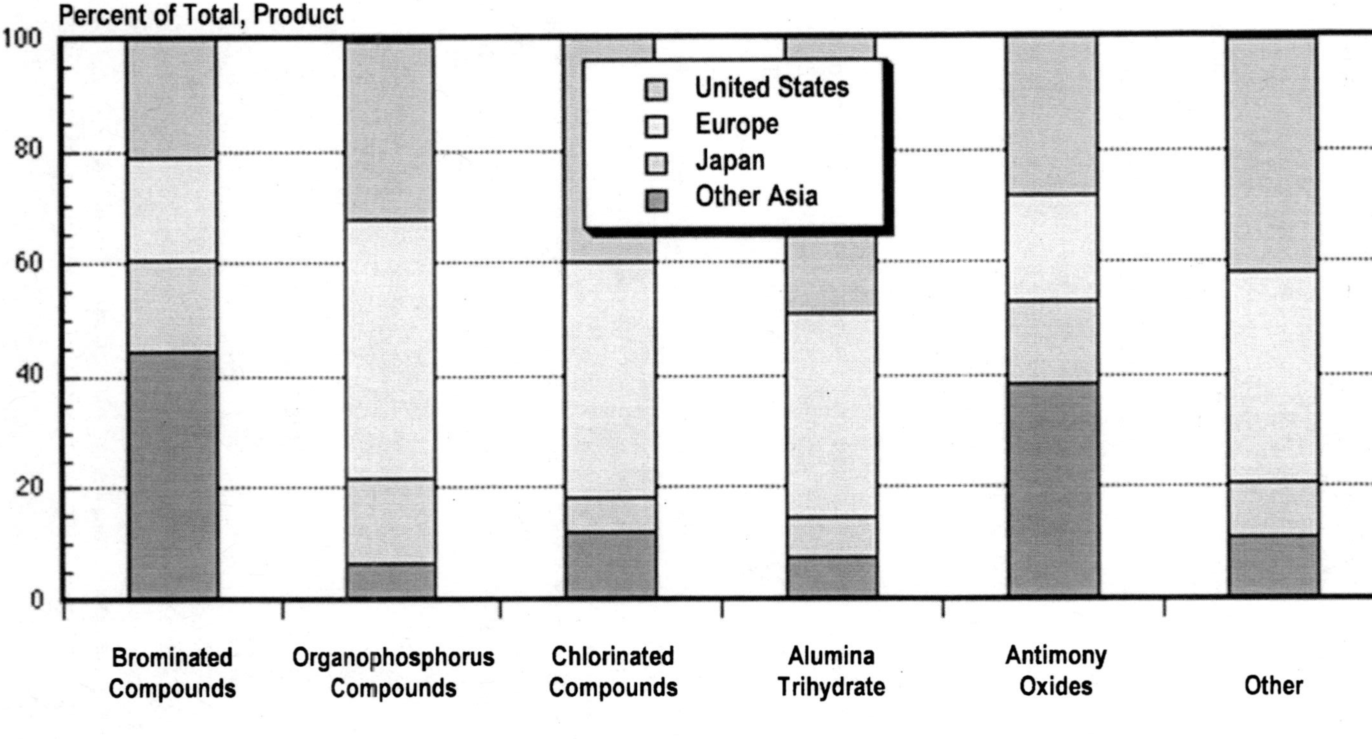


*Figure 2. Relative Consumption of Flame Retardants by Region - 2004*

With this background in place, one can see that research on fire safety of polymers has an important role to play in modern society. This volume is about the latest research at the intersection of the fields of fire and polymers with a strong focus on material chemistry. Within this volume, work is focused on improving the fire performance of polymers through a detailed understanding of polymer degradation chemistry. New and refined analytical techniques facilitate that analysis. Creative chemists continue to develop new approaches and new, more thermally stable flame retardant and polymeric structures. Mathematical fire models continue to become more sophisticated and the fire tests themselves are becoming better understood.

There are many diverse approaches to enhancing the fire stability of polymers. In the past, the most common approach involved the use of additives to the polymer via compounding approaches. Twenty years and more ago halogenated fire retardants (with antimony oxide) were the additives of choice to enhance the fire retardancy of many polymers. Over the past 10 years there has been a strong emphasis on non-halogenated fire retardants, and nano-scale additives in particular.

As one looks at previous Fire and Polymers volumes, topics have clearly changed over the years. In 1990 fire toxicity was the first section with six papers. In 1995 there again was a section on fire toxicity with seven chapters. In 2001 there was but one paper, in 2006, 3 and in this volume, only one. In 1990 there was a section on cellulosics, in 2001 only 1 chapter, and in the 2006 and this volume, none. In the 1995 volume there were twelve chapters on tests and regulations, in 2001, 2, and in this volume, none. In the 2006 volume, half of the papers were on nanocomposites and only two papers had a focus on halogen materials specifically. Once again, in this volume, the majority of the papers are concerned with nanocomposites and their affect on flammability, but there are nine papers dealing with either conventional or novel fire retardant systems. This peer-reviewed volume is designed to represent the state of the art in new approaches to flame retardant materials, and new understanding of polymer flammability phenomena (*9-12*).

## References

1. Karter, Jr., M. J. Fire Loss in the United States During 2007; National Fire Protection Association: Quincy, MA, August, 2008.
2. Fire Loss; National Fire Protection Association; http://www.nfpa.org
3. USFA State Fire Statistics; http://www.usfa.fema.gov/statistics/state/
4. Plastics in Europe, An analysis of plastics consumption and recovery in Europe; Association of Plastics Manufacturers; Summer 2004.

5. APC Year End Statistics for 2007; http://www.americanplasticscouncil.org
6. Toxicological Issues in Fires: Carbon Monoxide or Additives?, G.L. Nelson, Proceedings: Additives 2001: Plastics for the New Millennium, Executive Conference Management (Hilton Head, SC, March 18-21, 2001).
7. Fink, U.; Hajduk, F.; Ishikawa, Y.; Flame Retardants, http://www.sriconsulting.com/SCUP/Public/Reports/FLAME000
8. U.S. Demands for Flame Retardants to Grow; October, 2007; http://www.pcimag.com/copyright/BNP_GUID_9-5-2006_A_1000000...
9. Fire and Polymers, Hazards Identification and Prevention; Nelson, G.L.; Ed.; American Chemical Society; Washington, DC, 1990.
10. Fire and Polymers II, Materials and Tests for Hazard Prevention; Nelson, G. L.; Ed.; American Chemical Society; Washington, DC, 1995.
11. Fire and Polymers, Materials and Solutions for Hazard Prevention; Nelson, G.L. and Wilkie, C.A.; Eds.; Washington, DC, 2001.
12. Fire and Polymers IV, Materials & Concepts for Hazard Prevention; Wilkie, C.A.; and Nelson, G.L. ; Eds.; Washington, DC, 2006.

# Composites and Nanocomposites

# Chapter 2

# Thermal Behavior of Nanocomposites and Fire Testing Performance

**Alberto Fina, Sergio Bocchini, and Giovanni Camino**

**Polytechnic of Turin–Alessandria Site, Viale Teresa Michel 5, 15100 Alessandria, Italy**

A decade of research and development concerning polymer nanocomposites has shown the essential features of their combustion process. Nanocomposites actually display a fire retardant behavior because they avoid fire propagation by dripping of hot and flaming polymer particles and reduce the rate of combustion. The mechanism involved in nanocomposites fire retardance is based on formation on heating of a ceramic protective, insulating layer on the surface of the burning material resulting from coalescence of nanofillers enclosing char from surface polymer charring, catalysed by the nanofiller. The thermal evolution of nanocomposites to the fire protective structure is discussed in relation with dispersion and distribution of the nanofillers in the polymer matrix either as it results from nanocomposite preparation or from its thermal evolution. Evaluation of the fire retardant performance of nanocomposites in fire tests—representing different fire scenarios is discussed.

 

# Introduction

It is generally agreed that the combustion behavior of polymer nanocomposites strictly depends on the interface between polymer condensed phase and the gas phase. Since the first studies on the nanocomposites' behavior in fire (*1, 2*), it was pointed out that the behavior under forced combustion for different nanocomposites were quite similar: a reduction of the heat release rate consequent to a lower fuel feed rate often without substantial modifications of the polymer bulk degradation pathway. It is generally accepted that such a behavior is related to the formation of a physical shield built up by the inorganic nanoparticles left behind by polymer ablation, which acts as a barrier, slowing down the release of generated gas fuel. However, limited understanding of fundamental of physical and chemical process occurring in the condensed phase is available at present. Indeed, complex phenomena can take place in the surface mesophase during nanocomposite burning, affecting accumulation of inorganic particles and their interaction with the polymer while building of a surface structured ceramic phase takes place. Furthermore, the effectiveness of these phenomena on the fire performance of the nanocomposite strictly depends on the specific features of the considered test, such as geometrical setup, presence or absence of external heating source and the possibility of dripping.

This chapter addresses the mechanisms of thermal evolution of nanocomposites occurring during combstion, reviewing the present state of the art, and proposes a critical discussion of experimental results from different widely used fire tests.

# Thermal Evolution of Nanocomposites

## Physical effects

The currently available knowledge on the mechanisms for the formation of the physical shield during nanocomposite burning is the result of many years of work from different groups and went through the accumulation of evidence on re-organization of the inorganic nanoparticle inside the polymer during thermal degradation/combustion. Fundamental mechanisms of physical thermal evolution in nanocomposites are reviewed in this section.

Kashiwagi et al. (*2, 3, 4*) first examined the fire behavior of exfoliated polyamide 6 (PA6)-clay nanocomposites, showing that PA6/clay nanocomposite significantly reduced the peak heat release rate in cone calorimeter test compared to neat PA6 and explained this result with the formation of protective floccules on the sample surface acting as a thermal insulation layer. The analysis of the protective floccules evidenced that they are mainly formed by clay particles, in which lamellae are stacked with a d-spacing larger than the original clay without organic modifier. This higher interlayer distance was explained by the presence of thermally stable organic components possibly with graphitic-like structure, probably formed by the organic molecules entagled during clay re-assembling. Two mechanisms were proposed for accumulation of the initially well-dispersed clay particles. The first is the classical accumulation of clay particles on the surface of the burning material by polymer ablation. De-wetted clay particles, left behind by polymer pyrolysis, tend to aggregate and stack against each other because the degradation of the organic treatment of the clay surface, makes them more hydrophilic and less compatible with the polymer matrix. The second mechanism involves the thermal degradation of the clay organic treatment in the molten polymer and subsequent aggregation of particles. These aggregates of nanoparticles may be moved to the surface by the rising bubbles of volatile products obtained by polymer degradation and/or convection flow in the high temperature melt.

The mechanism was complimented by the studies of migration of the clay platelets inside the nanocomposites. Lewin (*5*) first pointed out the migration effect due to thermal decomposition of the organic modifier and subsequent migration to the surface driven by surface energy minimization.

The effect of filler dispersion in the polymer matrix on its surface coalescence during composite burning, was shown by Wang et al. (*6*). These authors found that surface ceramitization occurred with polystyrene-clay nanocomposites, either intercalated or exfoliated, whereas no surface structured accumulation was found in the case of corresponding microcomposite.

On the other hand, it has to be pointed out that the dispersion of nanoparticles, assessed in the bulk of the polymer matrix at room temperature, is not necessarily retained at the temperatures reached by the surface of the composite during burning. Indeed, the chemical interaction between the nanoclay and the thermally degrading polymer melt, which control the clay dispersion, may be significantly different from the interactions taking place with the polymer in the processing step at lower temperatures.

Pastore et al. (*7*) demonstrated a high temperature structural rearrangement of intercalated clay nanocomposites based on poly(ethylene-co-(vinyl acetate)) (EVA), using in-situ high temperature X-ray diffraction. Below 225 °C a migration of the layered silicate towards the nanocomposite surface and an increase of the interlayer distance were found both under nitrogen or air. At higher temperatures (up to 350 °C) inert or oxidizing atmosphere differently affect the structural modification of the nanocomposite. Indeed, deintercalation and the corresponding decrease of the silicate interlayer space were found under nitrogen as a consequence of thermal degradation of the organic modifier, whereas increase of the silicate interlayer space and exfoliation were the main processes found under air. It was proposed that at high temperature the role of the organic modifier lost by thermal decomposition is played under oxidative atmosphere by polar groups, such as ketones and esters created by surface polymer oxidation, that drive the platelets in intimate contact of the polymer in the oxygen-rich high polar surface of burning polymer nanocomposites.

In a series of articles on PP/nanoclay and PA6/nanoclay Lewin and coworkers accumulated evidence and completed the formation mechanism of the surface clay-rich barrier (*8,9,10,11*). Taking into account the morphological changes that take place when the nanocomposite temperature increases, it was proven that the organophilic layered silicate particles migrate and accumulate at the surface of the molten polymer at temperatures far below pyrolysis temperature. The causes of the migration were identified in the temperature and viscosity gradients of the melt with regard to heat source. The organomodified layered silicates were propelled to the surface by gas bubbles from the decomposing surfactant and polymer and/or by the difference in surface free energy between the polymer and the polymer/clay interface. The migration depends not only on temperature and time as expected for a diffusion process, but also on the extent of exfoliation of clay particles. In absence of polymer clay interaction as in the case of PP/nanoclay, the more exfoliated the structure, the more significant the migration effects, whereas in PA6, a polymer highly interacting with nanoclay, there are more factors that influence migration. Since the annealing in the presence of oxygen decreases migration, it was proposed that the presence of oxygen accelerated the degradation of the surfactant producing acidic clay that can strongly interact with PA6 by protonation and formation of strong linkages lowering clay mobility. Moreover, the extent of migration increases in the temperature range 250-275 °C whereas a further increase in temperature reduces the extent of migration. This phenomenon, which is contrary to expectations, is explained by the gradual conversion of the nanocomposite to a microcomposite due to an acceleration of organic modifier decomposition. The microcomposite, which is not a colloid, is not subject to surface potential forces and does not migrate. Instead, it aggregates in larger sediments and either stays suspended, or precipitates (*10*).

Although the discussion has been focused so far on clay nanocomposites, other nanoparticles have been shown to accumulate as an inorganic layer on the surface of the burning material, with beneficial effect on combustion properties (*12,13*). For example, fire retardance of polypropylene/multi-wall carbon nanotubes (MWNT) nanocomposites was studied in detail by Kashiwagi et al. (*14,15*) who observed that, during combustion, a nanotube network layer is formed which acts to insulate the PP from the external radiant flux, changing the transmission of heat from thermal conduction to radiative transfer. Schartel et al. obtained similar conclusions in PA6/MWNT (*16*), reporting the formation of MWNT interconnected network structure during burning, responsible for a melt viscosity increase preventing dripping and flowing of the nanocomposite during burning.

Hybrid organic/inorganic nanoparticles such as Polyhedral Oligomeric Silsesquioxanes (POSS) have been used as well in fire retardancy of polymers (*17,18,19*), taking advantage of their thermal ceramitization behavior. Indeed, the thermal degradation of silsesquioxanes leads to the formation of a ceramic phase, tipically silica or silicon oxycarbide, depending on POSS chemical structure and the degradation atmosphere (*20, 21, 22*). This ceramic phase can accumulate on the surface of the burning composite to produce a continuous thermally stable protective surface layer, similarly to what obtained with silicate nanoparticles.

## Chemical effects

In the cases described so far, a purely physical effect of the nanofiller has been reported on the structure of the burning composite, related to the formation of an inorganic layer by the accumulation and coalescence of nanoparticles on the surface, with no significant effect on the polymer degradation chemistry in the condensed phase. However, nanoparticles may also play a chemical role in specific cases. A detailed study of the flame retardant mechanism of polymer-clay nanocomposites (*23*) showed that during combustion a clay reinforced carbonaceous char is formed also for nanocomposites whose base polymer matrix normally produced little or no char when burned alone. Zanetti et al. (*24*) proposed a chemical oxidative dehydrogenation catalytic action of strong acid sites created by thermal decomposition of the organic modifier, in polypropylene/clay nanocomposites. The resulting conjugated polyenes are typical precursor of aromatic chars as it is well known for example from studies of thermal degradation of poly(vinyl chloride) and poly(vinyl acetate) (*25*). In EVA/montmorillonite nanocomposites, the acceleration of deacetylation by the catalytic effect of acidic clay sites was shown, with formation of unconjugated polyenes (*26*), which may evolve, through further dehydrogenation, to conjugated polyenes and then to polyaromatic compounds and carbonaceous char.

The catalytic effects are also suggested for other nanoparticles: it has recently been reported that linear low density polyethylene (LLDPE)/MWNT nanocomposites show the formation of a thin protective film of MWNT embedded in a carbon char generated by oxidative dehydrogenation of polyethylene on the surface of the nanocomposites, both during thermoxidative degradation and during cone calorimeter tests (*27, 28, 29*). The effect of POSS functionalised with different metals (Al, Zn, Ti, V) in polypropylene was also reported (*29,30,31*) to show different degrees of thermoxidative stabilisation, depending on the metal nature. With Al-POSS, the formation of conjugated unsaturation at relatively low temperature has been demonstrated, supporting the occurrence of oxidative dehydrogenation mechanisms in competition with well known degradation pathways for pure polypropylene.

# Combustion of Nanocomposites in Different Testing Scenarios

The described phenomena occurring during thermal degradation and combustion of nanocomposites directly affect the results of fire tests carried out on them. On the other hand, performance of polymer materials and nanocomposites in a fire test strongly depends on test scenario i.e. on the fire model.

At present, the fire retardance scientific community basically takes advantage of three different fire tests, namely vertical UL94, LOI and Cone Calorimeter. UL94 and LOI are generally referred to as flammability tests, in which the material behavior exposed to a small flame is addressed, in terms of capability to ignite and to self-sustain a flame, thus representing a scenario in which the material is at the origin of a fire. On the other hand, cone calorimeter test is representative of a forced combustion, in which the material is forced to burn under controlled heat flux. This test addresses the ignition time, the rate of combustion and the total heat released, modelling the contribution of the material to a fire started on other items. Moreover, these flammability and combustion tests also differ for the specimen positioning, the former being vertical tests, the latter most often being an horizontal test, although vertical configuration is provided for by the standard methods (*32, 33*). Considered these differences, it is certainly reasonable to expect different performance for a given fire retarded formulation compared with the reference material, when testing in different fire tests, expected to be representative of different fire scenarios. With polymer nanocomposites, the differences in performance obtained in flammability and forced combustion test are usually very significant, this having caused an ongoing discussion on the actual effectiveness of nanoparticles as fire retardants (*13, 34*).

The consequence of these facts is twofold: on the one hand, the relevance of different fire tests to real fire scenarios becomes crucial for the final application of polymer nanocomposites and, on the other hand, the scientific significance of standard tests must be carefully evaluated. In this section, a critical comment of phenomena behind the bare ranking results of some standard fire tests is proposed.

### Combustion tests

Cone calorimetry has been the most used test for the study of nanocomposite's fire behavior, since the pioneering work by Gilman et al. (*1*). Among the many parameters supplied by cone calorimetry, attention has been focused primarily on Peak of Heat Release Rate (pkHRR) during combustion because of its relevance to fire risk related to time to flashover. Typical nanocomposite's results reported in literature show a decreas in the pkHRR of about 50 to 70% compared to reference unfilled polymer, with either nanoclays (*1,4,23,34,35,36*), carbon nanotubes (*14,15,16,28*) or other inorganic nanoparticles (*37,38,39,40*). The general experimental observation is that the presence of dispersed nanoparticles switches the typical non-charring behavior of most thermoplastic polymers to that of charring materials, intended as materials which develop a surface protective layer when exposed to heat. Such surface layer generally grows in thickness thanks to progressive accumulation of nanoparticles upon polymer volatilization, often leading to the formation of a solid residue at the end of the test, with variable degree of compactness, ranging from isolated floccules (Figure 1a) to fully solid char with shape and size similar to the unburned specimen (Figure 1b). The after-burning-residue is the result of complex mechanisms occurring during combustion, involving the thermal evolution mechanism reviewed in the former section of this chapter, depending very much on polymer properties and nanofiller dispersion at room temperature. Also, the compactness of the residue during burning is generally related to the efficiency in HRR reduction, the higher the compactness, the lower HRR.

This "charring" behavior radically changes the profile of the HRR plot, as char-forming materials show the heat release rate peak at the early stage of the combustion because of the protective action of the char building up on the surface of the burning sample which slows down heat release. Conversely, non-charring materials show the peak of heat release rate at the end of the burning process, due to the so-called "thermal feedback effect" from standard thermally insulated sample holder (*41, 42, 43*). Since the physical phenomena behind the peaks in the two described cases are different, the simple comparison of pkHRR values to quantitatively assess relative fire risk of virgin polymer and corresponding nanocomposite may be misleading. It has been shown for PPgMA/clay nanocomposite that very different values for pkHRR reduction compared to PPgMA are obtained changing the specimen setup from insulating (70% reduction) to conductive (27% reduction) sample holder (*42*). This gives evidence of the different influence of the experimental setup on the pkHRR for charring (PPgMA/clay) and non-charring (PPgMA) materials. The elimination of thermal feedback contribution to pkHRR for non-charring polymers was proposed, thus affecting the calculated reduction of HRR: for PA6/montmorillonite nanocomposite, HRR reduction of about 33% was calculated when substracting thermal feedback contribution, compared with ~50% reduction obtained by bare calculation from pkHRR values (*43*). As a simple approach, the average value of HRR (i.e. the total heat evolved divided by burning time) appear to be more reliable parameter, compared to pkHRR decrease, when trying to quantify the reduction of combustion rate for nanocomposites.

In most of the cases, the reduction of combustion rate of polymer matrix in nanocomposites is explained with the barrier effect obtained upon nanoparticle and polymer char accumulation on the surface of the burning sample by the mechanisms described in the previous section of this chapter. The ceramic-char

(a)

(b)

*Figure 1. Residue obtained after cone calorimeter test, a) PP/5% organomodified montmorillonite, 50 kW/m², top view b) PA6/5% organomodified montmorillonite, 75 kW/m², side view. Both PP and PA6 burned in the same conditions left no significant residue at the end of test.*

barrier reduces the rate of fuel feed to the flame, either by the reduction of the effective incident heat flux onto the polymer, owing either to reradiation by the ceramic-char surface layer or by the slow diffusion of volatiles through the surface layer, by labyrinth effect, entrapment into porosity or adsorption. These phenomena are very effective in the horizontal confined sample configuration of the cone calorimeter test, because no material macroscopic flow occurs since the specimen is confined in the sample holder. Morover, the in-depth advancing of the flame front allows the ceramic-charred residue to be effective in protecting the underlying polymer, but this is a very specific conditions which does not relate, as an example, with lateral flame spread.

The use of other combustion tests, such as the cone calorimeter performed with vertical specimen setup and the radiant panel tests (*44, 45*), would certainly help in completing the assessment of nanocomposite behavior under forced combustion.

## Flammability tests

Among flammability tests, the most widely used ones are certainly the UL94 tests, either vertical or horizontal, and the Limiting Oxigen Index test. The LOI test gives a quantitative information about the material performance and is helpful in fire retardant materials development. Conversely, UL94 methods are pass/fail tests which aim at ranking the materials in terms of fire risk in selected scenarios but ranking by itself is useless in development of fire retardant materials. For example a burning sample extinguishing because of heat elimination by abundant dripping can be classified either as V-0 (most desirable) or V-2 (much less desired) depending on whether falling drops ignite the underlying cotton or not (*46,47*). Morover, this test was originally developed for devices and appliances, and the standards themselves state that the method is not intended to cover plastics when used as materials for building construction or finishing, whereas the test has become of general use in both industrial product specifications and in the scientific literature, because the test is extremely simple and cheap. However, from the scientific point of view, there is little knowledge on the phenomenological behavior of material in the simple pass/fail result, so that complementary description of polymeric materials burning process is highly advisable such as times of combustion, amount of material burned and rate of combustion.

Bartholmai and Schartel reported no reduction of flammability for maleic anhydride-grafted-polypropylene (PPgMA)/organoclay nanocomposites, the pure polymer being classed V-2 and the nanocomposites being non classified (NC) in the vertical test (*34*). This was explained by the different dripping behavior, as the unfilled PPgMA estinguishes thanks to dripping of flaming parts of the specimen, thus lowering heat flux to the polymer, whereas the nanocomposites showed decreased dripping, so that the specimens do not estinguish. However, the fire behavior observed was different in terms of burning velocity measured in horizontal UL94 tests, with reductions of about 50% for the nanocomposites compared to reference PPgMA. Similar effects have been reported for MWNT in PA6, where the interconnected network structure of CNTs hinders dripping, during vertical UL94 test (*16*).

Thus, in order to better describe the flaming behavior during the test, other information should be presented together with the final classification, such as the height of the flames, the burning rate and the end-of-test residue. A tentative example of this integrated approach is reported in the following for epoxy resin and epoxy resin/octaphenyl POSS nanocomposite (*48, 49*) (Table I). Unfilled epoxy resin (DGEBA/MDEA type) shows rapid burning up to the clamp after first ignition by the external flame, with dripping of flaming droplets (see images at burning times $t_a$=15÷90s in Table I), which ignite the cotton below the specimens. From the standard method, such a material cannot be classed accordingly to the vertical UL94. When adding 10% wt. octaphenyl POSS (ofePOSS), the burning behavior changes completely, with no droplets from the bottom of the specimen and a slow advancing of a weak flame (see Table I, $t_a$=15 s), before self estinguishing, which occurs at the average time $t_1$=34s. A similar behavior is observed after second ignition (see burning time $t_b$=15 and 30s in Table I) and self estinguishing occurs at the average time $t_2$=36s. Moreover, the residual weight after the end of the test is about 95% of the original specimen weight, evidencing that only surface burning occurred. This is most likely due to the ceramifying-charring self-protecting action of POSS on the specimen, when the surface is licked by the flame in the earlier stage of the test, by the mechanism reviewed in the former section of this chapter.

Despite these described differences, the UL94 standard method is unable to rank these materials in terms of fire retardance because both virgin epoxy and nanocomposite are classified NC, due to the fact the flame reached the top of the specimen or because of exceeding burning times, whereas it is obvious that there is a very important different fire hazard for the two materials in terms of fire propagation. From this example, the UL94 classification clearly appears inadequate, at least in some cases, as a single-test general method to evaluate flammability performance.

Complementary information may be obtained by LOI test, which represents the opposite configuration compared to vertical UL94, the specimen being ignited from the top. Relatively few papers reported LOI test for nanocomposite materials (*13,16,34,50,51,52*), generally showing either a decrease or no significant increase of the LOI value compared to their reference polymers. Indeed, the LOI setup condition appears to be critical for the production of an effective ceramic protective layer on the material surface, as the specimen below the flame front is not licked by the flame and there is no external irradiation. This prevents the formation of the protective char-ceramic surface layer on the material prior to being reached by the flame front.

**Table I. Comparison Epoxy and Composite Burning in UL94 Test**

| | | | | *Average Burning Times* | *UL94 class* | *Observations* |
|---|---|---|---|---|---|---|
| *Neat Epoxy resin* | 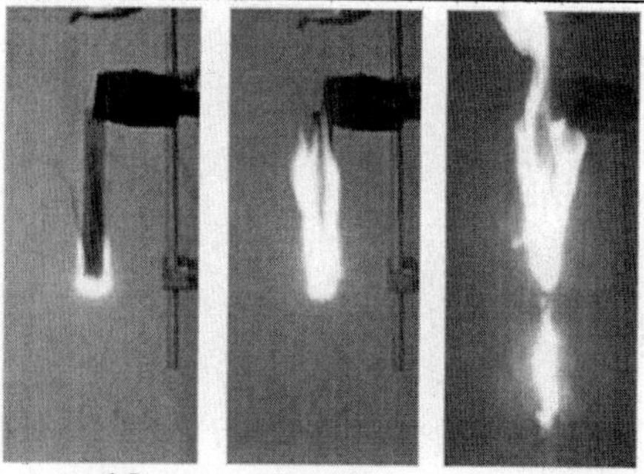 $t_a$=15 | $t_a$=60 | $t_a$=90 | No flameout | NC | Cotton ignited by flaming droplets. Flame reaches the clamp. End-of-test residue negligible. |
| *Epoxy resin with 10% ofePOSS* | 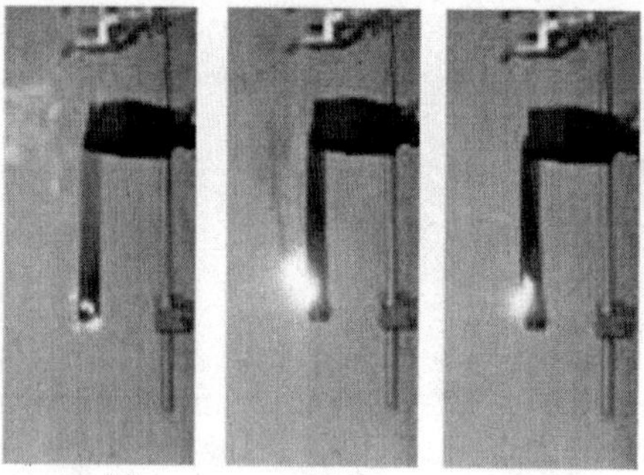 $t_a$=15 | $t_b$=15 | $t_b$=30 | Flameout occurring both on first and second ignition: $t_1$=34 s $t_2$= 36 s | NC | No flaming droplets. End of test residue: ~ 95% w/w |

NOTE: $t_a$: time after first ignition, (s)
$t_b$: time after second ignition, (s)
$t_1$ and $t_2$: average estinguishing time after first and second ignition, respectively, (s) (accordingly with the standard ASTM D3801)

## Conclusions

The combustion behavior of nanocomposites is one of their most attractive characteristics. Indeed, nanocomposites containing a few percent of well dispersed nanofiller burn at a much lower rate than the corresponding polymer without dripping of flaming particles, thus reducing the contribution of polymer materials to fire propagation. Nanocomposites are therefore effective fire retardant materials, which are defined as those materials which extend the time to flashover in fires.

The mechanism of fire retardance in nanocomposites is related to surface modification on heating, leading to a protective layer of a ceramic resulting from nanofiller coalescence, generally combined with char deriving from catalysed surface polymer charring.

The comprehensive assessment of fire retardant behavior of nanocomposites cannot be evaluated by a single test but should include different tests representing different fire scenarios, ranging from ignition to well developed fires. Each testing setup may show a different behavior of nanocomposites: as an example, flammability tests such as UL94 and LOI typically evidence for lower dripping, whereas cone calorimeter shows the reduction of burning rate during forced combustion. Moreover, from a material research perspective, test providing quantitative data on ignition and combustion behavior, such as the cone calorimeter, are essential for development of fire retardant materials, with progressive replacement of prescriptive codes with performance evaluation in materials selection for specific fire retardant applications.

Nanocomposites make thus a step forward towards reduction of fire risk and hazard for polymers because they avoid flame spreading by flaming dripping and reduce the rate of combustion. For the first time in polymer fire retardant records, we have polymeric materials available in which fire risk and hazard are reduced with simultaneous improvement of other polymer material properties and using an environmentally friendly technology. Most of the fire retardants used now have to be loaded into polymers in relatively large amount (10-70% by wt.) with negative effects on polymer physical and mechanical properties and on environmental issues.

Industrial development of nanocomposites seems now under way because since the appearance of the first clay-based nanocomposites two decades ago, a wide selection of nanofillers have become available on a commercial basis. A parallel extensive worldwide research effort has strongly improved our knowledge on preparation and characterization of nanocomposites, now available for industrial exploitation.

## Acknowledgements

The authors thanks Mr. Andrea Medici for Epoxy/POSS materials preparation and fire testing and Dr. Federica Bellucci for courtesy of Figure 1a.

## References

1. Gilman, J. W.; Lichtenhan, J. D. *SAMPE J.* **1997**, *33*, 40-46.
2. Kashiwagi, T.; Gilman, J. W.; Nyden, M. R.; Lomakin, S. M. In *Fire Retardancy of Polymers: The Use of Intumescence*; Le Bras, M.; Camino,

G.; Bourbigot, S.; Delobel, R. Ed.; The Royal Society of Chemistry: Cambridge, UK, 1998; pp. 175-202.

3. Gilman, J. W.; Kashiwagi, T.; Giannelis, E. P.; Manias, E.; Lomakin, S.; Lichtenhan, J. D.; Jones, P. In *Fire Retardancy of Polymers: The Use of Intumescence*; Le Bras, M.; Camino, G.; Bourbigot, S.; Delobel, R., Eds.; The Royal Society of Chemistry: Cambridge, UK, 1998; pp 203-221.
4. Kashiwagi, T.; Harris, R. H.; Zhang, X.; Briber, R. M.; Cipriano, B. H.; Raghavan, S. R.; Awad, W. H.; Shields, J. R. *Polymer* **2004**, *45*, 881-891.
5. Lewin, M. *Fire Mater.* **2003**, *27*, 1-7.
6. Wang, J.; Du, J.; Zhu, J.; Wilkie, C. A. *Polym. Degrad. Stab.* **2002**, *77*, 249-252.
7. Pastore, H. O.; Frache, A.; Boccaleri, E.; Marchese, L.; Camino, G. *Macromol. Mater. Eng.* **2004**, *289*, 783-786.
8. Tang, Y.; Lewin, M.; Pearce E. M. *Macromol. Rapid. Comm.* **2006**, *27*, 1545-1549.
9. Hao, J.; Lewin, M.; Wilkie, C. A.; Wang, J. *Polym. Degrad. Stab.* **2006**, *91*, 2482-2485.
10. Tang, Y.; Lewin, M. *Polym. Degrad. Stab.* **2007**, *92*, 53-60.
11. Lewin, M.; Tang, Y. *Macromolecules* **2008**, *41*, 13-17.
12. Castrovinci, A.; Camino, G. In *Multifunctional barriers for flexible structures: Textile, Paper and Leather;* Duquesne, S.; Magniez, C.; Camino, G., Eds.; Springer, Berlin, D, 2007; pp 87-108.
13. Bourbigot, S.; Duquesne, S. *J. Mater. Chem.* **2007**, *17*, 2283-2300.
14. Kashiwagi, T.; Grulke, E.; Hilding, J.; Harris, R.; Awad, W.; Douglas, J. F. *Macromol. Rapid Comm.* **2002**, *23*, 761-765.
15. Kashiwagi, T.; Grulke, E.; Hilding, J.; Groth, K.; Harris, R.; Butler, K. M.; Shields, J. R.; Kharchenko, S.; Douglas, J. F. *Polymer* **2004**, *45*, 4227-4239.
16. Schartel, B.; Potschke, P.; Knoll, U.; Abdel-Goad, M. *Europ. Polym. J.* **2005**, *41*, 1061-1070.
17. Devaux, E.; Rochery, M.; Bourbigot, S. *Fire Mater.* **2002**, *26*, 149-154.
18. Bourbigot, S.; Le Bras, M.; Flambard, X.; Rochery, M.; Devaux, E.; Lichtenhan, J.D. In *Fire Retardancy of Polymers: New Applications of Mineral Fillers.* Le Bras, M.; Wilkie, C.; Bourbigot, S., Eds; Royal Society of Chemistry; Cambridge, UK, 2005; pp 189-201.
19. Gupta, S.K.; Schwab, J.J.; Lee, A.; Fu, B. X.; Hsiao, B. S. In *Affordable Materials Technology-Plateform to global value and performance* Rasmussen, B. M.; Pilato, L. A.; Kliger, H.S. Eds; SAMPE pub.; Long Beach, CA; 2002 pp. 1517-1526.
20. Mantz, R.A.; Jones, P. F.; Chaffee, K. P.; Lichtenhan, J. D.; Gilman, J. W. *Chem. Mater.* **1996**, *8*, 1250-1259.
21. Fina, A., Tabuani, D.; Carniato, F.; Frache, A.; Boccaleri, E.; Camino, G. *Thermochim. Acta* **2006**, *440*, 36-42.

22. Fina, A.; Tabuani, D.; Frache, A.; Boccaleri, E.; Camino, G. In *Fire Retardancy of Polymers: New Applications of Mineral Fillers.* Le Bras, M.; Wilkie, C.; Bourbigot, S., Eds; Royal Society of Chemistry; Cambridge, UK, 2005; pp 202-220.
23. Gilman, J. W. In *Flame Retardant Polymer Nanocomposites* Morgan, A. B.; Wilkie, C. A. Eds; Wiley, Hoboken, NJ, 2007; pp 67-87.
24. Zanetti, M.; Camino, G.; Reichert, P.; Mülhaupt, R. *Macromol. Rapid Comm.* **2001**, *22*, 176-180.
25. McNeill, I. C. In *Comprehensive Polymer Science*, Allen, G.; Bevington, J.; Eastmond, G. C.; Ledwith, A.; Russo, S.; Sigwalt, P. Eds; Pergamon Press, Oxford, UK, Vol 6; 1989; pp 452-497.
26. Zanetti, M.; Camino, G.; Thomann, R.; Mülhaupt, R. *Polymer* **2001**, 42, 4501-4507.
27. Bocchini, S.; Frache, A.; Camino, G.; Claes, M. *Eur. Polym. J.* **2007**, *43*, 3222-3235.
28. Bocchini, S.; Annibale, E.; Frache, A.; Camino, G. *E-Polymers* **2008**, No 20 free access at www.e-polymers.org.
29. Fina, A.; Bocchini, S.; Camino G. *Polym. Degrad. Stab.* **2008**, *93*, 1647-1655.
30. Fina, A.; Abbenhuis, H. C. L.; Tabuani, D.; Camino, G. *Polym. Degrad. Stab.* **2006**, *91*, 2275-2281.
31. Carniato, F.; Boccaleri, E.; Marchese, L.; Fina, A.; Tabuani, D.; Camino, G. *Eur. J. Inorg. Chem.* **2007**, *4*, 585-591.
32. ISO standard 5660, 2002.
33. ASTM standard E 1354, 2003.
34. Bartholmai, M.; Schartel, B. *Polym. Adv. Technol.* **2004**, *15*, 355-364.
35. Zanetti, M.; Kashiwagi, T.; Falqui, L.; Camino, G. *Chem. Mater.* **2002**, *14*, 881-887.
36. Zheng, X.; Jiang, D. D.; Wang, D.; Wilkie,C. A. *Polym. Degrad. Stab.* **2006**, *91*, 289-297.
37. Kashiwagi T.; Gilman, J.W; Butler, K.M.; Harris, R.H.; Shields, J.R.; Asano, A. *Fire Mater.* **2000**, *24*, 277-289.
38. Lefebvre, J.; Le Bras M.; Bourbigot, S. In *Fire Retardancy of Polymers: New Applications of Mineral Fillers;* Le Bras, M.; Bourbigot, S.; Duquesne, S.; Jama C.; Wilkie, C. A., Eds.; Royal Society of Chemistry: Cambridge, UK, 2005; pp 42-53.
39. Du, M.; Guo, B.; Jia, D. Europ. Polym. J. **2006**, *42*, 1362-1369.
40. Kashiwagi, T. In *Flame Retardant Polymer Nanocomposites;* Morgan, A. B.; Wilkie, C. A., Eds.; Wiley: Hoboken, NJ, 2007, pp 285-324.
41. Schartel, B.; Hull T. R. *Fire Mater.* **2007**, *31*, 327-354.
42. Schartel, B.; Bartholmai, M.; Knoll, U. *Polym. Degr. Stab.* **2005**, *88*, 540-547.

43. Fina, A.; Canta, F.; Castrovinci, A.; Camino, G. In *Fire Retardancy of Polymers: New Strategies and Mechanisms;* Kandola, B.; Hull, R., Eds.; Royal Society of Chemistry: Cambridge, UK, 2009; Chapter 10, *in press.*
44. Standard EN ISO 9239, 2002 and standard ASTM E 648, 2008.
45. Gilman, J. W.; Bourbigot, S.; Shields, J. R.; Nyden, M.; Kashiwagi, T.; Davis, R. D.; Vanderhart, D. L.; Demory, W.; Wilkie, C. A.; Morgan, A. B.; Harris, J.; Lyon, R. E. *J. Mater. Sci.* **2003**, *38*, 4451-4460.
46. standard ASTM D 3801, 2000.
47. standard ASTM D 635, 1998.
48. Franchini, E.; Galy, J.; Gérard, J.F.; Medici, A.; Tabuani, D.; Camino, G. In *Proceedings of the 11th European Meeting on Fire Retardant Polymers (FRPM07)*, Bolton, UK, 3-6 July 2007.
49. Franchini, E.; Galy, J.; Gérard, J.F.; Tabuani, D.; Medici, A. *article in preparation.*
50. Kandola, B. K. In *Fire Retardant materials;* Horrocks, A. R.; Price D., Woodhead Publishing Limited: Cambridge, UK, 2001, pp 204-219.
51. Hartwig, A.; Putz, D.; Schartel, B.; Bartholmai, M.; Wendschuh-Josties, M. *Macromol. Chem. Phys.* **2003**, *204*, 2247-2257.
52. Bourbigot, S.; Duquesne, S.; Fontaine, G.; Bellayer, S.; Turf, T.; Samyn, F.; *Mol. Cryst. Liq. Cryst.* **2008**, *486*, 1367-1381.

## Chapter 3

# Functionalized-Carbon Multiwall Nanotube as Flame Retardant for Polylactic Acid

**Serge Bourbigot, Gaëlle Fontaine, Antoine Gallos, Caroline Gérard, and Séverine Bellayer**

**Procédés d'Elaboration des Revêtements Fonctionnels (PERF), LSPES–UMR/CNRS 8008, ENSCL, Avenue Dimitri Mendeleïev – Bât. C7a, B.P. 90108, 59652 Villeneuve d'Ascq Cedex, France**

In this work, we have investigated the benefit of combining the grafting of chemical function to improve the solubility and the dispersion in an organic matrix with the opportunity of grafting fireproofing chemicals (melamine-based compound) on multiwall carbon nanotube (MWNT). The functionalized MWNT (f-MWNT) is incorporated in polylactide (PLA) via melt blending. Transmission electron microscopy (TEM) reveals that a high level of nanodispersion is achieved in PLA with f-MWNT while the dispersion is poor with virgin MWNT. The PLA nanocomposite containing f-MWNT exhibit a significant reduction of peak of heat release rate (PkHRR) in cone calorimeter experiment but this effect is probably not optimized because we suspect an antithetic effect of the melamine-based compound grafted on the surface of the nanotube.

# Introduction

Polymeric materials are commonly used in everyday life increasing fire hazards and so flame retardants are very often incorporated into them to limit their flammability 0. Of particular interest is the developed nanocomposite technology consisting of a polymer and nanoparticles because they often exhibit remarkably improved mechanical properties and various other properties as compared with those of virgin polymer at loading as low as 3-5 wt.-% 0. Numerous nanocomposites containing different nanoparticles, including organomodified clays *0-5*), nanoparticles of $TiO_2$ 0, nanoparticles of silica 0, layered double hydroxides (LDH) *0-0*, carbon nanotubes (CNT) 0 or polyhedral silsesquioxanes (POSS) 0, have been prepared and characterized. All those materials exhibit low flammability associated with other properties such as enhanced mechanical or electrical properties.

Aliphatic polyesters, and particularly polylactic acid (PLA), currently deserve particular attention in the area of environmentally degradable polymer materials. They are well suited for the preparation of disposable devices because of their biodegradability *0-0*. In PLA nanocomposites it was reported that this family of composites exhibits improved properties including a high storage modulus both in the solid and melt states, increased flexural properties, a decrease in gas permeability, increased heat distortion temperature, an increase in the rate of biodegradability of pure PLA, etc… 0.

Kashiwagi et al. reported the first study on the flammability of polymer carbon nanotube nanocomposites 0. They showed significant flame retardant effectiveness of polypropylene (PP)/multi-walled carbon nanotubes (MWNT) (1 and 2% by mass) nanocomposites. Concurrently, Beyer demonstrated a small improvement in flammability properties of ethylene-vinyl acetate (EVA)/ MWNT (2.5 and 5% by mass) nanocomposites 0. Based on those results, one of the goals of this paper is to investigate the flame retardancy of PLA in a PLA/MWNT nanocomposite. An issue is to evenly disperse MWNT in the polymeric matrix to get the best flame retardancy properties. In previous work, we have shown that the nanodispersion (including MWNT) should be achieved to get the lowest peak of heat release rate (PkHRR) in a cone calorimeter 0.

The chemical functionalization of carbon nanotubes is one of the few methods used to enhance the interfacial adhesion between the nanotubes and the matrix 0. The chemical treatment permits to generate functional groups at the surfaces of the nanotubes. These functional groups could react with other chemicals, prepolymers, and polymers, thus enhancing the interfacial bond between the matrix and the tubes for their further application to polymer nanocomposites. Here our basic idea is to combine the benefit of grafting chemical function to improve the solubility and the dispersion in an organic matrix with the opportunity of grafting fireproofing chemicals on the nanotube.

This approach came out very recently in the literature with a reasonable success in acrylonitrile-butadiene-styrene (ABS) polymer using a phosphonate compound 0.

Melamine-based compounds are well known to be flame retardants and our strategy is to graft melamine onto the suface of nanotube. The synthesis of the functionalized carbon nanotube will presented in the first part of the paper. Nanodispersion of the nanotubes will be then examined. The third part of the paper will investigate the reaction to fire of the nanocomposites containing virgin nanotubes and functionalized nanotubes.

## Experimental

**Materials.** PLA (melt flow index (190 °C, 2.16 kg) = 6.61 g/10 min) was supplied by NatureWorks and dried overnight at 110 °C before use.

CNT are multiwall carbon nanotube (MWNT) supplied by Nanocyl (Nanocyl-7000 at 90% purity). Unless otherwise noted, all materials were obtained from commercial suppliers and used without further purification. All reactions involving air- or moisture-sensitive compounds were performed in a nitrogen atmosphere.

**Functionanized carbon nanotubes (f-MWNT).** Oxidation of MWNT by acid treatment was carried out by sonification for 2h (100W, 50kHz) of a suspension of 1g of MWNT in 250mL of sulfuric acid (97%)/nitric acid (65%) (3/1). The oxidized MWNT were washed with water until the pH was neutral, then filtered through 0.45μm Millipore membrane. The oxidized MWNT were dried at 80 °C for 20h. A mixture of 1g of oxidized MWNT and 250 mL of thionyl chloride was refluxed with stirring for 18h then unreacted thionyl chloride was removed by distillation and the activated MWNT (MWNT-COCl) were dried under reduced pressure. To a cold suspension of 2g of MWNT-COCl in 200 mL of dried dichloromethane were added dropwise 2mL of ethylene diamine; when no more acidic vapor evolved, the suspension was refluxed with stirring for 18h. The ethylene diamine functionalized MWNT were washed, filtered and dried at 80 °C. The sample of 6-chloro-2,4-diamino-1,3,5-triazine was obtained by the reaction of 2.0g of 2,3,6-trichloro-1,3,5-triazine (0.011 mol) with 1.47 mL of aqeous ammonia solution (28%) in a mixture of 25mL $H_2O$ and 25mL acetone at 45 °C overnight (this reaction is well known and can be found in textbooks about triazines). The precipitate was washed with water, filtered and dried under reduced pressure at ambient temperature leading to 1.5g of 6-chloro-2,4-diamino-1,3,5-triazine (95%). A suspension of 1g of ethylene diamine functionalized MWNT with an exess of 1g of 6-chloro-2,4-diamino-1,3,5-triazine in 200mL of water was refluxed for 18h. The f-MWNT were washed with water, filtered and dried under reduced pressure before use.

The acidic sites of the nanotubes were evaluated according to method described for single carbon nanotubes 0. The quantity of carboxylic acid function was evaluated to 1 mmol per gram of MWNT.

**Processing.** PLA was melt-mixed with MWNT and f-MWNT (2 wt.-%) at 185 °C (PLA) using a Brabender laboratory E350 mixer measuring head (roller blades, constant shear rate of 50 rpm) for 10 min in nitrogen flow to avoid hydrolysis.

**Fourier Transform InfraRed.** FTIR spectra were performed on Nicolet Impact 400D spectrometer in KBr pellets.

**Transmission electron microscopy.** All samples were ultra microtomed with a diamond knife on a Leica ultracut UCT microtome, at room temperature for PLA samples, to give sections with a nominal thickness of 70 nm. Sections were transferred to Cu grids of 400 mesh. Bright-field TEM images of nanocomposites were obtained at 300 kV under low dose conditions with a Philips CM30 electron microscope, using a Gatan CCD camera. Low magnification images were taken at 17,000x and high-magnification images were taken at 100,000x.

**Fire testing.** An FTT Mass Loss Calorimeter was used to carry out measurements on samples following the procedure defined in ASTM E 906. External heat flux of 35 kW/m² was used for running the experiments. The mass loss calorimeter was used to determine heat release rate (HRR).

## Results and Discussion

**Synthesis and characterization of the functionalized carbon nanotubes.** Functionalized MWNT (f-MWNT) were synthesized in four steps according to the protocol described in Figure 1. In the first step the MWNT are oxidized by treatment with a mixture of sulfuric acid/nitric acid. The carboxylic acid functions formed at the suface of the MWNT are then activate by thionyl chloride, this step was followed by the reaction with ethylene diamine. The 6-chloro-2,4-diamino-1,3,5-triazine, obtained by reaction of ammonia with 1,3,5-trichlorotriazine, react with the amine function on the MWNT and we obtained the melamine-based f-MWNT.

The purpose is twofold: (i) to increase the compatibility of MWNT with an organic polymeric matrix, and (ii) to nanodisperse flame retardant (FR) (here the melamine-based compound) in polymer.

Figure 2 shows FTIR spectra of pristine MWNT, oxidized MWNT, MWNT-CONH-$CH_2$-$CH_2$-$NH_2$, and f-MWNT. The spectrum of pristine MWNT exhibits two broad bands centered at 3430 $cm^{-1}$ (-OH) and 1080 $cm^{-1}$ (C-O) suggesting some impurities on the MWNTs. The oxidation of MWNT is clearly shown with the appearance of the band at 1635 $cm^{-1}$ assigned to carbonyl group

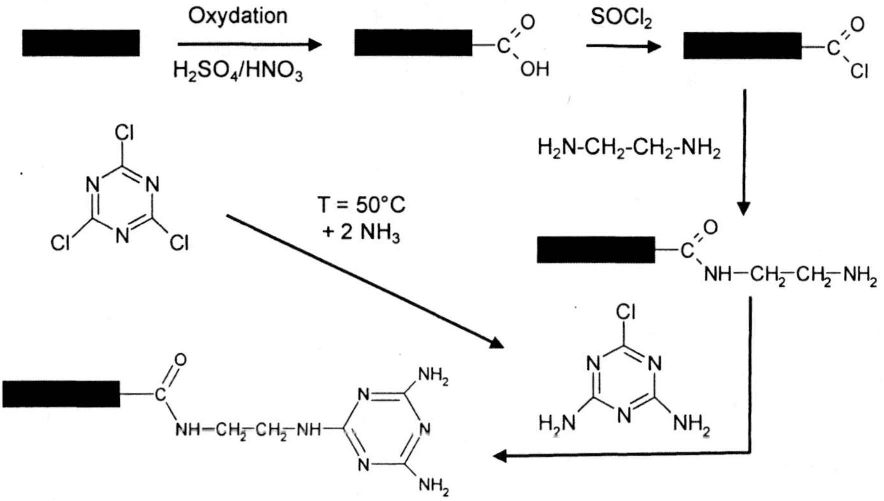

*Figure 1. Synthetic steps for the preparation of f-MWNT*

in carboxylic function and of the broad and intense band centered at 3430 $cm^{-1}$ assigned to hydroxyl group. The reaction of oxidized MWNT with the ethylene diamine leads to the formation of amide group. Amide and carboxylic groups exhibits characteristic bands at very close wavenumbers (intense band of carbonyl in -CONH and –COOH at 1635 $cm^{-1}$ and a broad band of –OH /-NH groups centered at 3430 $cm^{-1}$). Evidence of the formation of an amide bond is given by the band at 1550 $cm^{-1}$ characteristic of a monosubstitued amide. The last step of the functionalization of the MWNTs is evidenced by the shift of the broad band centered at 3430 $cm^{-1}$ to 3330 $cm^{-1}$ corresponding to the transformation of $-NH_2$ to -NH. Additional bands observed on the spectrum are assigned to melamine-based MWNT compound.

**Characterization of the nanodispersion.** Figure 3 shows TEM images at low and high magnification of PLA/MWNT. Large bundles of nanotubes can be distinguished on the picture at low magnification (Figure 3-a) and numerous agglomerates are observed at high magnification (Figure 3-b). Using non-functionalized nanotubes the dispersion is poor and must be enhanced to expect good flame retardancy properties.

f-MWNTs are much better dispersed in PLA than MWNT as revealed by TEM at low and high magnification (Figure 4). Tiny agglomerates can be distinguished on the picture at low magnification (Figure 4-a) but almost all nanotubes are single and relatively well separated from each other (Figure 4-b). So, it is demonstrated that the functionalized nanotubes with a melamine-based compound permit one to achieve a high level of nanodispersion.

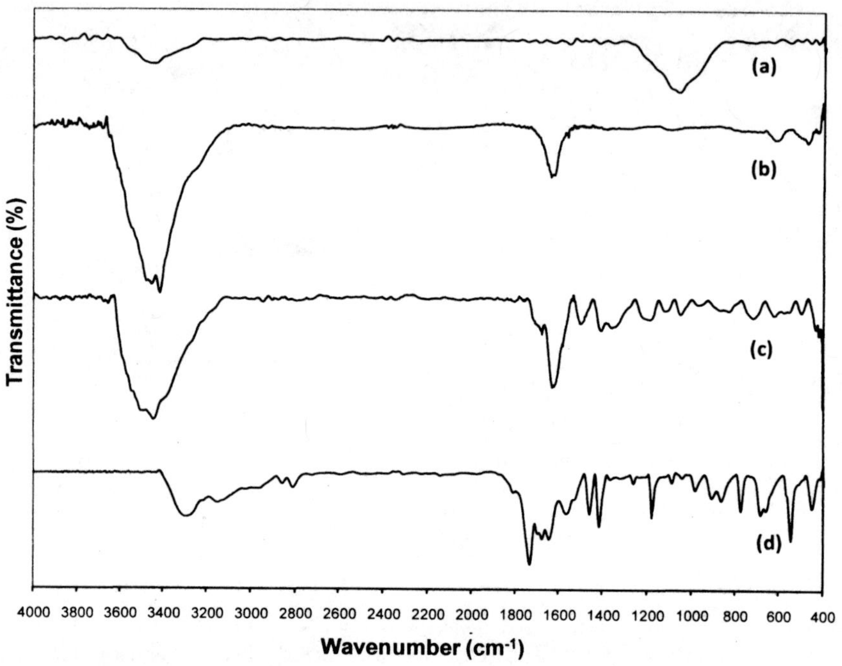


*Figure 2. FTIR spectra of (a) pristine MWNT, (b) MWNT-COOH, (c) MWNT-CONH-$CH_2$-$CH_2$-$NH_2$, and (d) f-MWNT*

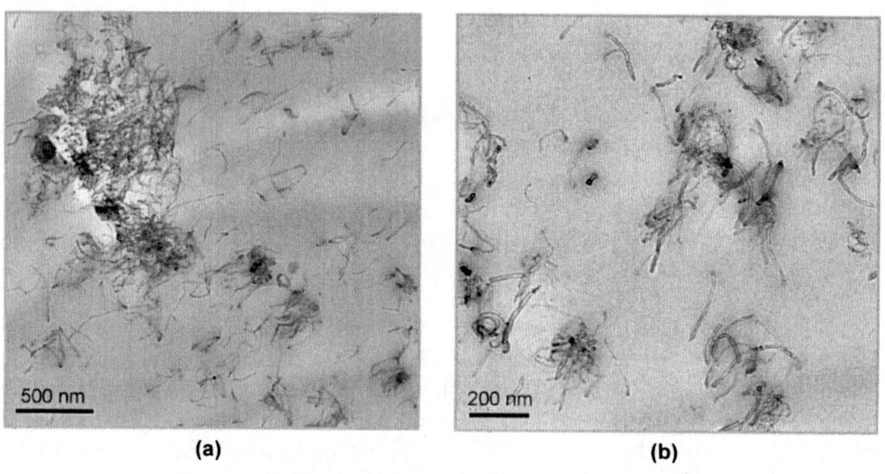


*Figure 3. TEM images of PLA/MWNT at low magnification (a) and at high magnification (b).*

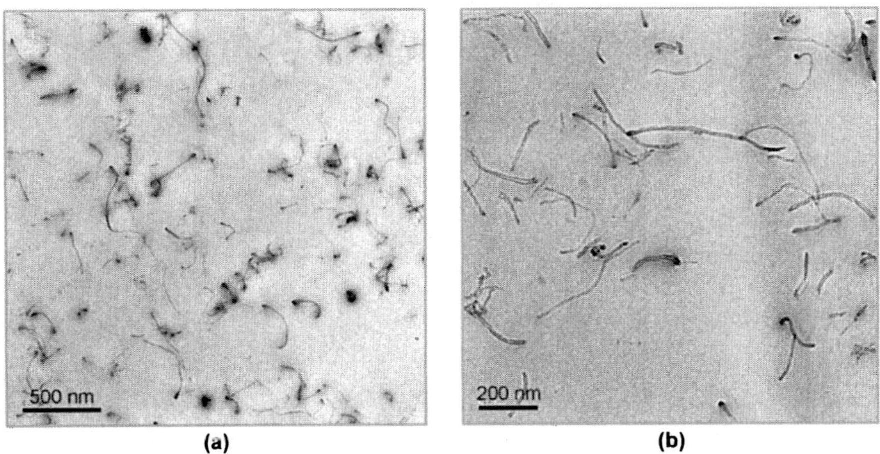


*Figure 4. TEM images of PLA/f-MWNT at low magnification (a) and at high magnification (b).*

**Reaction to fire.** PLA/f-MWNT exhibits lower HRR values than pure PLA and PLA/MWNT (reduction by 28% of PkHRR compared to virgin PLA) while PLA/MWNT does not exhibit any flame retardancy (Figure 5). This might be assigned at least partially, to the excellent dispersion of f-MWNT in PLA 0. In the final residues, the char of PLA/f-MWNT is relatively compact and cohesive while that with MWNT exhibits islands of char without any cohesion. It is also noteworthy that the times to ignition of PLA/MWNT and of PLA/f-MWNT are not enhanced compared to virgin PLA. According to the previous work of Kashiwagi et al. 0, we may assume this is because of an increase in the radiation in-depth absorption coefficient by the addition of carbon nanotubes. We also know that the incorporation of nanotubes in a polymer increases its viscosity in the melt compared to virgin polymer 0 upon heating. In the same conditions, virgin PLA melts and 'boils' creating strong convection due to mass transfer. At this time, heat conductivity of the nanocomposite becomes lower than that of the virgin polymer while it is the reverse before heating. Radiation absorption and lower heat conductivity causes accumulation of heat at the surface of the material upon heating and the nanocomposite reaches its ignition temperature more rapidly.

We have also examined if melamine is a flame retardant for PLA. The incorporation of 30 wt.-% (the typical amount for conventional FR) of melamine in PLA provides a high LOI value (35 vol.-%) but it increases PkHRR and shortens the time to ignition, which means that melamine is not an efficient flame retardant for PLA in terms of cone calorimetry. Visual observations suggests that melamine decreases the viscosity of the materials upon heating

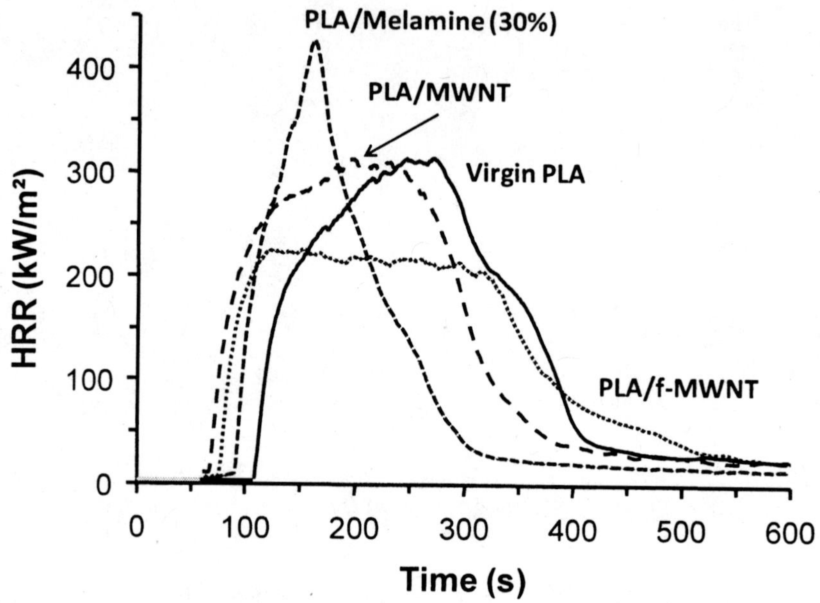


*Figure 5. HRR curves as a function of time of PLA/MWNT and PLA/f-MWNT compared to virgin PLA and PLA/Melamine.*

(LOI test conditions). During an LOI test, the burning materials flows away along the bar and the combustion is stopped, which might explain in such conditions why LOI value is quite high while cone performance is poor. This discussion might also explain why the reduction of PkHRR of PLA/f-MWNT is not high as expected while the nanodispersion level is excellent. Further work is in progress with other FR compounds to be grafted onto the surface of nanotube to investigate this.

## Conclusion

In this work, we have investigated the synthesis of functionalized carbon nanotubes containing flame retardant groups (melamine-based compounds) and the reaction to fire of PLA nanocomposites containing MWNT and f-MWNT. High level of nanodispersion is achieved with f-MWNT while the dispersion is poor with virgin MWNT. The nanocomposite containing f-MWNT exhibits significant reduction of PkHRR but this effect is probably not optimized because it is suspected that the melamine-based compound grafted on the surface of the nanotube creates an antithetic effect.

## Acknowledgment

This work was partially supported by the European project Interreg IV 'NANOLAC'. The authors are indebted to Mr. Pierre Bachelet from our group for skilful technical assistance and for helpful discussion. Mr. Michael Claes from Nanocyl (Sambreville, Belgium) is gratefully acknowledged for supplying carbon nanotubes, for helpful collaboration and discussion.

## References

1. Bourbigot, S.; Le Bras, M., *In: Flammability handbook,* Troitzsch, J. Ed., Hanser Verlag Pub., Munich, **2004**, p. 133.
2. Alexandre, M. ; Dubois. P. *Mat. Sci. Eng. R.* **2000**, *28*, 13.
3. Gilman, J.W.; Kashiwagi, T.; Lichtenhan, J.D. *SAMPE J.* **1997**, *33*, 40.
4. Bourbigot, S.; VanderHart, D.L.; Gilman, J.W.; Bellayer, S.; Stretz, H.; Paul, D.L. *Polymer* **2004**, *45*, 7627.
5. Song, L.; Hu, Y.; Tang, Y.; Zhang, R.; Chena, Z.; Fa W. *Polym. Deg. Stab.* **2005**, *87*, 111.
6. Laachachi, A.; Leroy, E.; Cochez, M.; Ferriol, M.; Lopez-Cuesta J.M. *Polym. Deg. Stab.* **2005**, *89*, 344.
7. Kashiwagi, T.; Morgan, A.B.; Antonucci, J.M.; VanLandingham, M.R.; Harris Jr., R.H.; Awad, W.H.; Shields, J.R. *J. Appl. Polym. Sci.* **2003**, *89*, 2072.
8. Lefebvre, J.; Le Bras, M.; Bourbigot S. *Fire Retardancy of Polymers: New Applications of Mineral Fillers*, Le Bras, M.; Wilkie, C.A.; Bourbigot, S.; Duquesne, S.; Jama C. (Eds.), p. 42, The Royal Society of Chemistry: Cambridge, **2005**.
9. Zammarano, M.; Gilman, J.W.; Franceschi, M.; Meriani S. *Proceedings of the 16th BCC Conference on Flame Retardancy*. Lewin, M. BCC: Norwalk, CT, **2005**.
10. Kashiwagi, T.; Du, F.; Winey, K.I.; Groth, K.M.; Shields, J.R.; Bellayer, S.P.; Kim, H.; Douglas, J.F.. *Polymer* **2005**, *46*, 471.
11. Bourbigot, S.; Le Bras, M.; Flambard, X.; Rochery, M.; Devaux, E.; Lichtenhan, J.D. Fire Retardancy of Polymers: New Applications of Mineral Fillers, Le Bras, M.; Wilkie, C.A.; Bourbigot, S.; Duquesne, S.; Jama C. (Eds.), p. 189, The Royal Society of Chemistry: Cambridge, 2005.
12. Vinka, E.T.H.; Rabagob, K.R.; Glassnerb, D.A.; Gruberb. P.R. *Polym. Deg. Stab.*, **2003**, *80*, 403.
13. Auras, R.; Harte, B.; Selke, S. *Macromol. Biosci.*, **2004**, *4*, 835.
14. Lunt, J. *Polym. Deg. Stab.*, **1998**, *59*, 146.
15. Kimura, K.; Horikoshi, Y. *Fujitsu Sci. Tech. J.*, **2005**, *41*, 173.
16. Ray, S.S.; Okamoto, M. *Macromol. Rapid Commun.*, **2003**, *24*, 815.

17. Kashiwagi, T.; Grulke, E.; Hilding, J.; Harris Jr., R.H.; Awad, W.H.; Douglas, J. *Macromol. Rapid Commun.*, **2002**, *23*, 761.
18. Beyer, G. *Fire Mater.*, **2002**, *26*, 291.
19. Bourbigot, S.; Duquesne, S.; Jama C. *Macromol. Symp.*, **2006**, *233*, 180.
20. Sun, Y.P.; Fu, K.; Lin, Y.; Huang, Y. *Acc. Chem. Res.*, **2002**, *35*, 1096.
21. Ma, H.Y.; Tong, L.F.; Xu, Z.B.; Fang, Z.P. *Adv. Funct. Mater.* **2008**, *18*, 414.
22. Vogel, A.I.; Tatchell, A.R.; Furnis, B.S.; Hannaford, A.J.; Smith, P.W.G. Vogel's Textbook of Practical Organic Chemistry, 5th Edition; Pearson Education Ltd: Harlow, **1989**.
23. Afremow, L.C. Infrared spectroscopy: its use in the coating industry, Federation of Socities for Paint Technology: Phliadephia, PA, **1969**.
24. Hu, H.; Bhowmik, P.; Zhao, B.; Hamon, M.A.; Itkis, M.E.; Haddon R.C. *Chem. Phys. Letters*, **2001**, *345*, 25.
25. Kuan, H.C.; Ma, C.C.; Chang, W.P.; Yuen, S.M.; Wu, H.H.; Lee, T.M. *Comp. Sci. Tech.* **2005**, *65*, 1703.

# Chapter 4

# Use of Layered Double Hydroxides as Polymer Fire-Retardant Additives: Advantages and Challenges

**Charles Manzi-Nshuti, Liying Zhu, Calistor Nyambo, Linjiang Wang, Charles A. Wilkie*, and Jeanne M. Hossenlopp***

**Department of Chemistry and Fire Retardant Research Facility, Marquette University, P.O. Box 1881, Milwaukee, WI 53201-1881**

Layered Double Hydroxides (LHDs) have been identified as a promising new additive class for generating polymer nanocomposites with enhanced thermal stability and improved flammability properties. An advantage of these materials over structurally similar smectite clays is the ability to tune physical and chemical properties via simple synthetic strategies that can modify the metal hydroxide layer and/or the identity of the charge-balancing interlayer anion. Recent advances in development of LDHs for polymer fire retardancy applications are reviewed here and a discussion of future challenges is provided.

Organically modified smectite clays have been extensively studied as polymer nanocomposite additives and the have been shown to improve thermal stability, flammability, and mechanical properties (1). Layered double hydroxides, LDHs, can be categorized as anionic analogs of clays. In an LDH, the metal hydroxide layer consists of a combination of divalent and trivalent cations and have the general formula $[M^{2+}{}_xM^{3+}{}_{1-x}(OH)_2]^{x+}(A^{n-}{}_{x/n})\cdot mH_2O$ where $A^{n-}$ represents the intercalated anion (2). Due to the abilility to alter the identity and relative stoichiometry of the metals and/or the interlayer anion, LDHs and related anionic clays have been proposed for a number of applications, including fire retardancy of polymers (3).

The most common LDH used in polymer nanocomposite applications is hydrotalcite, $Mg_6Al_2(CO_3)(OH)_{16}\cdot 4(H_2O)$, or a modified Mg/Al LDH where the carbonate is replaced by an organic species to promote dispersion. In literature reports, these additives were found to improve thermal stability and some flammability properties of a variety of polymers (4). The addition of iron to hydrotalcite was found to provide improvement in fire properties of an ethylene-vinyl acetate copolymer (4j). While Mg/Al LDH provides some promising improvements in selected polymer fire properties, use of these additives with other fire retardants as also begun to attract attention as a strategy for producing effective formulations (5).

While LDHs have shown some promise as fire retardant additives, a number of issues remain in characterizing and optimizing LDHs for this purpose. Recent work in our laboratories has focused on identifying how factors such as the LDH structure (metal ion composition and anion size and structure), loading, and dispersion quality affect the thermal stability and flammability of selected model polymers. Key results are reviewed here.

## Experimental Methods

LDHs were synthesized via coprecipitation or by exchange of the desired anion with a nitrate-containing LDH precursor. Syntheses were carried out under nitrogen in order to exclude $CO_2$ and thus minimize carbonate contamination, following literature methods (6). Relative amounts of divalent to trivalent metal ions in the product were controlled by the initial concentrations of metal nitrates. LDH products were characterized via powder x-ray diffraction (XRD) and Fourier transform infrared spectroscopy (FTIR).

Polymer (nano)composites can be formed via melt blending, *in situ* polymerization, or solution blending methods; the work reviewed here focuses on melt blended samples. The LDH was mixed at different mass ratios with commercial polymer samples in a Brabender Plasticorder for approximately 10 minutes at 60 rpm and 185 ° C. Dispersion and nanocomposite formation were assessed using XRD and transmission electron microscopy (TEM). Thermal stability was assessed via thermogravimetric analysis (TGA). TGA experiments were performed on a SDT 2960 machine at the 15 mg scale under air or a flowing nitrogen atmosphere at a scan rate of 20 °C/min. Flammability was assessed using cone calorimetry. Cone calorimeter measurements were performed on an Atlas CONE-2 according to ASTM E 1354 at 35 or 50 $kW/m^2$ incident flux using a cone shaped heater; the exhaust flow was set at 24L/sec. The specimens for cone calorimetry were prepared by the compression molding of the sample (about 30 g) into 3 x 100 x 100 $mm^3$ square plaques. Typical results from cone calorimetry are reproducible to within about ± 10%; these uncertainties are based on many runs in which thousands of samples have been combusted.

## Results and Discussion

The advantage of using an LDH is the wide range of tunability – one can vary the stoichiometry, the identity of the divalent and/or trivalent metal ion, and the identity of the anion and all of these have been studied. The initial investigation was to evaluate whether an LDH could work with either polar or non-polar polymers. Accordingly, a magnesium-aluminum undecenoate LDH was melt blended with four different polymers, polyethylene (PE), polypropylene (PP), polystyrene (PS) and poly(methyl methacrylate) (PMMA) and the morphology and fire performance were evaluated[7]. From TEM, only in the case of PMMA is the LDH well-dispersed in the polymer; for all three of the non-polar polymers, only immiscible systems are obtained. From cone calorimetry, minimal changes (<10%) in the peak heat release rate (PHRR) are seen for both PE and PP while a reduction of about 20% is observed for the poorly-dispersed PS composite. With PMMA, in which the LDH is well-dispersed, one obtains a PHRR reduction of 52%. Thus PMMA can be identified as the most likely candidate for further study and the next section of this paper addresses this polymer.

PMMA was melt-blended with a series of undecenoate –containing LDHs in order to examine the effect of systematic variations in the divalent metal (8). A series of LDHs containing $Zn^{2+}/Al^{3+}$, $Co^{2+}/Al^{3+}$, $Ni^{2+}/Al^{3+}$, and $Cu^{2+}/Al^{3+}$ metal hydroxide layers with $M^{2+}$:$M^{3+}$ ratios of 2:1 were tested, along with a $Zn^{2+}$:$Al^{3+}$ 3:1 stoichiometry. The $Co^{2+}/Al^{3+}$ melt-blended composite showed evidence for intercalation in x-ray diffraction analysis and the best dispersion in TEM analysis while all of the other samples were best characterized as microcomposites. Surprisingly, the 2:1 Zn:Al system was noticeably better dispersed then the 3:1 system. Since the increased ratio of Zn to Al will decrease the amount of aluminum and thus decrease the anion exchange capacity, one might have guessed that the 3:1 system would be better dispersed.

None of the composite samples exhibited improvement in mechanical properties (tensile strength, elongation at break, Young's modulus) when compared with PMMA alone. Note that all reference samples were treated under the same conditions in the Brabender mixer to provide an appropriate comparison.

Thermal stability (as determined by the temperature required for 50% mass loss in thermogravimetric analysis (TGA)) was improved in the presence of all additives except the copper-containing LDH which was determined to be unstable under processing conditions. The nanocomposite $Co^{2+}/Al^{3+}$ loaded at 10% provided the best improvement in cone calorimetry evaluation of peak heat release rate (PHRR), with a 41% improvement over the pure polymer. This is better than the 30% improvement reported for PMMA melt-blended with organically-modified MMT (9). In addition, it is important to note that even the LDH microcomposite samples exhibited some improvement in PHRR with 26% and 35% improvement for 2:1 and 3:1 $Zn^{2+}:Al^{3+}$ LDHs at 10% loading. Recalling that the 2:1 Zn:Al system was better dispersed than the 3:1, it is suprising to see that the 3:1 gave a larger reduction in the PHRR, since this is normally connected to dispersion. It is clear that the connection between dispersion and selected fire properties is not as tight with the LDH family as it is with the montmorillonite type clays.

A magnesium – aluminum undecenoate LDH has also been blended with PMMA and it appears to be fairly well-dispersed, certainly at the micro level and perhaps at the nanometer level; the reduction in the PHRR is 52% when 10% of the LDH is present[7]. When montmorillonite type clays have been used with PMMA, the best reduction that has been seen in the PHRR is about 30%. It is clear based upon these few observations that the LDH family is a new family that must be developed to understand how dispersion is affected and the connection between dispersion and fire retardant properties. A plot of the heat release rates for zinc, magnesium, nickel and cobalt all with aluminum as the trivalent anion, are given in Figure 1.

Anion effects have also been explored for LDHs with PMMA (10). The effect of alkyl chain length in a series of *n*-alkyl carboxylates [$CH_3(CH_2)_xCOO^-$ with x = 8, 10, 12, 14, 16, and 20] was examined using $Mg^{2+}/Al^{3+}$ LDH. The best dispersion is obtained when a C-10 anion is used and this is described as exfoliated. As the chain length increases, intercalation becomes the dominant mode of nano-dispersion. TGA analysis of thermal stability indicated no significant chain length effect. However, PHHRs were found in to increase with increasing chain length, with C-10 and C-12 additives providing the best results. A plot of the heat release rates for selecetd anions is shown in Figure 2.

In addition to carboxylate anions, anions of phosphorus and sulfur have also been used; the anions under study are bis(2-ethylhexyl)phosphate, 2-ethylhexylsulfate and dodecylbenzenesulfonate (11). Both the phosphate and the sulfonate are well-dispersed in PMMA and can be described as exfoliated while the sulfate is a micro-composite. The best reduction in PHRR, 45%, is seen for the sulfonate at 10% loading while the phosphate gives a reduction of

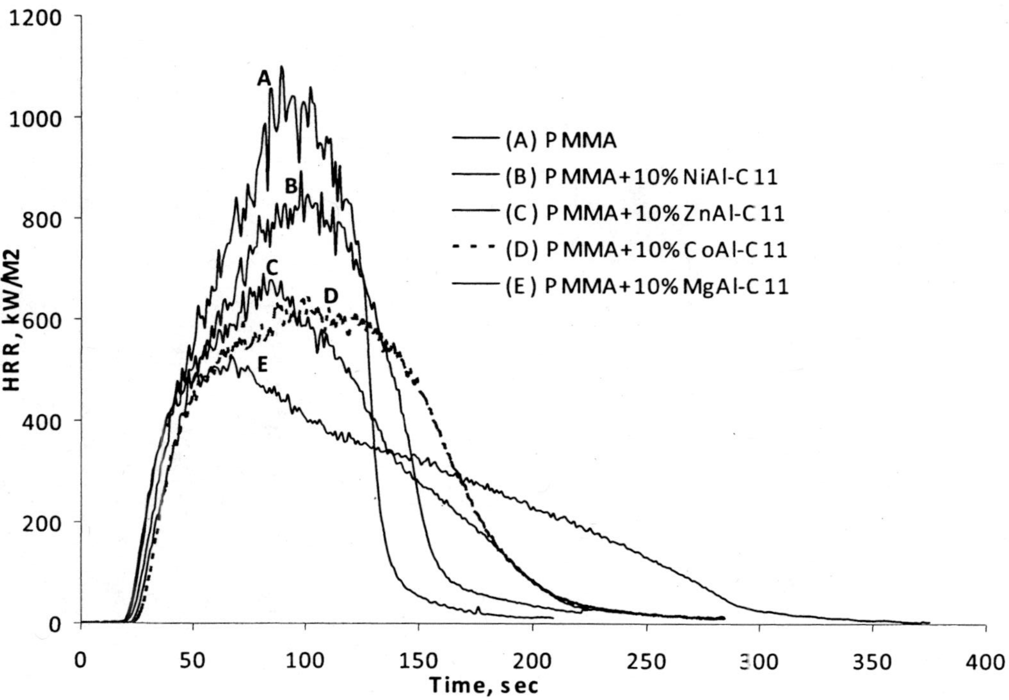


*Figure 1. Heat release rate plots for the various divalent metal-containing LDHs with PMMA.*

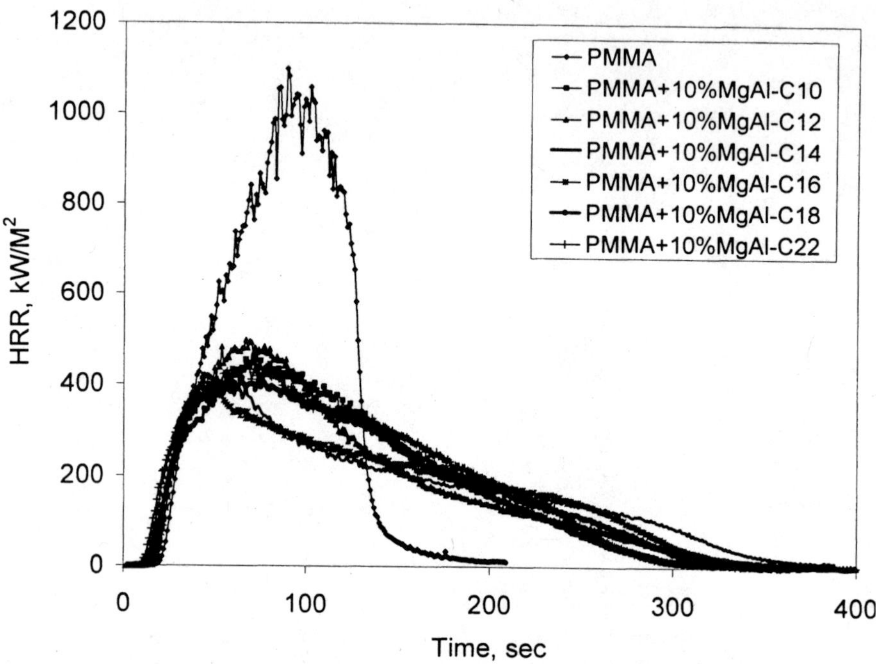


*Figure 2. Heat release rate curves for PMMA nanocomposites with selected carboxylates*

37%. It is of interest to note that there is very little dependence on the amount of LDH for the phosphate; the reduction is 29% at 3% loading, 33% at 5% loading and 37% at 10% loading. While the sulfonate significantly increases as the laoding increases, going from 17% at 3% loading, 29% at 5% loading and 45% at 10% loading. Figure 3 provides the heat release rate curves for all three of these additives at 10% loading. Once again, there are more questions raised then answers from these results.

The trivalent metal ion has also been varied – both aluminum and iron have been used with calcium as the divalent metal ion (12). When the morphology is evaluated by the combination of XRD and TEM, one concludes that the Ca-Al system is nano-dispersed while the Ca-Fe system is not. The reduction in the PHRR is 54% for Ca-Al and 34% for Ca-Fe. Two important conclusions can be drawn from this, aluminum is probably a better choice than iron as the trivalent metal ion in an LDH but, even though the Ca-Fe system is not well-dispersed, there is still a very substantial reduction in the peak heat release rate. This makes one wonder if better dispersion can be obtained, will the fire properties be improved? The heat release rate plots for both Ca-Al and Ca-Fe at 10% loading in PMMA are shown in Figure 4.

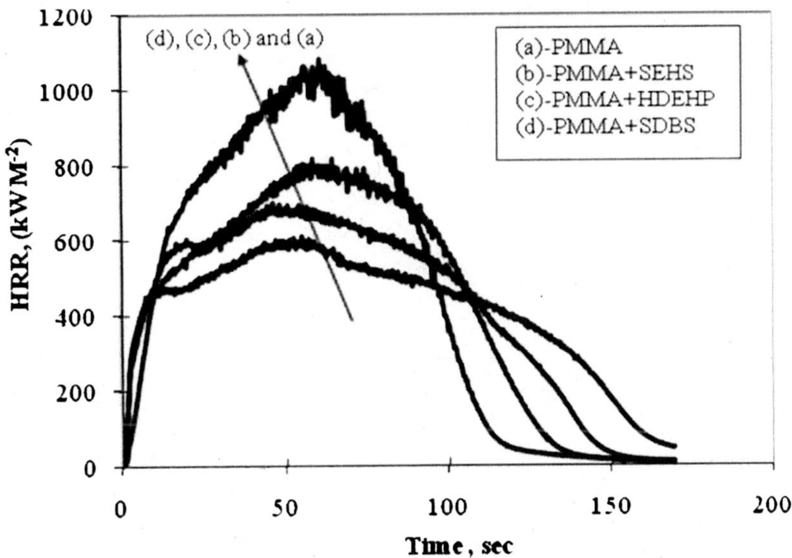


*Figure 3. Heat release rate curevs for PMMA (nano)composites with 10% of the sulfate, sulfonate and phosphate LDHs*

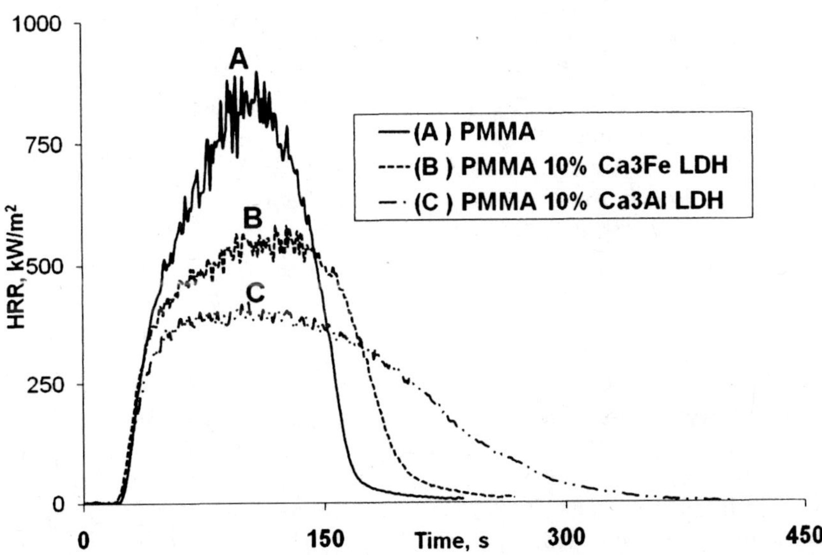


*Figure 4. Heat release rate curves for PMMA (nano)composites containing 10% of CaFe and CaAl LDHs.*

$Zn^{2+}/Al^{3+}$ and $Mg^{2+}/Al^{3+}$LDHs, both with 2:1 divalent to trivalent metal ion stoichiometries, have been prepared with oleate $[CH_3(CH_2)_7CH{=}CH(CH_2)_7COO^-]$ anions. The Zn/Al LDH loaded at 10% provided a 28% reduction in PHRR, similar to the analogous undecenoate-containing LDH. The $Mg^{2+}/Al^{3+}$ oleate additive produced a 48% reduction in PHRR when loaded at 10%, the best result observed to date in these laboratories for melt-blended PMMA. Work is in progress to complete characterization of the polymer/additive composites; TEM and x-ray data for the $Zn^{2+}/Al^{3+}$ suggest a disordered system with tactoid formation.

Another series of additives containing the dodecyl sulfate anion $[C_{12}H_{25}OSO_3^-]$ has been investigated (13). $Zn^{2+}/Al^{3+}$ dodecyl sulfate LDH loaded at 10% resulted in a similar reduction in PHRR (25%) as the corresponding undecenoate and oleate systems. However, the improvement in PHRR for $Mg^{2+}/Al^{3+}$ dodecyl sulfate was only 22%, significantly lower than found for the oleate anion. An HDS containing Zn and Ni with dodecyl sulfate was also prepared (Zn:Ni = 2:1) and a 40% reduction in PHRR was observed at 10% loading.

Key results for the melt-blended PMMA system include:

- Improvements in PHRR (40-48%) that exceed those found with modified MMTs have been observed for different combinations of metals and anions. Further work is necessary to determine whether the chemical composition or other physical property is the most important factor.
- Nanocomposite formation is not required for PHRR reductions but the role of dispersion and nanocomposite formation requires further investigation.
- Time-to-ignition in cone calorimetry is largely unperturbed by LDHs melt-blended in PMMA (not shown above).
- The effects of loading on PHRR are inconsistent and require further investigation, particularly if combinations of additives are being evaluated for synergy.

## Nonpolar Polymers

While the initial results with PE and PP were not encouraging, that with PS was since even though the dispersion was not good, there was still a 20% reduction in the PHRR – thus polystyrene can be a model system for use of LDHs with nonpolar polymers. The *n*-alkyl carboxylate $Mg^{2+}/Al^{3+}$ LDHs described in the preceding section were bulk polymerized with PS (10) and the morphology was evaluated. One might anticipate that as the chain length increases, the potential for compatability between the anions and a non-polar polymer will increase so one might expect to see better dispersion with longer chain lengths. Indeed this is the case but one never achieves nano-dispersion of any of the these LDHs in PS. The reduction in PHRR decreases as the chain length increases, which is not a suprise because this will also increase the fire load. What is a surprise is that one obtains a 56% reduction in PHRR for the C-10 system, which is poorly dispersed. This reduction is comparable to what is seen for well-dispersed montmorillonite type materials in PS. A plot of the heat release rate for selected carboxylates is shown in Figure 5.

The dodecyl sulfate series was also tested with PS with magnesium and zinc as the divalent metals and aluminum as the trivalent metal. With both sets of metals, the time to ignition drops dramatically when 3% LDH is added but does not further drop as the amount is increased. There is a significantly larger reduction in the PHRR for magnesium (37%) compared to zinc (17%). With both metals, the change in the PHRR does not depend significantly on the amount of the LDH.

Preliminary investigations of LDHs in other nonpolar polymers such as polypropylene (PP) and polyethylene (PE) are underway. It is very suprising to find that a MgAl LDH gives good dispersion in PMMA but does not sem to be well-dispersed in PE or PP while the ZnAl LDH may be just the opposite of this. From cone calorimetry, the reduction in the PHRR in PE is 28% with MgAl and 58% with ZnAl. For comparison, in PMMA with MgAl the reduction is 48% but 29% with ZnAl.

We have observed similar results for a related metal hydroxide additive, copper hydroxy dodecyl sulfate in PE (14). Mg/Al and Zn/Al oleate were also blended with PE. The Mg-containing oleate LDH resulted in 29% improvement in PHRR when loaded at 10%, similar to the results above. In contrast, Zn/Al oleate led to a striking 58% reduction for 10% loading in PE. These preliminary results again suggest that LDHs hold promise as components of fire retardant additives but the role of chemical structure and addive dispersion are not easily predicated and require further exploration.

## Combinations of LDHs with Other Additives

Combinations of melamine with Zn/Al undecenoate LDH were added to PMMA via in situ polymerization (15). Melamine alone served to increase the time to peak heat release rate (tPHRR) and decrease the average specific extinction area (ASEA), while the LDH alone led to a better improvement in PHRR. A combination of 10% melamine and 5% LDH resulted in the best overall performance in cone calorimetry testing, based on PHRR and fire performance and growth indices. While combinations of the two additives at higher loadings also performed well in cone testing, it is expected that mechanical properties will be negatively impacted (16).

Another study has combined an LDH with ammonium polyphosphate (APP) in PS. Based on TGA data, there is clear synergy between the two

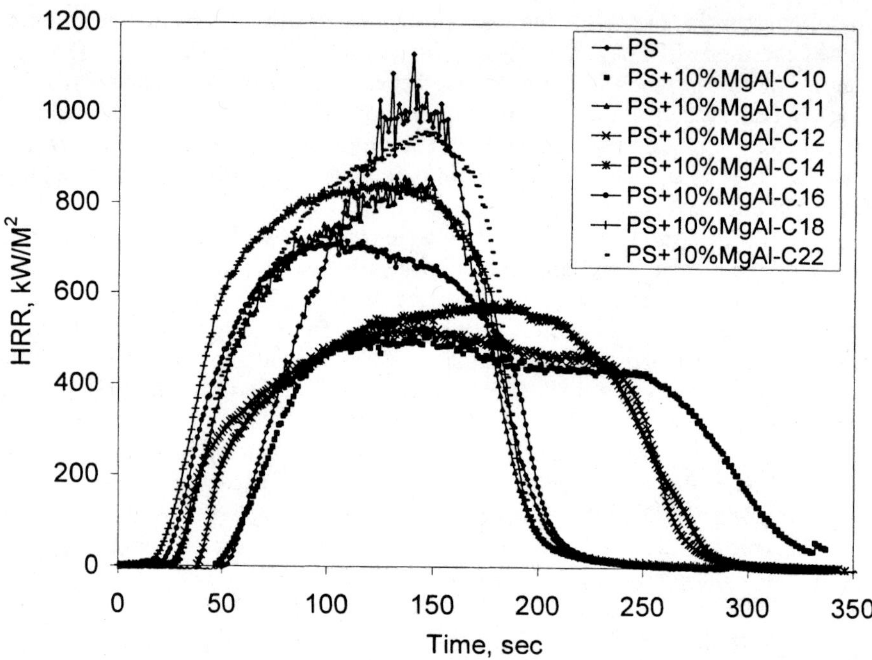


*Figure 5. Heat release rate curves for PS-LDH system with various carboxylates.*

additives. Likewise, from cone calorimetry the experimental reduction in the PHRR is almost twice the value that might be expected. We are actively pursuing other examples of potential synergistic combinations of LDHs with conventional fire retardancts and we feel that these systems have the potential to be very useful.

## Conclusions

LDH additives can improve thermal stability (TGA) and PHRR in cone calorimetry measurements, even without nanocomposite formation or good dispersion. Further work in necessary to: characterize LDH and HDS morphologies, evaluate dispersion, analyze chars in more detail and characterize the evolution of the metal species during thermal/fire degradation, identify the source of improvements in thermal/fire properties (i.e. elaborate on the whether the physical or chemical properties of the additives are key factors in determining PHRR reduction), and explore additive combinations.

## Acknowledgments

This work was performed under the sponsorship of the US Department of Commerce, National Institute of Standards and Technology, Grant 60 NANB6D6018.

## References

1. See for example, (a) Alexandre, M.; Dubois, P., *Mater. Sci. Eng.* **2000**, *R28*. (b) Tyan, H.; Leu, C.; Wei, K., *Chem. Mater.* **2001**, *13*, 222. (c) Yeh, J.-M.; Liou, S.-J.; Lai, C.-Y.; Wu, P.-C., *Chem. Mater.* **2001**, *13*, 1131. (d) Gilman, J.W.; Kashiwagi, T.; Nyden, M.; Brown, J.E.T.; Jackson C.L.; Lomakin, S.; Gianelis, E.P.: Manias, E, in *Chemistry and Technology of Polymer Additives*; Al-Malaika, S.; Golovoy, A.; Wilkie, C.A, Eds.; Blackwell Scientific, **1999,** 249. (e) Xie, W.; Gao, Z.; Pan, W.-P.; Hunter, D.; Singh, A.; Vaia, R., *Chem. Mater.* **2001**, *13*, 2979.
2. Leroux, F.; Besse, J.-P. *Chem. Mater.,* **2001**, *13*, 3507.
3. Morioka, H.; Tagaya, H.; Kadokawa, J.-I.; Chiba, K. *Recent Res. Devel. In Mat. Sci.,* **1998**, *1*, 137-188 and extensive references therein.
4. (a) Chen, W.; Qu, B., *Chem. Mater.* **2003**, *47*, 3208. (b) Zammarano, M.; Franceschi, M. ; Bellayer, S. ; Gilman, J.W. ; Meriani, S., *Polymer* **2005**, *46*, 9314. (c) Shi, L.; Li, D.; Wang, J.; Li, S. ; Evans, D.G. ; Duan, X., *Clays Clay Mater.* **2005**, *53*, 294. (d) Jiao, C.M.; Wang, Z.Z.; Ye, Z.; Hy, Y.; Fan, W.C., *J. Fire Sci.* **2006**, *24*, 47. (e) Leroux, F., *J. Nanosci. Nanotech.* **2006**, *6*, 303. (f) Taviot-Gueho, C.; Leroux, F., *Structure Bond.* **2006**, *119*, 121. (g) Longchao, D.; Qu, B.; Zhang, M., *Polym. Degrad. Stab.* **2007**, *92*, 497. (h) Costa, F.R.; Wagenknecht, U.; Heinrich, G., *Polym. Degrad. Stab.* **2007**, *92*, 1813. (i) Costache, M.C.; Heidecker, M.J.; Manias, E.; Camino, G.; Frache, A.; Beyer, G.; Gupta, R.K.; Wilkie, C.A., *Polymer* **2007**, *48*, 6532. (j) Jiao, C.-M.; Wang, Z.-Z.; Chen, X.-L.; Hu, Y., *J. Appl. Polym. Sci.* **2008**, *107*, 2626.
5. (a) Zhang, G.; Ding, P.; Zhang, M.; Qu, B. *Polym. Degrad. Stab.* **2007**, *92*, 1715. (b) Ye, L.; Ding, P.; Zhang, M.; Qu, B., *J. Appl. Polym. Sci.* **2008**, *107*, 3694.
6. Wang, G. -A.; Wang, C. C.; Chen, C.Y. *Polymer* **2005**, *46*, 5065.
7. Nyambo, C.; Wang, D.; Wilkie, C.A. *Polym. Adv. Tech.*, in press. DOI 10.1002/pat1272.
8. Manzi-Nshuti, C.; Wang, D.; Hossenlopp, J.M.; Wilkie,C.A., *J. Mater. Chem.* **2008**, *18*, 3091.
9. Jash, P.; Wilkie,C.A., *Polym. Degrad. Stab.* **2005**, *88*, 401.
10. Nyambo, C.; Songtipya, P.; Manias, E.; Jimenez-Gasco, M.M.; Wilkie, C.A., *J. Mater. Chem.*, in press. DOI 10.1039/b806531d

11. Wang, L.; Su, S.; Wilkie, C.A. manuscript in preparation.
12. Manzi-Nshuti, C.; Wang, D.; Hossenlopp, J.M.; Wilkie, C.A. submitted.
13. Zhu, L.; Su, S.; Hossenlopp, J.M., manuscript in preparation.
14. E. Kandare, E.; Hossenlopp, J.M. "Effects of Hydroxy Double Salts and Related Nanodimensional Layered Metal Hydroxides on Polymer Thermal Stability," ACS Symposium Series, Polymer Degradation, Optimization and Performance, in press.
15. C. Manzi-Nshuti, C.; Hossenlopp, J. M.; Wilkie, C. A. *Polym. Degrad. Stab.*, in press. DOI: 10.1016/j.polymdegradstab.2008.07.005.
16. Nyambo, C.; Kandare, E.; Wang, D.; Wilkie, C.A. *Polym. Degrad. Stab.*, in press. DOI: 10.1016/j.polymdegradstab. 2008.05.029

## Chapter 5

# Effect of Flame Retardants on the Thermal, Burning, and Char Formation Behaviour of Polypropylene–Nanoclay Compounded Polymers

**Baljinder K. Kandola[1], A. Yenilmez[1], Richard A. Horrocks[1], G. Smart[1], W. Kun[2], and Yuan Hu[2]**

**[1]Centre for Materials Research and Innovation, University of Bolton, Bolton BL3 5AB, United Kingdom**
**[2]State Key Laboratory of Fire Science, University of Science and Technology of China, 96 Jinzhai Road, Hefei, Anhui 230026, People's Republic of China**

The effect of different flame retardant types on the thermal stability, flammability and char formation tendency of polypropylene (PP) nano/micro composite is studied. PP, compatabiliser, nanoclay, UV-stabiliser and different flame retardants have been compounded in a twin screw extruder to produce polymers with improved thermal and flame retardant properties. Thermal analysis has been used to study the thermal properties, and limiting oxygen index (LOI) and slightly modified UL-94 test for flammability of the samples. All flame retardants acting in the condensed phase (phosphorus- and nitrogen- containing) lowered the rate of decomposition, whereas halogenated flame retardants had a little effect. The addition of nanoclay with or without flame retardants increased the thermal stability of all samples and helped in char formation. All samples with flame retardants and no clay, burnt completely, which is not unexpected, given the low levels (5%) of flame retardants used here. However, the flame spread was low. On addition of clay to the compounded polymer, a change in burning behaviour was observed, flame spread was reduced and samples self-extinguished, except for the one containing melamine phosphate. A tube furnace was used for char formation at different temperatures and the charred structures have been examined with digital images, optical and scanning electron microscopy, and FTIR. This information has been used to understand the mechanisms of thermal degradation of different flame retarded PP - nano/micro composite samples.

## Introduction

Polypropylene or polypropene (PP) is a useful commodity polymer mainly used in apparel, upholstery, floor coverings, medical, geotextiles and automotive applications, due to its low cost, light weight, good mechanical properties and low reactivity towards other chemicals. The main advantage is that PP being an addition polymer made from the monomer propylene, is unusually resistant to many chemical solvents, acids and bases. However, being wholly aliphatic hydrocarbon structure, burns very rapidly with a relatively smoke-free flame and without leaving any char residue *(1)*. Lack of polar groups in the structure makes it difficult to react with reactive flame retardants. Additive flame retardants if used, are required in large amounts (>20% w/w) to provide the required fire protection to products *(1)*. However, such high levels of additives cause polymer processing problems, in particular for their extrusion into thin films or fibres. The recent developments in the area of polymer – nanocomposites have suggested that by the addition of just a small quantity (<5%) of organically modified layered silicate nanoclay (montmorillonite) to a polymer matrix could enhance many of the properties of that polymer, including the fire performance *(2)*. In polypropylene the lack of polar groups in the polymer chain makes direct intercalation or exfoliation of the nanoclays almost impossible *(3)* without the use of a compatibiliser *(4-7)*. Maleic anhydride - grafted polypropylene (PP-g-MA) can be used as a compatabiliser, which enhances the interaction between the clay and polymer with strong hydrogen bonding between -OH or -COOH and the oxygen groups of clay *(4-7)*. PP-g-MA or PP containing polar monomers also enhance the char formation during combustion of PP *(8,9)*, which may be due to improvement in nanoclay dispersion. The nanocomposites thus formed exhibit higher thermal stability and lower flammability than virgin PP *(10,11)*. Polymer - nano/micro composites usually are not flame retardant enough to pass stringent flammability tests and still require some amount of conventional flame retardants. When nanoclays and flame retardants are concurrently interspersed into the polymer, their interactions may be additive, synergistic or antagonistic *(12)*. It must be noted, however that with this approach of using flame retardants as additives, the polymer content in the formulation is reduced compared to the unmodified polymer, hence less combustible material is present in the former. The present study is the part of a larger project exploring the production of fire retardant synthetic nano/micro composite fibres. In our earlier publications we have demonstrated that nanoclays can be nanodispersed in polypropylene with proper choice of compatibilser, and the compounded polymer can be extruded into fibres *(4,6)*. Nanoclays, although increasing the thermal stability of polypropylene and helping in char formation, do not reduce the flammability of PP fibres to a large extent *(4)*. Nanoclay and a small amount of flame retardant (5%), when added together to PP containing certain compatibilsers, the extruded fibres could self extinquish. In our previous work we have only used ammonium polyphosphate *(6),* whereas in this study different phosphorus- , nitrogen- and halogen containing flame retardants are studied. The main aim of this work is to understand the mechanism of combined the action of nanoclay and different types of flame retardants on thermal stability and flammability of PP. A number of polypropylene samples containing compatibiliser, nanoclay, UV-stabiliser and different flame retardants have been compounded in a twin screw extruder and their thermal and fire retardant behaviour studied using different techniques. The chars formed during combustion have been characterised to understand the mechanism of flame retardancy of different systems.

## Experimental

### Materials

Following materials were obtained from commercial sources:
*Polypropylene* (PP): Moplen HP516R, Basell Polyelefins.
*Grafted PP* (PB): Polybond 3200, maleic anhydride grafted PP (PP-g-MA) with a maleic anhydride level 1% (w/w), Crompton Corp.
*UV- Stabiliser* (NOR): Flamstab NOR 116FF, Ciba SC.
*Nanoclay* (E): Bentone Elementis E-107; montmorillonite modified with dimethyl, dehydrogenated tallow quaternary ammonium ion.

*Flame retardants (FR)*:
APP : Ammonium polyphosphate, Albemarle Corporation.
NH: Melamine phosphate, Rhodia Specialities Ltd.
NH 1197 : Pentaerythritol phosphate, NH1197, Great Lakes.
FR 245 : Tris(tribromophenyl) cyanurate, FR 245, DSBG, Israel.
FR 372 : Tris(tribromoneopentyl) phosphate, FR 372, DSBG, Israel.

### Polymer Preparation

The polypropylene and additives were hand-mixed in a plastic container prior to compounding. In PP, the level of PP-g-MA (PB) has been kept 1 wt % ; clay 3 wt %; UV stabiliser 1 wt %; and FR 5 wt % (see Table 1). Sample details are given in Table 2. A Thermoelectron Prism Eurolab 16 twin screw extruder with a temperature profile over six heating zones between 179–190°C was used for compounding. Polymer was pelletised after cooling in the water bath. For LOI and burning tests, coarse mono filaments (strands) of compounded polymers (diameter 1.8 ± 0.2 mm) were collected before pelletising.

**Table I. Mass percentages of various components in the formulations**

| *Sample* | *PP (%)* | *NOR (%)* | *PB (%)* | *E (%)* | *FR* (%)* |
|---|---|---|---|---|---|
| PP-NOR | 99.0 | 1.0 | - | - | - |
| PP-NOR-PB-E | 95.2 | 0.9 | 0.9 | 2.9 | |
| PP-NOR-FR* | 94.3 | 0.9 | - | - | 4.7 |
| PP-NOR-PB-E-FR* | 90.9 | 0.9 | 0.9 | 2.7 | 4.5 |

NOTE : * FR = APP, NH, NH 1197, FR245 or FR 372.

## Characterisation and Testing

X-ray diffraction (XRD) studies were carried out on compounded polymer samples using a Bruker D8 Advance X-ray Diffractometer with CuKα radiation (40kV, 40mA) monochromatised with a graphite sample monochromator over the 2θ range 2-60° with a step size of 0.02° and a count time of 12.2 s per step.

Transmisssion electron microscopy (TEM) was carried out using a Hitachi H-800 microscope at an acceleration voltage of 100 kV. High resolution electron microscopic (HREM) images were obtained with a transmission electron microscope (JEOL – 2010, Japan) at an acceleration voltage of 100 kV.

Thermogravimetric analysis and differential thermal analysis (DTA) were performed on a TA instruments SDT 2960 simultaneous DTA–TGA instrument under flowing air. Limiting oxygen index (LOI) tests were conducted using a standard procedure *(12)*. Flame spread tests were done by a modified UL-94 test on strands (length = 110 mm, diameter = 1.8 ± 0.2 mm) to observe their burning behaviour in both horizontal and vertical orientations. The first 10 mm of sample burning was not taken into account and so times of burning were recorded once the flame had reached a line drawn 10mm from the edge against which flame of 30mm was applied for 10s as specified in the test. A video film was taken of the burning of each sample from which times to reach 50 ($t_1$) and 100 mm ($t_2$) marks and/or to achieve flameout were noted. Three replicates of each sample were burnt and results averaged. The burning behaviour of each sample was observed and noted.

A Carbolite tube furnace was used to prepare chars of polymers at different temperatures (200 - 500°C) under isothermal heating conditions for 5 minutes.

The chars were examined for their surface characteristics and morphology by scanning electron microscopy using a Cambridge Stereoscan 200 SEM. A Nicolet, Magna-IR Spectrometer 550 was used to study any structural changes in the chars, using KBr discs containing 2% (w/w) char.

## Results and Discussion

### X-ray Diffraction Studies

The XRD curves for all compounded samples, which contain E-107 clay, compatibiliser PB and flame retardants are shown in Figure 1.

Although the XRD study was carried out in the range $2\theta = 2 - 60^{\circ}$, only the regions between $2\theta = 2 - 8^{\circ}$ are analysed in detail and shown in Figure 1 and the intergallery d-spacings are given in Table 2. For the clay $d_{001}$ value is 2.53nm (Table 2). For sample PP-NOR-PB-E, no peak could be observed, which could indicate exfoliation of the clay, although the TEM results discussed in a later section show that the nanoclays are neither exfoliated nor intercalated, but only nanodispersed. In all flame retardant containing samples the clay peak could be observed, but at slightly lower $2\theta$ angle (see Figure 1), hence higher d- spacings (Table 2). FR245 containing sample has a greater d-spacing (3.46 nm), followed by NH and NH1197 (3.32, 3.31 nm), APP (3.23nm) and FR372 (3.20nm). The differences are due to effect of different flame retardants on dispersion of the clay as seen by the microscopic studies.

### Transmission and High Resolution Electron Microscopies

The transmission electron micrographs of different samples in Figure 2 show the effect of two types of flame retardants on clay dispersion. Careful observation of enlarged sections of the more highly resolved TEM micrographs, in spite of their rather poor definition, shows that in Figure 2(a) larger particles are appear to be the flame retardant NH1197, whereas, the smaller and more structured particles appear to be clay platelets that have started to delaminate and separate, thereby indicating that delamination is taking place. This delamination is also apparent in higher resolution electron microscopic images of this sample in Figure 3. The dispersion appears to be much better for the sample PP-NOR-PB-E-FR372 (Figure 2(b), indicating that the flame retardant FR372 is much better dispersed as is the nanoclay. In general it can be concluded that although structures observed here are not obviously intercalated or exfoliated microcomposites, dispersion is indeed at the nanolevel, i.e., particle size $\ll 1\mu m$. The clay particle widths appear to be $< 0.1\mu m$ with lengths up to $0.5\mu m$.

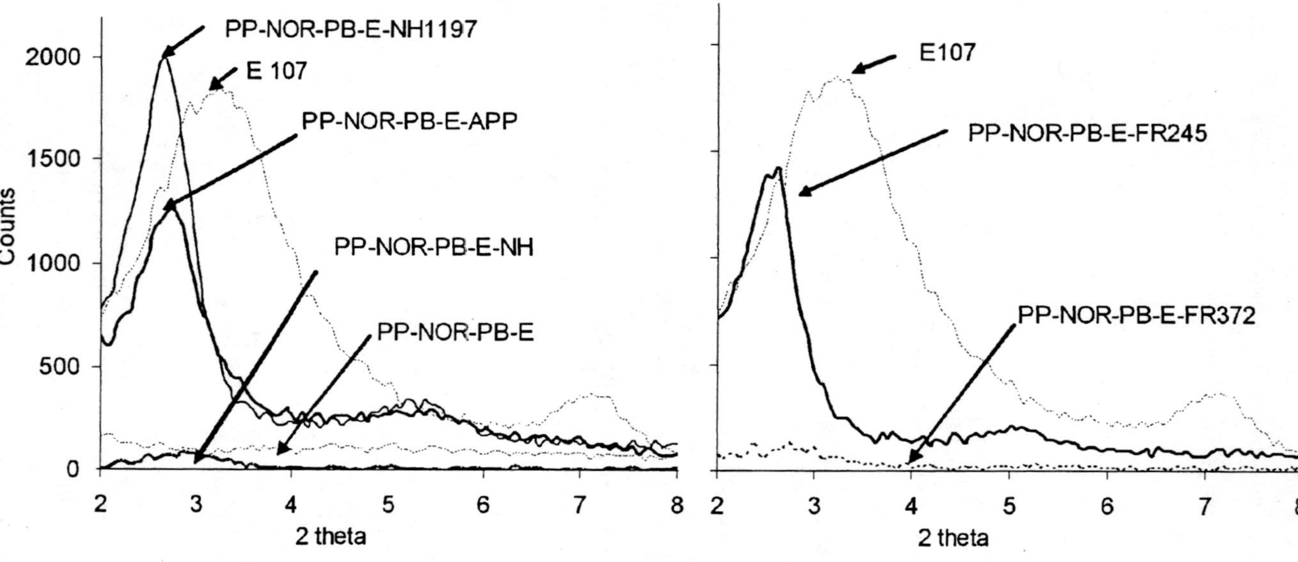


*Figure 1. XRD curves for compounded clay-containing polymer samples showing the effect of clay and flame retardants on dispersion.*

**Table 2. XRD and thermal analytical results of compounded polymers**

| *Sample* | *XRD results, d - spacing (nm)* | *DTA melt-ing peak max. (°C)* | *Decomposition temp from DTG curves (°C)* | | *Char yield from TGA curves (%) at* | | | *Char yield from Isothermal heating in furnace, (%) at* | |
|---|---|---|---|---|---|---|---|---|---|
| | | | *Onset* | *Maxima* | *300 °C* | *400 °C* | *500 °C* | *300 °C* | *350 °C* |
| PP- NOR | | 166 | 222 | 318 | 58 | 3 | 1 | 95 | 77 |
| PP-NOR-PB-E | -, - | 165 | 217 | 379 | 75 | 4 | 3 | 96 | 85 |
| PP- NOR-APP | | 165 | 243 | 299; 344 | 79 | 19 | 11 | 99 | 77 |
| PP-NOR-NH | | 166 | 229 | 312; 349 | 82 | 15 | 8 | 99 | 81 |
| PP-NOR-1197 | | 166 | 247 | 336 | 75 | 7 | 5 | 99 | 76 |
| PP-NOR-FR245 | | 165 | 239 | 318; 334 | 66 | 3 | 1 | 98 | 75 |
| PP-NOR-FR372 | | 166 | 224 | 292; 311 | 58 | 6 | 3 | 98 | 76 |
| PP-NOR-PB-E-APP | 3.23, 1.62 | 168 | 225 | 256; 363 | 86 | 9 | 6 | 99 | 86 |
| PP-NOR-PB-E-NH | 3.32 | 168 | 239 | 304; 345 | 78 | 10 | 6 | 97 | 79 |
| PP-NOR-PB-E-1197 | 3.31, 1.61 | 168 | 229 | 381 | 83 | 6 | 4 | 98 | 93 |
| PP-NOR- PB-E-FR245 | 3.46, 1.72 | 167 | 215 | 376; 390 | 82 | 6 | 3 | 97 | 89 |
| PP-NOR-PB-E-FR372 | 3.20 | 166 | 223 | 300; 365 | 85 | 4 | 3 | 97 | 85 |

NOTE : d-spacing for clay E 107 = 2.53 nm

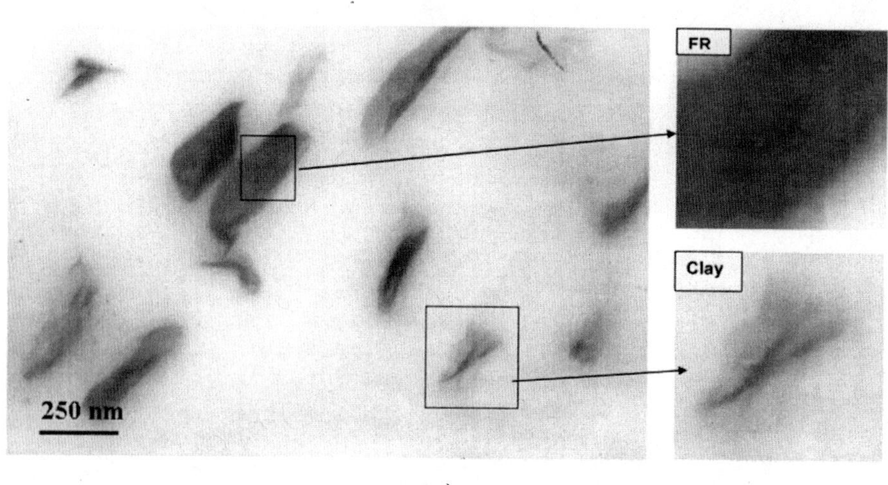


a)

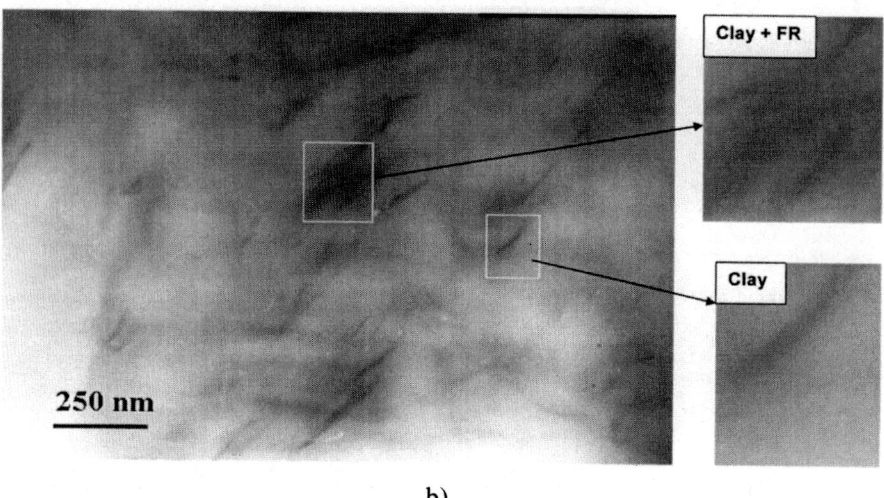


b)

*Figure 2. TEM of a)PP-NOR-PB-E-NH1197 and b) PP-NOR-PB-E-FR372 with enlarged section of higher resolution micrograph.*

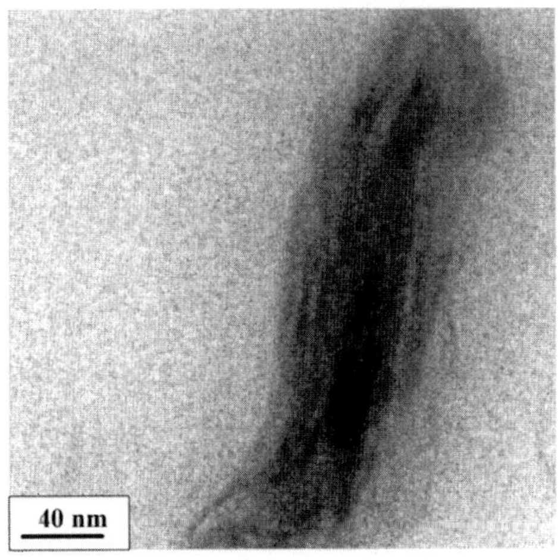


*Figure 3. HREM of sample PP-NOR-PB-E-NH1197.*

## Thermal Analysis

Table 2 lists the fusion minima temperature or the melting temperatures of all the polymers obtained from DTA curves. It can be seen that addition of clay, graft and all flame retardants has a minimal effect on the melting point of the polypropylene. In Figure 4 the TGA curves of selective samples are given and analyses of all TGA responses of all samples are given in Table 2. The onset of decomposition temperature (temperature where the mass loss starts, measured from DTG curves (not shown here)) for PP containing UV-stabiliser is 222°C, which has been slightly lowered by the presence of clay (217 °C). For flame retardant - containing samples (PP-NOR-FR), values varied between 224 – 247°C depending upon the flame retardant type. Presence of clay in PP-NOR-PB-E-FR samples reduced the onset of decomposition temperature in each sample compared to the respective PP-NOR-FR samples, the only exception are the samples containing melamine phosphate. This observation is consistent with other studies in the literature where it has been demonstrated that clays initiate decomposition and lower the ignition time of different polymers *(9)*. All flame retardants acting in the condensed phase (phosphorus- and nitrogen - containing) lowered the rate of decomposition as can be seen from Figure 4(a), higher decomposition temperature values (second DTG maxima in Table 2) and increased char residue values as reported in Table 2.

Halogenated flame retardants FR 245 and 372 had little effect on the rate of decomposition (Figure 4(b) and Table 2).The addition of nanoclay with or without flame retardants increased the thermal stability of all samples as indicated by respective DTG maximum temperature values in Table 2. Nanoclay

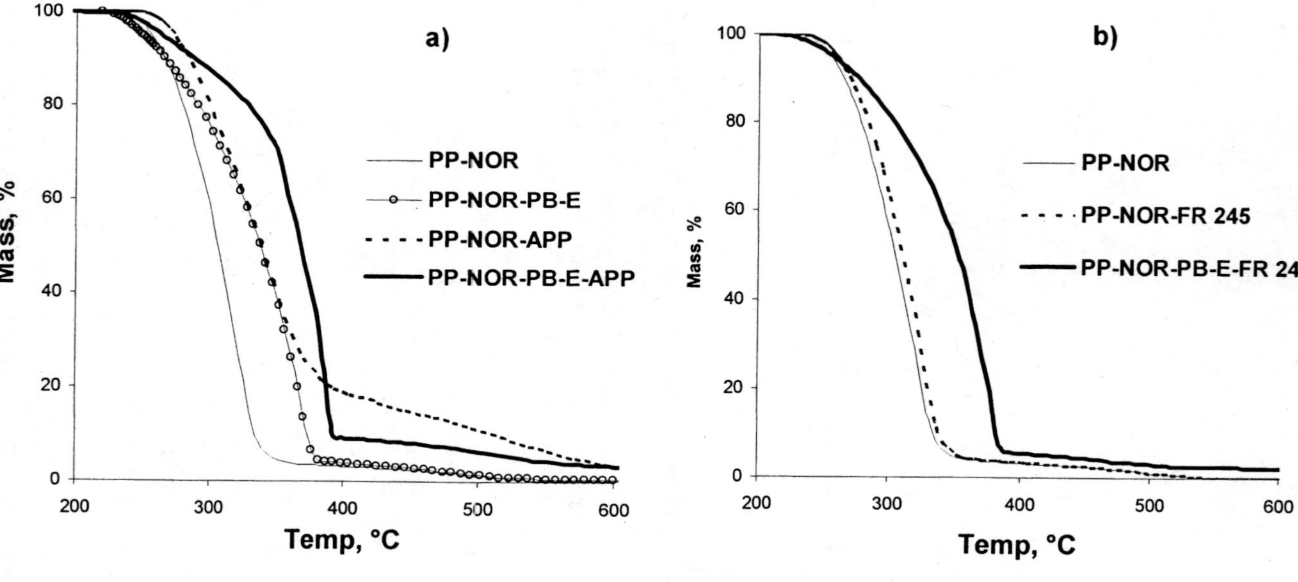


*Figure4. (a) TGA responses of PP-NOR, PP-NOR-PB-E, PP-NOR-FR, PP-NOP-PB-E-FR, where FR is (a) APP and (b) FR 245.*

together with melamine phosphate had no further effect on thermal stability of the flame retarded sample (see Table 2), showing the incompatibility of melamine phosphate with the organically modified clay. This has also been observed in our previous studies for nylon *(12)*. Thus addition of a nanoclay together with conventional flame retardants enhances both the overall thermal stability and char formation in otherwise non-char - forming PP polymer.

## Charaterization of the Char Residue

All samples were heated in a carbolite furnace at different temperatures from 250 – 500°C, which were then studied by scanning electron microscopy and FTIR. Below 300°C, the degraded residues were still largely polymeric and not charred enough to make KBr discs. At 375 °C, control sample completely volatilized, and in others char left was not enough to carry out all the studies. Hence, only results for 300 and 350 °C – heated samples are discussed here. The percentage of residues left for all samples are given in Table 2 and the digital images of residues at 350 °C of selected samples are shown in Figure 5. The presence of maleic-annhydride grafted polypropylene, PB and clay in PP (sample PP-NOR-PB-E) has encouraged char formation (see Figure 5(d)) whereas no char could be seen for the control sample, PP-NOR, which when still a molten polymer, starts volatilizing after depolymerization. PP-g-MA not only improves the clay dispersion in PP, but also enhances the char formation. Gilman *(9)* has shown that by adding 15% MA-grafted PP with 2 or 5% organically modified clay, the char formation in PP is increased considerably, whereas no char could be observed for samples containing clays only. Similarly others *(13)* have used about 7.5% MA-grafted PP and 5% clay to show increase in char formation. However, in this work we have only used 1% PB and 3% clay and the results are promising. PP-g-MA acts as compatibilser and helps in improving dispersion of the clay in polymer *(4-6)*. The stability of the dispersion during combustion prevents aggregation or phase separation of the nanoparticles, thus promoting continuous coverage of surface by residue *(14)*. This reduces the rate of gas escape from polymer melt affecting the viscosity and hence, prevents the melt dripping of the polymer during vertical UL-94 testing and holds the polymer together *(15)*. As can be seen from Figure 5 (b) APP in PP helps in some char formation, but in presence of PB and clay, the char formation is considerabley enhanced (Figure 5(e)). Even with the halogenated sample, which is a non char-former, the clay has helped in char formation (sample PP-NOR-PB-E-FR372, Figure 5(f)).

Scanning electron microscopic images of these heated samples of polymers are shown in Figure 6 and corroborate the analyses of images in Figure 5. For example, the PP control sample in Figure 6(a) shows the absence of any charred

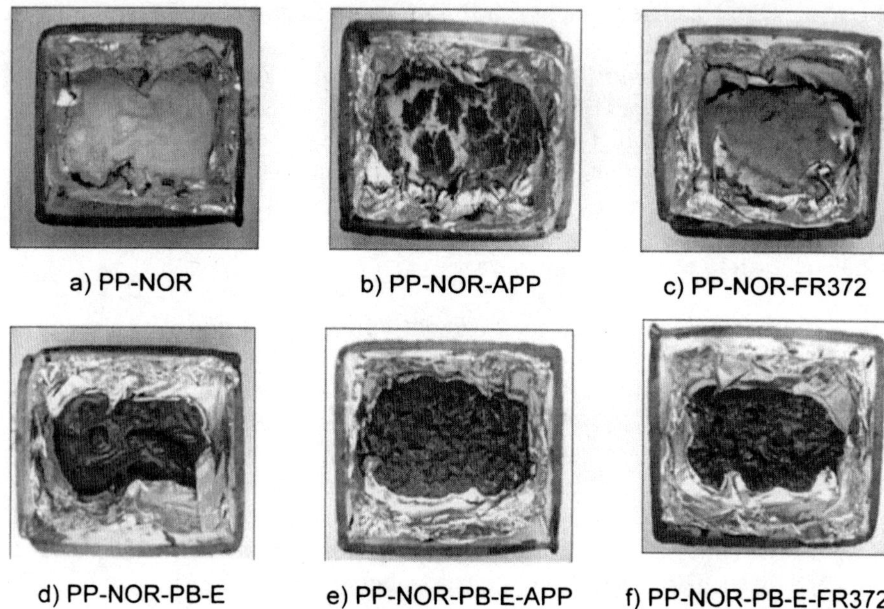


*Figure 5. Digital images of charred samples of polymers obtained by heating the polymer in a tube furnace at 350 °C for 5 minutes.*

structure. The nanoclay presence (Figure 6(b)) changes the morphology of the polymer where it is seen to be more textured. Obvious char formation is not apparent until both nanoclay and flame retardant are present together (Figures 6 (c) – (e)) and respective char structures appear to differ albeit in an ill-defined way.

All these chars were also studied by FTIR and spectra of selected sample heated at 350 °C are shown here in Figure 7. The PP heated at 300 and 350 °C showed the characteristic peaks at 1455 and 1355 $cm^{-1}$ due to C-H bending and the broad peak between 2800 and 3000 $cm^{-1}$ due to C-H stretching vibration (see Figure 7). The peaks at 1633 and 1719 $cm^{-1}$ correspond to the absorption of C=C and C=O stretching. In case of PP-NOR-PB-E sample, prominent peaks at 1035 and 1085 $cm^{-1}$ due to Si-O stretching vibration of montmorillonite clay could be observed. Similar peaks were also observed in a recent work by Qin et al *(11)*. These peaks were more prominent in all flame retarded samples containing nanoclays. For flame retarded PP samples, no new peaks due to the flame retardant chemicals could be observed, which may be due to the low percentages present or the peaks (eg, for PP-NOR-APP sample, 1190 $cm^{-1}$ corresponding to

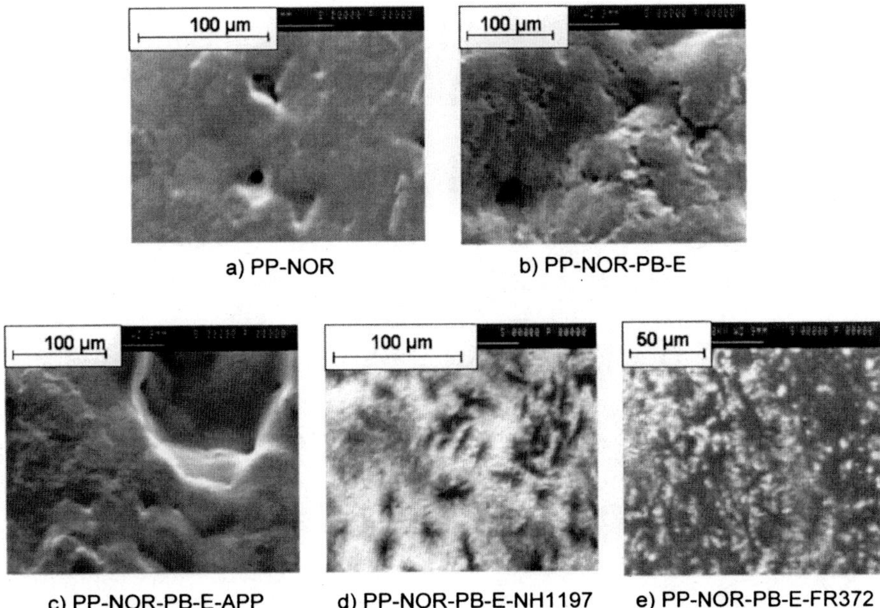


a) PP-NOR b) PP-NOR-PB-E

c) PP-NOR-PB-E-APP d) PP-NOR-PB-E-NH1197 e) PP-NOR-PB-E-FR372

*Figure 6. SEM images of charred samples of polymers obtained by heating the polymer in a tube furnace at 350 °C for 5 minutes*

P-O-C stretching , 1250 and 1350 $cm^{-1}$ due to P=O stretching) are superimposed by other peaks in that region. Similarly no new peak in the respective samples containing nanoclay was observed, which could have indicated possible chemical interaction between the flame retardant and the decomposing PP polymer.

## Limiting Oxygen Index

Limiting oxygen index (LOI) values for various samples are listed in Table 3, where it can be seen that nanoclay presence reduces them compared to either control or respective flame retardant containing samples. This behaviour is not unexpected as nanoclay presence changes the burning behaviour of polymer by reducing its thermoplasticity *(12)*. In presence of clay, the polymer does not melt and so cannot lose energy via melt dripping. The polymer thus burns more easily and consistently as is also evident from burning test results given in Table 3.

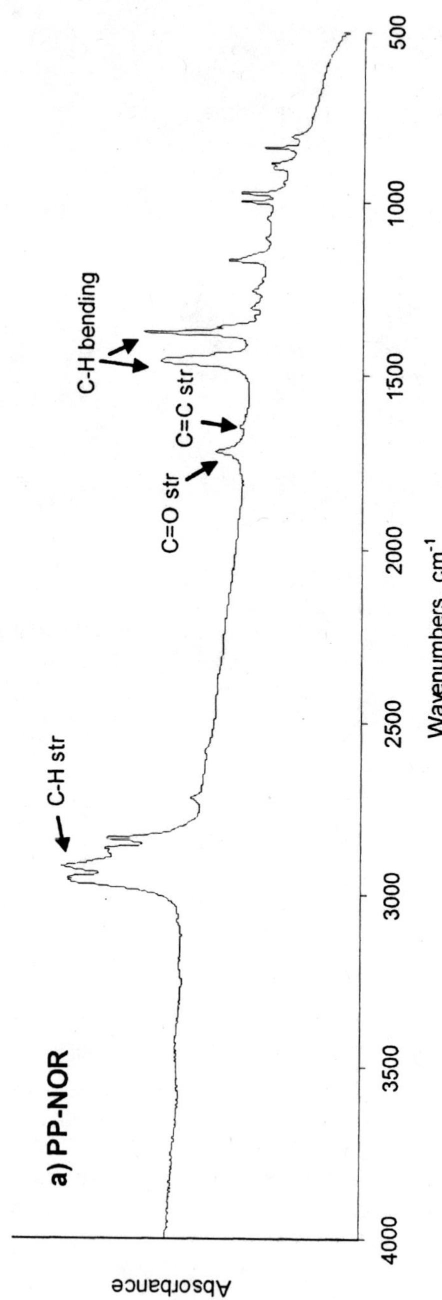
a) PP-NOR
C-H str
C=O str
C=C str
C-H bending
Absorbance
4000
3500
3000
2500
2000
1500
1000
500
Wavenumbers, cm-1

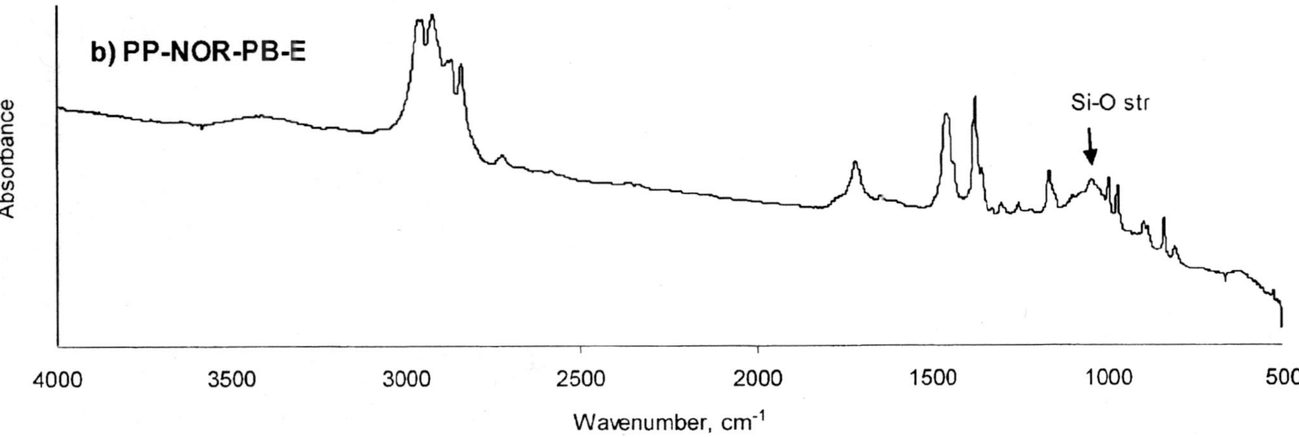


*Figure 7. FTIR spectra of charred samples of polymers obtained by heating the polymer in a tube furnace at 350 °C for 5 minutes.* *Continued on next page.*

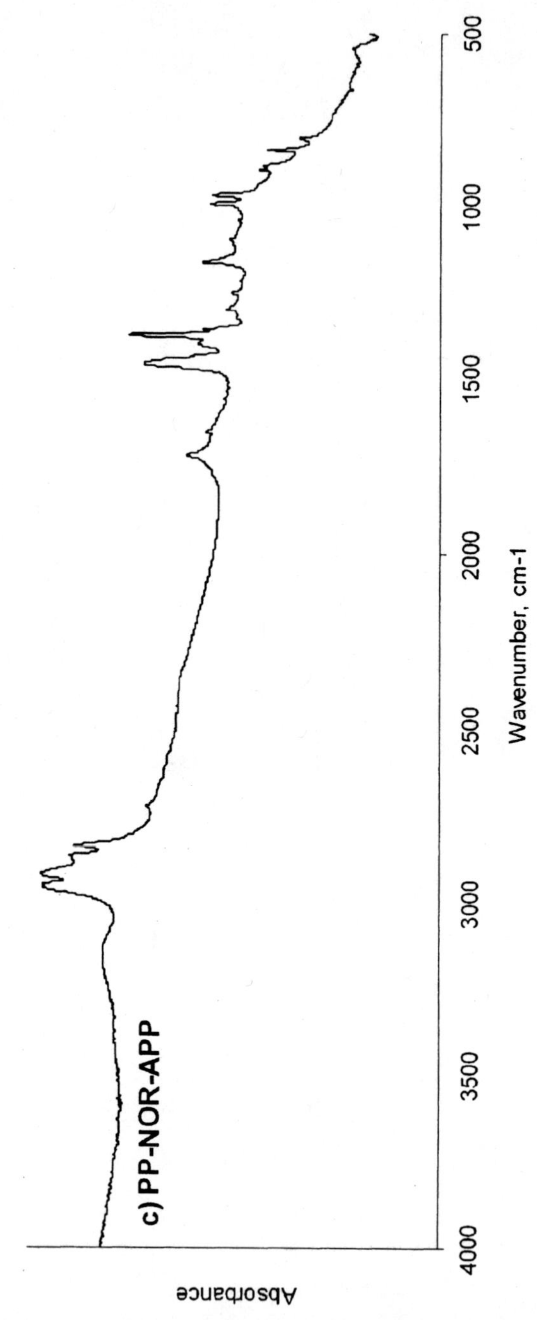
c) PP-NOR-APP
Absorbance
4000
3500
3000
2500
2000
1500
1000
500
Wavenumber, cm-1

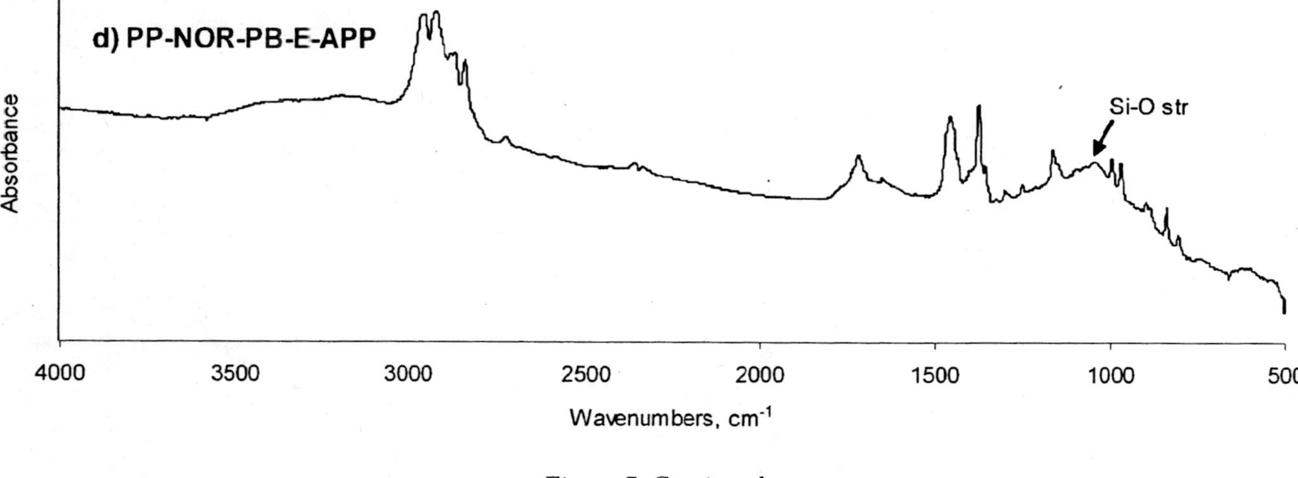


*Figure 7. Continued.*

## Flame Spread

Flammability of the compounded polymer strands was studied using a UL-94 test rig, igniting the sample and recording times taken to reach 50 ($T_1$) and 100 mm ($T_2$) marks and to extinguish (flame out, FO) in both horizontal and vertical modes. Selected images are shown in Figures 8 and 9. In the horizontal mode PP sample melted and burned with flaming drips. All samples with flame retardants and no clay, burnt completely, which is not unexpected, given low levels (5%) of flame retardants used here. However, the flame spread was low as seen from Table 3. On addition of clay to the compounded polymer, a change in burning behaviour was observed, samples self-extinguished, except for the one containing melamine phosphate (Table 3 and Figure 8). In some samples (eg containing APP) the flame flickered, which was probably was due to poor dispersion of the flame retardant in the polymer *(6)*.

In the vertical mode burning of the control sample was more vigorous with melting drips (see Table 3 and Figure 9). With flame retardants there was not much effect on burning, except for the sample containing FR372 which self exinquished. The clay presence in PP reduced melt-dripping, but had minimal effect on burning, however when present with NH 1197 or FR 372, the effect was promising and the samples self-extinquished (Table 3 and Figure 9).

## Flame Retardant Mechanism

Polypropylene undergoes decomposition via random chain scission involving formation of some unsaturated end groups, eg $-C(CH_3)=CH_2$. Accompanying oxidation via hydroperoxide formation yields further scission products comprising alcohols, ketones and acidic functionalities. There is a little evidence of any tendency to cross-link and form char during this process, thereby presenting a significant flame retardant challenge.

It can be seen from the discussion in previous sections that nanoclays act as char promoter in PP containing PP-g-MA. It has been reported in the literature *(11,16)* that thermal degradation of organoclays can create acidic sites on the clay, which catalyse the initial decomposition stages of a polymer, resulting in a lowering of ignition time and LOI values for many polymer-nanocomposite samples. The acidic sites however, catalyse formation of a protective surface coating on the surface of polymer, which acts as a thermal barrier leading to delay of thermo-oxidative degradation. In some cases clays can also catalyse dehydrogenation and crosslinking of polymer chains, thus reducing combustion.

Phosphorus- and nitrogen- containing flame retardants acting in condensed phase produce char, which acts as a thermal barrier for the decomposing polypropylene. There is no evidence of chemical interaction between the flame retardants and PP in our work or from the literature. Halogenated flame

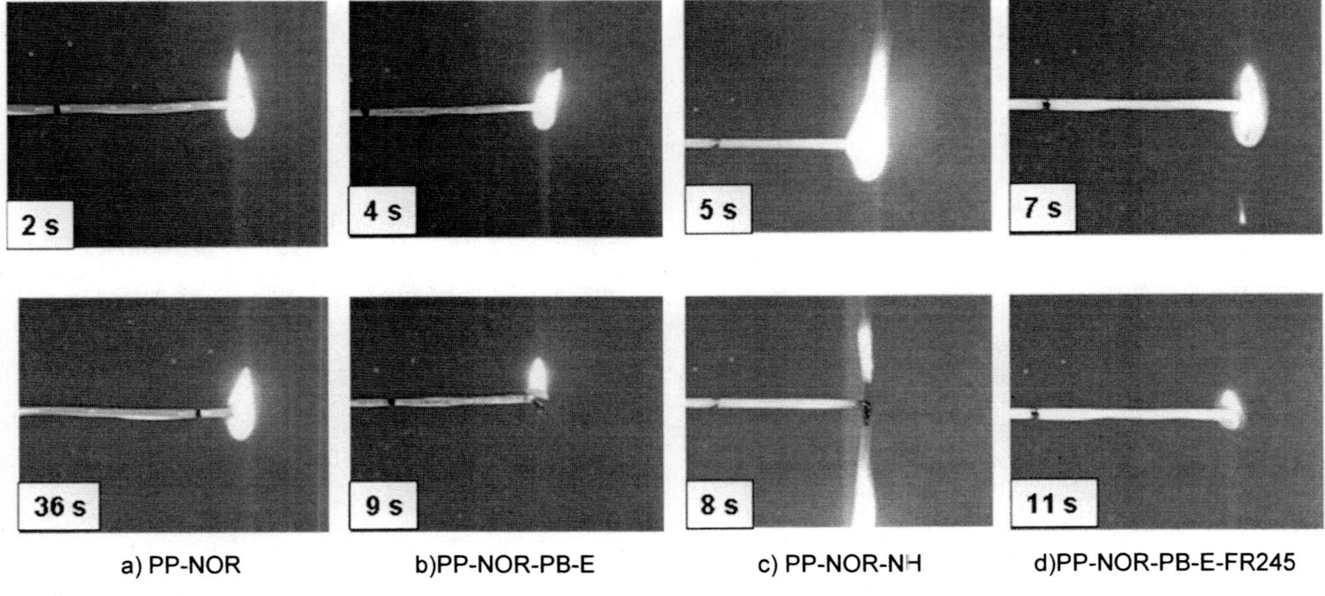


*Figure 8. UL-94, horizontal burning of strands.*

**Table 3. LOI and flame spread results of compounded polymers**

| *Sample* | *LOI (%)* | *Horizontal burn* | | | *Vertical burn* | | |
|---|---|---|---|---|---|---|---|
| | | *$T_1$ (s)* | *$T_2$ (s)* | *FO (s)* | *$T_1$ (s)* | *$T_2$ (s)* | *FO (s)* |
| PP- NOR | 18.5 | 41 | 82 | (M) | 22 | 42 | (BD) |
| PP-NOR-Pb-E | 17.5 | - | - | 10 (SE) | 15 | 30 | * |
| PP- NOR-APP | 19.0 | 33 | 71 | * | - | - | 5 (BD, SE) |
| PP-NOR-NH | 19.9 | - | - | 8 (BD) | 18 | 36 | * |
| PP-NOR-1197 | 18.6 | 45 | 89 | * | 23 | 44 | * |
| PP-NOR-FR245 | 18.9 | 29 | 57 | * | 17 | 38 | (BD,*) |
| PP-NOR-FR372 | 21.9 | - | - | 5 (BD, SE) | 24 | - | 30 (BD, SE) |
| PP-NOR-Pb-E-APP | 17.8 | - | - | 32 (SE) | 17 | 33 | * |
| PP-NOR-Pb-E-NH | 17.9 | 27 | 51 | * | 14 | 26 | * |
| PP-NOR-Pb-E-1197 | 17.8 | - | - | 10 (SE) | - | - | 7 (SE) |
| PP-NOR- Pb-E-FR245 | 17.2 | - | - | 12 LBD, SE) | 20 | 40 | * |
| PP-NOR-Pb-E-FR372 | 18.7 | - | - | 11 (BD, SE) | - | - | (BD, SE) |

NOTE : $T_1$ = time to 50mm; $T_2$ = time to 100mm; FO = flame out time ; M = melting ; BD = burning drips ; LBD = low burning drips SE = self extinguished ; * = burnt completely.

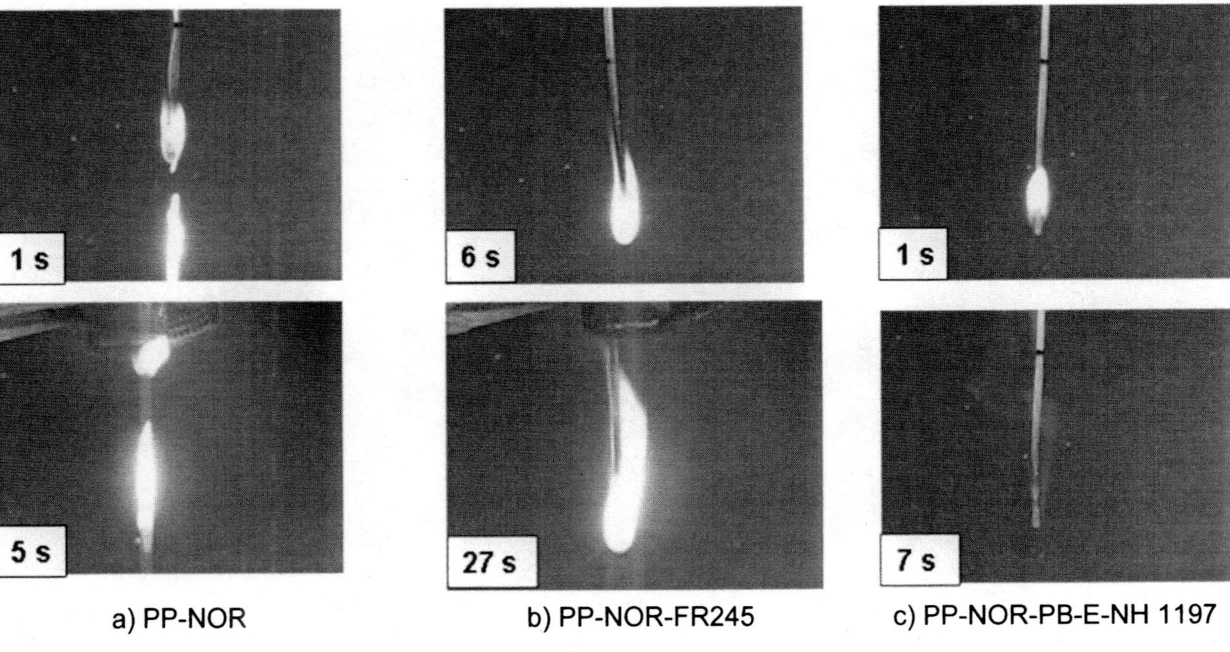


*Figure 9. UL-94, vertical burning of strands*

retardants acting in vapour phase help in removing active free radicals from the combustion zone, thus reducing combustion of the polymer. In PP-PB-clay-FR samples, char formation has been generally enhanced for both phosphorus- and halogen-containing flame retardants, with the former yielding greater increases. Hu et al *(13)* has suggested that in case of phosphorus – containing flame retardants, phosphoric acid released can react with quaternary ammonium salt of the clay, thus creating acidic sites on the clay as observed by Qin et al *(11,16)*, which enhance char formation. Similarly HBr released from halogenated flame retardants can react with the quaternary ammonium salt of the clay *(13)*. Although supporting this mechanism, we have not found any evidence of these reactions in this work.

## Conclusions

The dispersion of nanoclay can be improved by compounding polymer-clay samples and by adding compatibilser. LOI values are not changed significantly following addition of clay and even in presence of other flame retardants. However, the presence of nanoclay alone changes the thermal stability and burning behaviour of the polypropylene. With additional flame retardant presence, the polymer can show self extinguishing properties.

## Acknowledgements

The authors wish to thank the Engineering and Physical Sciences Research Council and the Ministry of Defence, UK, for funding; and Rhodia Consumers Specialities Ltd and Camira Ltd., UK for their collaboration and advice.

## References

1. Zhang, S.; Horrocks, A.R. *Prog. Polym. Sci*, **2003**, *28*, 1517.
2. Zhu, J.; Uhl, F.M.; Morgan, A.B.; Wilkie, C.A. *Chem. Mater,* **2001**, *13(12)*, 4649.
3. Százdi, L.; Pukánszky Jr. B.;J.; Vansco, G.J.; Pukánszky, B. *Polymer,* **2006**, *47(13)*, 4638.
4. Horrocks, A.R.; Kandola, B.K.; Smart, G.; Zhang, S.; Hull, T.R. *J Appl. Polym. Sci.*, **2007**, *106 (3)*, 1707.
5. Kandola, B.K.; Smart, G. Horrocks, A.R.; Joseph, P.; Zhang, S.; Hull, T.R.; Ebdon, P.; Hunt, B.; Cook, A. *J Appl. Polym. Sci.*, **2008**, *108 (2)*, 816.

6. Smart, G.; Kandola, B.K.; Horrocks, A.R.; Marney, D. *Polym. Adv. Techn.*, **2008**, *19*, 658.
7. Nam, P.H.; Maiti, P.; Okamoto, M.; Kotaka, T.; Hasegawa, N; Usuki, A. *Polymer,* **2001**, *42(23)*, 9633.
8. Manias, E.; Touny, A.; Wu, L.; Strawhecker, K.; Lu, B.; Chung, T.C. *Chem. Mater,* **2001**, *13(10)*, 3516.
9. Gilman, J.W. Chapter 3 in *Flame Retardant Polymer Nanocomposites,* Morgan, A.B.; Wilkie, C.A., Eds.; Wiley-Interscience: New Jersey, 2007.
10. Qin, H.L.; Zhang, S.M.; Zhao, C.G.; Feng, M.; Yang, M.S.; Shu Z.H. *Polym. Degrad. Stab.,* **2004**, *85*, 807.
11. Qin, H.L.; Zhang, S.M.; Zhao, C.G.; Hu, G.; Yang, M.S. *Polymer,* **2005**, *46*, 8386.
12. Horrocks, A.R.; Kandola, B.K.; Padbury S. *J Text. Inst.*, **2003**, *94 (3)*, 46.
13. Hu, Y.; Song, L. Chapter 8 in *Flame Retardant Polymer Nanocomposites,* Morgan, A.B.; Wilkie, C.A., Eds.; Wiley-Interscience: New Jersey, 2007.
14. Kashiwagi, T.; Harris, R.H.; Zhang, X.; Briber, R.M.; Cipriano, B.H.; Raghavan , S.R.; Awad, W.H.; Shields, J.R. *Polymer,* **2004**, *45 (3)*, 881.
15. Kashiwagi, T.; Du, F.M.; Douglas, J.F.; Winey, K.I.; Harris, R.H.; Shields, J.R. *Nat. Mater.,* **2005**, *4*, 928.
16. Xie, W.; Gao, Z.M.; Pan, W.P; Hunter, D.; Singh, A.; Vaia, R. *Chem Mater.,* **2001**; *13*, 2980.

# Chapter 6

# Flammability of Polymer–Inorganic Nanocomposites

**Feng Yang, Irina Bogdanova, and Gordon L. Nelson***

**College of Science, Florida Institute of Technology, 150 West University Boulevard, Melbourne, FL 32901 6975**

Polymer/alumina and polymer/silica nanocomposites were prepared by using a single-screw extrusion approach. A systematic investigation of thermal degradation and flame retardant mechanisms of these polymer nanocomposites (PNC) was conducted. It was found that the nanoparticles play a significant role in the thermal degradation processses of polymer nanocomposites. Metal oxide nanoparticles (alumina) have the ability to catalyze the degradation of carbonyl containing polymers, which accelerates the degradation process. On the other hand, nanoparticles may also affect the thermal degradation pathway by restriction mechanisms as well as a free radical trapping effect. The thermal stability of PNC is the combined result of the above factors. The flammability of PNCs was evaluated by Oxygen Index, UL 94 and Cone Calorimetry. The relationship between thermal stability and melt flow, versus flammability is used to explain the flame retardant mechanisms involved. Maximium reduction in Peak Rates of Heat Release (PHRR) depended upon resin with the greatest reduction observed for Polycarbonate (PC), 54%, versus 43% for polystyrene (PS) and 17% for poly(methyl methacrylate) (PMMA).

 

In general, a remarkable reduction in peak HRRs has been reported for nanocomposites, as well as enhanced physical properties (*1-6*). However, nanocomposites cannot be used alone as flame retardant materials, because they generally cannot meet the criteria of regulatory tests like UL-94 and DIN. As is widely accepted, the cone calorimeter measures flammability quantitatively to provide a scientific understanding of the fire behavior of a material instead of resembling the real fire scenario. In most cone calorimeter tests, samples are positioned horizontally to the cone heater, which facilitates the formation of a thermal barrier on the surface of the sample, and results in a relatively slower burning process. In contrast, many regulatory tests require the fire to travel through the samples vertically, which leads to a short time for the formation of such a thermal barrier on the surface of the nanocomposites (*7, 8*). Therefore, a complete flammability evaluation by both cone and regulary tests helps one understand the full picture of flame retardancy of nanocomposites.

In the past, several approaches had been attempted to explain the improved flame retardancy of polymer / clay nanocomposites (*9-11*), including the formation of a protective surface barrier/insulation layer, radical trapping by paramagnetic iron within clay, as well as char formation enhancement when nanoparticles are present. However, less effort has been put on additive size, shape, chemical composition and degradation process effects on the flammability of PNCs. It is important to understand the basic changes in the pyrolysis process and how it will affect the flammability of materials. In the present research, the impact of nanofiller chemical composition and their shapes on the decomposition process of PNCs is studied, and the relationships between degradation processes and flammability of PNCs are discussed.

## Experiment

### Preparation of PNCs

Poly(methyl methacrylate) (PMMA) for silica PNCs was provided by ICI Acrylics, Inc. and PMMA for alumina PNCs was from Ineos. LEXAN® 103 and LEXAN® 3360 polycarbonate (PC) were from GE Plastics. Styron 685DW and Styron 685P polystyrene for PS / silica nanocomposites were obtained from Dow Plastics, Inc.. The difference between Styron 685DW and 685P is that the latter resin contains flow control additives, which makes the melt flow of 685P thicker than 685DW. A developmental grade of polycarbonate was provided by Dow Plastics, Inc., which has no additives in the resin. Nanoscale silica material, Aerosil® 972 (16 nm, BET surface area is 110±20 $m^2/g$) (an organo-silica where the surface is covered by methyl groups), was provided by Degussa Co. and was used as received. Alumina, alpha form (40-80 nm, irregular) and aluminum hydroxide (15 nm, spherical) were purchased from Nanostructured and Amorphous Materials Co., USA. Alumina whiskers (2800 nm/2-4 nm) was purchased from Sigma-Aldrich, USA. Aluminasol 200® is from Nissan Chemical Inc., which is an alumina hydrate rod with average particle size of 100 nm/10 nm. All alumina nanoparticles were surface treated with 3-methacryl-oxypropyltrimethoxysilane (MAP) or methyltrtimethoxysilane (MT) before use.PNCs were obtained by a single-screw extrusion technique developed in our laboratory using a 3/4" CW Brabender Table Top Independent Extruder, and a 2" ribbon die was used at the orifice of the extruder (*12, 13*). Polymer pellets were pre-dried under vacuum at 100 °C for 24 hours to eliminate moisture and solvent.

### Flammability Evaluation

Oxygen Index testing (OI) was performed on all samples according to ASTM D2863. The flame spread rate of all the samples was investigated by the Horizontal Burning Test according to ASTM D635. All PC nanocomposites were subjected to a Vertical Burning Test based on ASTM D3801 (UL94). Heat and visible smoke release rates of nanocomposites were evaluated at 35 kW/m$^2$ with a Custom Scientific Cone Calorimeter according to modified ASTM E1354. Test samples were made of two conjunctive 2×4 inches strips with a 1/16 inch thickness instead of 4×4×1 inch samples. Three to five tests were conducted for each sample. The error of testing is approximately 5%.

## Results and Discussion

### Cone Calorimetry Study

Previous reports by the authors showed that PNCs generally exhibit higher thermal stability than pristine polymers (*14, 15*). However, it was found that the degradation mechanisms for polymers are very different for metal oxides and nonmetal oxides nanofillers. A nonmetal oxide, such as silica, may enhance the thermal stability of polymers by radical trapping and chain mobility restriction effects, while the heterogeneous catalysis effects on the formation of carboxylic acid and reactivity of surface trapped radicals during the degradation process caused by metal oxides additives may have a negative impact on the thermal stability of a given polymer (*16*). As a result, although PNCs with metal oxides, alumina, as nanofillers generally exhibited enhanced thermal stability, the improvement was not as significant as silica based PNCs. Shapes of alumina have the less impact on the thermal stability of PNCs. However, the particle size of alumina may greatly affect the degradation temperature of polycarbonate, and this is because further interaction between polycarbonate molecular chain and nanoparticles occurs principally when the particles size matchs the segmental chain length of polycarbonate.

With the above comments in mind, significant reduction in peak heat release rates was found for alumina based PNCs when they were subjected to Cone Calorimetry evaluation as shown in Table I to Table III. A remarkable 54% reduction in PHHR was achieved for 100nm rod shaped alumina at 2 wt% in polycarbonate, while an average of 30% reduction in PHHRs was seen for PS/alumina PNCs. It is noticible that the reduction in PHHRs is strongly affected by the polymer type, as the highest reductions were found for PC PNCs in Table I and least for PMMA PNCs in Table III. Besides the chemistry difference of PC, PS and PMMA, the apparent difference among these polymers is the melt flow viscocity at test conditions. PMMA usually forms a liquid flow with very low viscocity at Cone conditions, while extremely thick melts were formed for PC during the tests. According to the thermal barrier protective layer mechanism, the viscocity of the polymer melts could significantly affect the the efficacy of the formation of such a protective layer. Moreover, both the times to ignition and times of peak of polycarbonate PNCs are shorter than pristine polymers, which suggests that decomposition of polymer happens at an earlier stage in the burning process.

As for PC PNCs, it is noted that the amounts of residue at the end of the tests are less or the same as pristine PC, which indicates that no enhancement in char formation was achieved. Enhancement of residue was previously found for PC/silica PNCs (*17*). According to the discussion in the previous section, the presence of alumina could catalize the cleavage of carbonyl groups, which as well leads the pyrolysis of PC to become faster and more complete. It is also found that the reduction in PHRRs happened at very low loading level of nanofillers. For example, a 47% reduction was found at only 0.5 wt% of 100nm MAP modified rod alumina PC PNC as shown in Table I, while 1 wt% of the same nanofiller could reduce the PHRR of PS to 36% according to Table 2. This phenomenon strongly suggested that the flame retardant mechanism of nanoalumina is totally different than the traditional alumina or aluminum trihydrate in micron size that follows a "heat sink" mechanism, because for the later one needs high amounts to impact the flammability of polymers. Moreover, it also shown that the reduction in peak heat release rates reached its highest at 2-3% of alumina loading level in most cases, and further addition of alumina to the PNCs made slight or no changes as expected, sometimes even negative. As discussed in the previous section, the presence of alumina in PNCs may accelerate the decomposition of polymers by catalyzing effect, as well as enhance the thermal stability by radical trapping and chain mobility restriction. When the amount of alumina reaches a certain level in the polymer, the catalysis effect becomes dominant over the enhancement effect, which leads to less reduction in PHRRs of PNCs at higher alumina concentrations.

**Table I. Cone Calorimetry of PC/Alumina PNCs.**

| Material | PC (Dow) | PC-40-0.5-MAP | PC-40-1-map | PC-40-2-map | PC-40-5-map | PC-40-0.5-mt | PC-40-1-mt | PC-40-2-mt |
|---|---|---|---|---|---|---|---|---|
| PHRR (kW/m²) | 1049 | 950 (-9%) | 807 (-23%) | 689 (-34%) | 732 (-30%) | 918 | 916 | 616 |
| TTI(s) | 111 | 80 | 75 | 83 | 39 | 90 | 86 | 77 |
| Time of Peak (s) | 143 | 107 | 108 | 112 | 112 | 109 | 99 | 98 |
| HR (MJ/m²/g) | 1.79 | 1.73 | 1.82 | 1.68 | 1.86 | 1.84 | 1.88 | 1.71 |
| MLR (g/m²s) | 33.3 | 32.1 | 26.0 | 24.8 | 26.8 | 30.0 | 32.3 | 16.8 |
| Residue (%) | 24.0 | 18.8 | 21.6 | 20.3 | 24.2 | 21.2 | 24.7 | 16.0 |

| Material | PC (Dow) | PC-100-0.5-map | PC-100-2-map | PC-100-5-map | PC-100-0.5-mt | PC-100-1-mt | PC-100-2-mt |
|---|---|---|---|---|---|---|---|
| PHRR (kW/m²) | 1049 | 558 | 487 | 560 | 518 | 616 | 532 |
| TTI(s) | 111 | 65 | 54 | 59 | 61 | 79 | 67 |
| Time of Peak (s) | 143 | 88 | 78 | 97 | 102 | 109 | 90 |
| HR (MJ/m²/g) | 1.79 | 1.91 | 1.79 | 1.70 | 1.58 | 1.70 | 1.75 |
| MLR (g/m²s) | 33.3 | 17.6 | 12.3 | 11.8 | 18.0 | 20.6 | 12.5 |
| Residue (%) | 24.0 | 26.9 | 10.8 | 9.0 | 15.5 | 22.0 | 19.8 |

| Material | Lexan® 3620 | PC-15-0.5-map | PC-15-1-map | PC-15-2-map | PC-15-5-map | PC-15-1-mt | PC-15-2-mt | PC-15-5-mt |
|---|---|---|---|---|---|---|---|---|
| PHRR (kW/m²) | 918 | 691 | 791 (-14%) | 591 | 656 (-29%) | 679 | 645 | 512 |
| TTI(s) | 83 | 85 | 56 | 65 | 47 | 72 | 77 | 66 |
| Time of Peak (s) | 103 | 107 | 75 | 95 | 84 | 97 | 97 | 85 |
| HR (MJ/m²/g) |  | 1.75 |  | 1.74 | 1.70 | 1.77 | 1.80 | 1.64 |
| MLR (g/m²s) | 31 | 25.7 | 26 | 23.4 | 24 | 26.2 | 25.4 | 20.6 |
| Residue (%) | 22.7 | 20.5 | 21.7 | 20.3 | 22.8 | 22.8 | 15.6 | 25.0 |

| Material | Lexan® 3620 | PC-wh-1-map | PC-wh-3-map | PC-wh-5-map | PC-wh-1-mt | PC-wh-2-mt |
|---|---|---|---|---|---|---|
| PHRR (kW/m²) | 918 | 666 | 566 | 540 | 670 | 655 |
| TTI(s) | 83 | 96 | 76 | 80 | 78 | 75 |
| Time of Peak (s) | 103 | 116 | 96 | 101 | 106 | 100 |
| HR (MJ/m²/g) |  | 1.70 | 1.65 | 1.64 | 1.71 | 1.76 |
| MLR (g/m²s) | 31 | 30.3 | 21.2 | 22.0 | 27.2 | 20.5 |
| Residue (%) | 22.7 | 25.4 | 26.8 | 24.3 | 23.8 | 23.1 |

**Table II. Cone Calorimetry of PS/Alumina PNCs.**

| Material | PS 685DW | PS-40-1-map | PS-40-5-map | PS-40-5-mt |
|---|---|---|---|---|
| PHRR (kW/m²) | 1528 | 1007 | 1079 | 878 (-43%) |
| TTI(s) | 36 | 24 | 27 | 24 |
| Time of Peak (s) | 46 | 53 | 55 | 50 |
| HR (MJ/m²/g) | | 3.11 | 2.98 | 2.92 |
| MLR (g/m²s) | 32.5 | 16.2 | 30.0 | 20.6 |
| Residue (%) | 1.8 | 0 | 3.3 | 15.8 |

| Material | PS 685DW | PS-100-map-1 | PS-100-map-3 | PS-100-map-5 | PS-100-1-mt | PS-100-3-mt | PS-100-5-mt |
|---|---|---|---|---|---|---|---|
| PHRR (kW/m²) | 1528 | 975 (-36%) | 1097 (-28%) | 1157 (-24%) | 1080. | 1144 | 940 |
| TTI(s) | 36 | 31 | 24 | 25 | 24 | 24 | 29 |
| Time of Peak (s) | 46 | 78 | 46 | 41 | 47 | 37 | 63 |
| HR (MJ/m²/g) | | 3.00 | | | -- | -- | 3.04 |
| MLR (g/m²s) | 32.5 | 28.1 | 27 | 27.2 | 26.5 | 29.0 | |
| Residue (%) | 1.8 | 4.8 | | | | 4.7 | 6.0 |

| Material | PS 685DW | PS-15-1-map | PS-15-3-map | PS-15-5-map | PS-15-1-mt | PS-15-3-mt | PS-15-5-mt |
|---|---|---|---|---|---|---|---|
| PHRR (kW/m²) | 1528 | 1096 | 1046 | 1016 | 1090 | 1068 | 1027 |
| TTI(s) | 36 | 27 | 27 | 24 | 27 | 28 | 24 |
| Time of Peak (s) | 46 | 57 | 58 | 56 | 52 | 58 | 51 |
| HR (MJ/m²/g) | | 3.16 | 3.24 | 3.10 | 3.29 | 3.20 | 3.40 |
| MLR (g/m²s) | 32.5 | 26.2 | 26.6 | 28.1 | 28.9 | 27.3 | 25.9 |
| Residue (%) | 1.8 | 5.0 | 8.8 | 6.8 | 1.7 | 6.0 | 4.7 |

| Material | PS 685DW | PS-wh-1-map | PS-wh-3-map | PS-wh-5-map | PS-wh-1-mt | PS-wh-3-mt | PS-wh-5-mt |
|---|---|---|---|---|---|---|---|
| PHRR (kW/m²) | 1528 | 1073 | 1043 | 1053 | 1033 | 1011 | 991 |
| TTI(s) | 36 | 32 | 31 | 28 | 31 | 27 | 28 |
| Time of Peak (s) | 46 | 72 | 68 | 53 | 67 | 64 | 55 |
| HR (MJ/m²/g) | | 3.21 | 3.08 | 3.43 | 3.25 | 3.33 | 3.28 |
| MLR (g/m²s) | 32.5 | 25.2 | 29.2 | 29.1 | 27.6 | 25.8 | 26.5 |
| Residue (%) | 1.8 | 1.9 | 2.2 | 5.2 | 2.7 | 6.7 | 10.7 |

**Table III. Cone Calorimetry of PMMA/Alumina PNCs.**

| Material | PMMA | PMMA-40-1-map | PMMA-40-map-3 | PMMA-40-map-5 | PMMA-40-3-mt | PMMA-40-5-mt |
|---|---|---|---|---|---|---|
| PHRR ($kW/m^2$) | 851 | 831 | 719 | 730 | 841 | 765 |
| TTI(s) | 23 | 22 | 19 | 20 | 18 | 16 |
| Time of Peak (s) | 48 | 42 | 43 | 46 | 47 | 43 |
| HR ($MJ/m^2/g$) | 2.60 | 2.64 | 2.37 | 2.38 | 2.47 | 2.30 |
| MLR ($g/m^2s$) | 29.1 | 29.0 | 25.0 | 27.1 | 30.0 | 26.3 |
| Residue (%) | 1.1 | 1.1 | -- | 4.9 | 4.9 | 5.3 |

| Material | PMMA (Ineos) | PMMA-100-1-map | PMMA-100-3-map | PMMA-100-5-map | PMMA-100-1-mt | PMMA-100-3-mt | PMMA-100-5-mt |
|---|---|---|---|---|---|---|---|
| PHRR ($kW/m^2$) | 851 | 765 | 705 | 720 | 818 | 719 | 688 |
| TTI(s) | 23 | 24 | 25 | 25 | 20 | 24 | 19 |
| Time of Peak (s) | 48 | 46 | 45 | 55 | 44 | 54 | 44 |
| HR ($MJ/m^2/g$) | 2.60 | 2.52 | 2.49 | 2.46 | 2.53 | 2.35 | 2.38 |
| MLR ($g/m^2s$) | 29.1 | 27.8 | 25.4 | 28.4 | 26.8 | 26.4 | 24.7 |
| Residue (%) | 1.1 | 3.3 | 5.7 | -- | -- | 4.3 | 6.5 |

| Material | PMMA (Ineos) | PMMA-15-map | PMMA-15-map-3 | PMMA-15-map-5 | PMMA-15-5-mt |
|---|---|---|---|---|---|
| PHRR ($kW/m^2$) | 851 | 730 | 741 | 712 | 745 |
| TTI(s) | 23 | 22 | 20 | 22 | 22 |
| Time of Peak (s) | 48 | 49 | 45 | 51 | 50 |
| HR ($MJ/m^2/g$) | 2.60 | 2.27 | 2.56 | 2.20 | 2.37 |
| MLR ($g/m^2s$) | 29.1 | 25.9 | 27.8 | 27.2 | 27.4 |
| Residue (%) | 1.1 | 5.1 | 8.1 | 4.7 | 8.1 |

| Material | PMMA (Ineos) | PMMA-wh-map-1 | PMMA-wh-map-3 | PMMA-wh-map-5 |
|---|---|---|---|---|
| PHRR ($kW/m^2$) | 851 | 906 | 906 | 791 |
| TTI(s) | 23 | 20 | 22 | 19 |
| Time of Peak (s) | 48 | 43 | 43 | 41 |
| HR ($MJ/m^2/g$) | 2.60 | 2.74 | 2.67 | 2.62 |
| MLR ($g/m^2s$) | 29.1 | 31.6 | 32.9 | 24.0 |
| Residue (%) | 1.1 | 11.6 | 7.1 | -- |

**Table IV. UL 94 of PC PNCs.**

| Material | OI | Vertical Burning* | | |
|---|---|---|---|---|
| | | $T_f$ (s) | $T_1$ avg., (s) | $T_2$ avg., (s) |
| PC (Dow) | 23.3 | 110.0 | 6.8 | 15.2 |
| PC-Al-100-MAP-1 | 24.3 | 135.0 | 9.2 | 17.8 |
| PC-Al-100-MAP-2 | 27.1 | 156.0 | 10.8 | 20.4 |
| PC-Al-100-MAP-5 | 25.3 | 215.0 | 16.4 | 26.6 |
| PC-Al-40-MAP-1 | 23.7 | 158.0 | 13.0 | 18.6 |
| PC-Al-40-MAP-2 | 24.2 | 183.0 | 12.8 | 23.8 |
| PC-Al-40-MAP-5 | 25.3 | 151.0 | 13.0 | 17.2 |
| PC (Lexan 3620) | 22.1 | 128.0 | 5.6 | 18.9 |
| PC-Al-15-MAP-1 | 22.5 | 114.0 | 8.0 | 15.3 |
| PC-Al-15-MAP-2 | 23.4 | 143.0 | 16.9 | 11.6 |
| PC-Al-15-MAP-5 | 22.5 | 130.0 | 12.5 | 11.6 |
| PC-Al-15-MT-1 | 23 | 152.0 | 14.0 | 16.2 |
| PC-Al-15-MT-2 | 23 | 142.0 | 16.9 | 12.9 |
| PC-Al-15-MT-5 | 22 | 147.0 | 21.7 | 7.0 |
| PC-Al-wh-MAP-1 | 22 | 90.0 | 3.9 | 12.7 |
| PC-Al-wh-MAP-3 | 23 | 75.0 | 2.7 | 10.9 |
| PC-Al-wh-MAP-5 | 22 | 76.0 | 2.9 | 12.3 |
| PC-Al-wh-MT-1 | 23 | 91.0 | 3.4 | 14.8 |
| PC-Al-wh-MT-3 | 23 | 102.0 | 3.1 | 16.1 |
| PC-Al-wh-MT-5 | 22 | 82.0 | 3.7 | 12.7 |
| PC (Lexan 103) | 22.5 | 99.6 | 7.2 | 17.7 |
| PC-Si-16-1 | 22.5 | 56.7 | 5.2 | 6.2 |
| PC-Si-16-3 | 22.5 | 125 | 15.7 | 9.3 |
| PC-Si-16-5 | 22.1 | 147 | 7.2 | 22.2 |

* all samples drip but not ignite cotton.

## UL 94 and OI Testing of PNCs

All PNCs were subjected to Oxygen Index and small flame testing as reported in Tables IV-VI. All polycarbonate PNCs listed in Table IV are V-2 rated materials according to the UL94 Standard. Because polycarbonate is an inherent V2 rated material, no horizontal burning test was performed. As the VB results in Table IV suggest, the shape and size of additive can significantly impact the vertical burning behavior of polycarbonate. Specifically, PNCs from alumina whisker with 2-4 nm in diameter showed remarkable reduction in total after flame time from 128 seconds to 75 seconds at 3-5% loading level, which is promising for the development of higher fire rating polycarbonates. However, the OI indices of larger size alumina based PNCs showed that OI of 100 nm rod alumina gave higher values than polycarbonate itself, while little to no change in oxygen indices were found for smaller size alumina filled PNCs. It is noted that 100 nm rod PNCs gave the largest PHRR reductions in Cone testing. Further comparison between PNCs from 15 nm spherical alumina and 16 nm spherical silica suggested that chemical composition has some impact on vertical burning behavior of polycarbonate. The largest reduction in total time of vertical burning test (Tf) was for 16nm silica PNCs (*18*) at 1% loading.

In Table V, the UL 94 results of PS PNCs are listed. Vertical burning tests were also performed to gather further information on the flammability of PS PNCs, although they are all HB to "no-rating" materials. As the OI data indicated, only slight increases were achieved by PS nanocomposites, while no distinguishable difference can be found when comparing samples by their shape and chemical composition. However, burning rates measured by the HB test decreased with the increase of alumina concentration. In another words, PS/alumina burns slower than PS. However, PS/alumina nanocomposites burned faster at the VB test than PS. All PS/alumina samples dripped heavily in the HB test which removed heat from the burning samples. In contrast, the burning rates for PS/silica (20 nm) nanocomposites were increased in both HB and VB tests when compared to pristine PS, while no dripping happened during the HB evaluation among these samples. When comparing the horizontal burning rates of alumina PNCs at similar nanofiller size range, i.e. 100 nm rod shaped versus 40-80 nm spherical/irregular and 15 nm spherical shaped versus whiskers, it is found that 1-dimensional rod/whisker shaped alumina reduced the burning rate of PS more than their 0-dimensional spherical/dot shaped counterparts.

All PMMA based PNCs are "no-rating" materials in UL 94 testing as shown in Table VI. Little to no change in OI was observed, which suggests that sufficient fuels were fed to the gas phase, while the presence of alumina did not significantly slow down the pyrolysis process of polymers. Horizontal burning rate for PMMA/alumina nanocomposites generally decreased with increased concentration of alumina, i.e. PMMA/alumina nanocomposites burned slower than PMMA, which is different than silica based PMMA nanocomposites. As shown in Table VI, the PMMA/silica nanocomposites exhibit substantially

**Table V. UL 94 of PS PNCs.**

| Sample | OI | Horizontal Burning Rate (cm/min) | Vertical burning test* | |
|---|---|---|---|---|
| | | | $T_1$, (sec) | $T_2$, (sec) |
| PS (STYRON® 685DW) | 16.9 | 6.1 | 76 | - |
| PS-Al-100-MAP-1 | 16.9 | 5.8 | 69 | - |
| PS-Al-100-MAP-3 | 17.2 | 5.2 | 53 | - |
| PS-Al-100-MAP-5 | 17.5 | 4.4 | 66 | - |
| PS-Al-100-MT-1 | 16.9 | 5.5 | 59 | - |
| PS-Al-100-MT-3 | 17.2 | 5.1 | 54 | - |
| PS-Al-100-MT-5 | 17.5 | 5.1 | 55 | - |
| PS-Al-40-MAP-1 | 17.2 | 5.8 | 58 | - |
| PS-Al-40-MAP-3 | 17.5 | 5.3 | 62 | - |
| PS-Al-40-MAP-5 | 17.2 | 5.4 | 51 | - |
| PS-Al-40-MT-1 | 17.1 | 6.2 | 61 | - |
| PS-Al-40-MT-3 | 17.1 | 6.3 | 63 | - |
| PS-Al-40-MT-5 | 16.9 | 5.4 | 61 | - |
| PS-Al-15-MAP-1 | 17.5 | 4.4 | 60 | - |
| PS-Al-15-MAP-3 | 17.2 | 5.4 | 55 | - |
| PS-Al-15-MAP-5 | 17.2 | 5.2 | 47 | - |
| PS-Al-15-MT-1 | 17.5 | 5.4 | 53 | - |
| PS-Al-15-MT-3 | 17.2 | 5.1 | 43 | 7 |
| PS-Al-15-MT-5 | 17.5 | 4.7 | 21 | 11 |
| PS-Al-wh-MAP-1 | 17.5 | 6.1 | 41 | - |
| PS-Al-wh-MAP-3 | 17.5 | 4.4 | 55 | - |
| PS-Al-wh-MAP-5 | 17.2 | 4.5 | 61 | - |
| PS-Al-wh-MT-1 | 17.5 | 4.7 | 59 | - |
| PS-Al-wh-MT-3 | 17.2 | 4.7 | 49 | - |
| PS-Al-wh-MT-5 | 17.2 | 3.9 | 58 | - |
| PS(685p)-Si-20-PTCS-1 | 17.0 | 5.7 | 92 | - |
| PS(685p)-Si-20-PTCS-3 | 17.3 | 5.8 | 83 | - |
| PS(685p)-Si-20-PTCS-5 | 17.5 | 6.1 | 68 | - |

* ignites cotton and burns up to the clamp

**Table VI. UL 94 of PMMA PNCs.**

| Sample | OI | Horizontal Burning(cm/min)[a] |
|---|---|---|
| PMMA (INEOS) | 16.9 | 5.1 |
| PMMA-Al-100-MAP-1 | 16.9 | 4.8 |
| PMMA-Al-100-MAP-3 | 17.2 | 4.9 |
| PMMA-Al-100-MAP-5 | 17.5 | 4.6 |
| PMMA-Al-40-MAP-1 | 17.2 | 5.1 |
| PMMA-Al-40-MAP-3 | 17.5 | 5.2 |
| PMMA-Al-40-MAP-5 | 17.2 | 4.5 |
| PMMA-Al-15-MAP-1 | 17.5 | 4.3 |
| PMMA-Al-15-MAP-3 | 17.2 | 4.9 |
| PMMA-Al-15-MAP-5 | 17.2 | 5.1 |
| PMMA-Al-wh-MAP-1 | 17.5 | 4.3 |
| PMMA-Al-wh-MAP-3 | 17.5 | 4.1 |
| PMMA-Al-wh-MAP-5 | 17.2 | 4.2 |
| PMMA (ICI) | 17.5 | 4.7 |
| PMMA-silica-90-1 | 17.5 | 6.9 |
| PMMA-silica-90-5 | 21.2 | 7.2 |

[a] Burning rate=450/(t- t1), where t1 is the burning time from the beginning to 25 mm, and t is the burning time from the beginning to 100 mm.

higher burning rates compared to virgin PMMA. However, all the nanocomposites burn without dripping, which is a very different behavior from the pure PMMA that drips significantly during the test.

## Conclusion

Silica and alumina nanocomposites were obtained by a single-screw extrusion approach from PC, PS and PMMA, and their thermal stability and flammability were investigated. In general, PNCs exhibit higher thermal stability than pristine polymers. The degradation mechanisms for polymers are different for metal oxides and nonmetal oxides nanofillers. A nonmetal oxide, such as silica, may enhance the thermal stability of polymers by radical trapping and chain mobility restriction effects, while the heterogeneous catalysis effect on the formation of carboxylic acid and reactivity of surface trapped radicals during the degradation process caused by metal oxides additives may have a negative impact on the thermal stability of a given polymer. As a result, although PNCs with metal oxides, alumina, as nanofillers exhibited enhanced thermal stability, the improvements were not as significant as silica based PNCs.

Shapes of alumina have less impact on the thermal stability of PNCs. However, the size of alumina greatly affects the degradation temperatures of polycarbonate, and this is because further interaction between polycarbonate molecular chain and nanoparticles occurs better when the particle size matchs the segmental chain length of polycarbonate.

The Cone Calorimetry results of PNCs showed significant reduction in PHRR. Maximum reduction of PHRR was highest for PC-54% versus 43% for PS and only 17% for PMMA. Viscosity of polymer melts plays an important role in determining the flammability of PNCs under the Cone test conditions. Data also showed that the reduction in peak heat release rates was highest at 2-3% of alumina loading level in most cases, and further addition of alumina to the PNCs made slight or no changes as expected, sometimes even negative, which could be caused by the heterogeneous catalysis effects becoming the dominant factor in the pyrolysis process when alumina content reaches a certain level. The burning behavior of silica based and alumina based PNCs are very different when subjected to UL 94 evaluation. For instance, faster burning rates were found for silica based PMMA and PS PNCs, while alumina based PNCs showed slower burning rates in Horizontal Burning tests.

## References


1. Lee J., T. Takekoshi; Giannelis E. *Mater. Res. Soc. Symp. Proc.*, **1997**, 457-513.
2. Zanetti M.; Camina G.; Mulhaupt R. *Polym. Degrad. Stab.*, **2001**, *74*, 413-417.
3. Kashiwagi T.; Shields J.; Harris R., R. H. Jr.; Davis R. D. *J. Appl. Polym. Sci.* **2003**, *87*, 1541-1553.
4. Vacatello M. *Macromolecules*, **2001**, *34*, 1946-1952.
5. Austin P. J; Buch R. R.; Kashiwagi T. *Fire Mater,* **1998**, *22*, 221-237.
6. Yang F.; Yngard R.; Nelson G. L. *J. Fire Sci.*, **2005**, *23*(3), 209-226.
7. Gilman J. W. *Appl. Clay Sci.*, **1999**, *15*(1-2), 31-49.
8. Gilman J. W.; Kashiwagi T.; Lichtenhan J. D. *SAMPE J.*, **1997**, *33*, 40-46.
9. Gilman J. W. *App. Clay Sci.*, **1999**, *15*(1-2), 31-49.
10. Morgan A. B; Harris R. H. Jr.; Kashiwagi T.; Chyall L. J.; Gilman J. W. *Fire Mater.*, **2002**, 26, 247-253.
11. Zhu J.; Morgan A. Uhl.; Wilkie C. A. *Chem. Mater.*, **2001**, *13*, 4649-4654.
12. Yang F.; Nelson G. L. *J. Appl. Polym. Sci.*, **2004**, *91*, 3844-3850.
13. Yang F.; Nelson G. L. In *Proceedings of The 11th International Conference, ADDITIVES 2002,* Clearwater Beach, FL, March 25, **2002**.
14. Yang F.; Bogdanova I.; Nelson G. L. *Polym. Adv. Tech.*, **2008**, *19*, 602-608.
15. Yang F.; Bogdanova I.; Nelson G. L. In: *Fire Retardancy of Polymers: New Strategies and Mechanisms, Chapt. 7*, R. T. Hull editor, RSC Publishing, Cambridge , **2008**, 95-109, in press.

16. Ravve A., In: *Principles of Polymer Chemistry, Chapt. 9,* Ravve A. editor, Kluwer Academic / Plenum Publishers, New York, **2000**.
17. Yang F.; Nelson G. L. *Polym. Adv. Tech.*, **2006**, *17*, 320-326,.
18. Yang F.; Bogdanova I.; Nelson G. L. In Proceedings of *The 19th Annual BCC Conference on Flame Retardancy,* Stamford, Connecticut, June 9-11, **2008**.

## Chapter 7

# Flame-Retardant PET–PC Blends Compatibilized by Organomodified Montmorillonites

**B. Swoboda[1], E. Leroy[1,2], F. Laoutid[1,3], and J.-M. Lopez-Cuesta[1,*]**

**[1]CMGD, Ecole des Mines d'Alès, 6 av. de Clavières, 30100 Alès, France**
**[2]Current address: GEPEA (UMR CNRS 6144), 37 Bd de l'Université, B.P. 420, 44606 St-Nazaire Cedex, France**
**[3]Current address: Materia Nova–SMPC, University of Mons-Hainaut, Place de Parc 20, B–7000 Mons, Belgium**
***Corresponding author: Jose-Marie.Lopez-Cuesta@ema.fr**

Flame retardant PET nanocomposites and compatibilized PET/PC blends were obtained using a montmorillonite modified by a methyl-triphenoxyphosphonium salt. The organomodified MMT was shown to exfoliate into the PET matrix and to act as a compatibilizer between PET and PC in the blends. The combination of this nanofiller with triphenyl phosphite gives good mechanical properties and reaction to fire performance, including V-0 rating in the UL94 test. These results illustrate the advantage of combining various formulation strategies to obtain engineering polymeric materials, even from recycled plastics having intrinsically low performances (recycled PET and PC in this case).

## Introduction

The present main applications of reclaimed PET bottles after regeneration are as fibers, packaging, strips and films (*1*). Actually, the re-use of PET as an higher added value engineering plastic is limited due to several issues: when recycled PET is melt processed, various degradations mechanisms (thermal, mechanical, oxidative or hydrolytic) take place, generally promoted by remaining impurities, leading to a decrease of the average chain length, and thus to a decrease of rheological and mechanical properties (*2*).

In a two recent papers (*3,4*), we studied the reaction to fire of polyethylene terephtalate/polycarbonate (PET/PC) blends made from recycled polymers, with the aim to increase the fire retardancy of recycled PET.

Both the morphology and the compatibilization of the blends were shown to have a strong influence on final properties. For example, while the reaction to fire improved almost linearly for low percentages of added PC, blends containing more than 50% w./w. of PC reacted to fire like pure PC. This change of behavior was correlated with the formation of a continuous PC phase in the blend. Concurrently, the compatibilization of the blend by a transesterification reaction between PET and PC (*3*), leading to the formation of copolymers at the interface, was shown to decrease the overall fire performances, despite a strong increase of charring. This was ascribed to the fact that the transesterification reaction, as a side effect, decreases the average PET chain length, which results in a strong decrease of the viscosity of the blends and of the temperature at which mass loss begins.

The introduction of triphenylphosphite (TPP), after the compatibilization of the blends resulted in significantly improved reaction to fire, including V-0 rating in the UL94 test *(4).* TPP actually has a triple role: acting as a flame retardant due to its phosphorus content (9% by weight), but also as a chain extender able to promote bridging reactions between PET chain ends, and as an inhibitor of the transesterification reaction between PET and PC. The combination of these three mechanisms results in a strong increase of both the thermal stability and charring of the PET/PC blends (*4*).

Recent research on PET/PC blends (*5*) shows the possibility to achieve blend compatibilization by adding specifically designed organomodified montmorillonites which locate at the interface between PET and PC due to thermodynamic driving forces in the melt.

In the present article, we will show that such compatibilization can be obtained using TPP derived phosphonium modified montmorillonites, and that the combination of such nanoparticles with pure TPP enables one to obtain recycled PET/PC blends with substantially improved mechanical and fire behavior.

# Experimental

## Materials and processing

The recycled PET (post-consumer bottles flakes) was supplied by VALORPLAST (France) (ηPET = 0.76 dl/g, in 2-chlorophenol at 25°C, average molecular weight: 26300 g/mol. calculated using the Mark-Houwink equation (K = $3.8x10^4$; a = 1.3)).

Recycled PC (compact disc production waste, without metallic coating) was supplied by MPO (France) (ηPC = 0.37 dl/g, in 2-chlorophenol at 25°C, average molecular weight calculated using the Mark-Houwink equation (K = $6.0x10^4$ ; a = 1.22): 18000 g/mol). The CD's have been ground using a rotary cutter mill in order to obtain PC flakes of typically 8 mm.

Purified Na Montmorillonite (MMT-$Na^+$, Nanofil® EXM 757) was purchased from Süd-Chemie.

Triphenylphosphite (TPP, CAS: 101-02-0) and methyltriphenoxyphosphonium iodide salt (Figure 1, CAS: 17579-99-6) were purchased from Acros Organics.

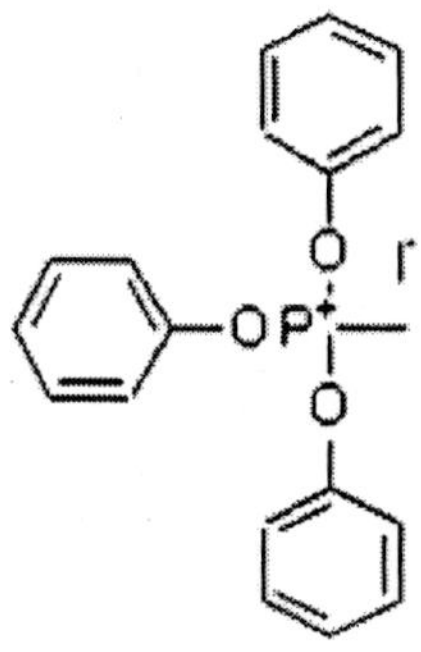

*Figure 1. Chemical structure of methyltriphenoxyphosphonium iodide salt.*

### *Organic modification of MMT-$Na^+$ and properties of MMT-P*

The ion exchange reaction was performed in aqueous medium at low temperature (typically between 273 and 277 K): 50 g of MMT-$Na^+$ were dispersed with stirring in 3 liters of distilled water for 24h. Concurrently the phosphonium salt was dissolved with stirring for 12 hours in an 80/20 w./w. mixture of distilled water and acetone with a concentration of 25g of salt per 3 liters of solvent. The two solutions containing the MMT-$Na^+$ and the phosphonium salt, respectively, were then mixed with stirring for 16 hours. The resulting organomodified MMT which will be called MMT-P in the following, was then filtered, washed by centrifugation using a mixture of water and acetone, and finally lyophilized in order to obtain a fine powder. The modified montmorilonite MMT-P has the following characteristics:

- According to X-ray diffraction analysis, the gap between clay layer increased from 1.24 nm (MMT-$Na^+$) to 1.60 nm due to the intercalation of phosphonium ions.
- Infrared spectroscopic analysis of the MMT-P powder showed the presence of absorption bands at 1586 $cm^{-1}$, 1483 $cm^{-1}$, and 686 $cm^{-1}$, which are typical of the initial phosphonium salt (Figure 1).
- The weight percentage of organic phase present in MMT-P after washing and lyophilization was 9.5 wt.%. This value was determined by thermogravimetric analysis (TGA, 5°C/min, 25-700°C, samples of typically 25 mg, using a Netzsch STA 409 device). Another important characteristic of MMT-P revealed by TGA is that the onset of the mass loss due to the phosphonium ion takes place around 300°C, while pure TPP starts to evaporate around 200°C (Figure 2). This increase of thermal stability is of particular interest since the melt mixing with PET will be done at 270°C.
- Finally, the electrophoretical mobility (Zeta Potential) of MMT-$Na^+$ and MMT-P were measured using a Malvern Nanosizer NANOZS: in water, it decreases from 40.9 mV to 22.9 mV; while in tetrahydrofuran (THF) it increases from 11.2 mV to 44.5 mV. This indicates that MMT-P particles are strongly organophilic, as opposed to initial hydrophilic MMT-$Na^+$.

The different formulations were compression molded at 270 °C at a pressure of 100 bars for 30 seconds, directly after blending, in order to obtain $100 \times 100 \times 4$ $mm^3$ sheets, which were then cut to the requisite size for fire and mechanical testing.

*Polymer compounding*

Due to the high sensitivity of PET to moisture, the following protocol was used before melt mixing: PET and PC flakes were dried separately in vacuum at 120 °C for 16h. The choice of these conditions was the result of a detailed study of PET drying kinetics and sensitivity to moisture during melt processing (*3*).

Melt blending was performed using a Haake internal mixer (Rheomix) at 270 °C and 60 rpm for 10 minutes (except for formulation 9, see Table I).

Samarium acetylacetonate catalyst was used for the compatibilization of the formulation # 10 during melt mixing. Catalyst was directly dispersed in PETr/PCr pellets in order to obtain catalyst weight contents of 0.05%.

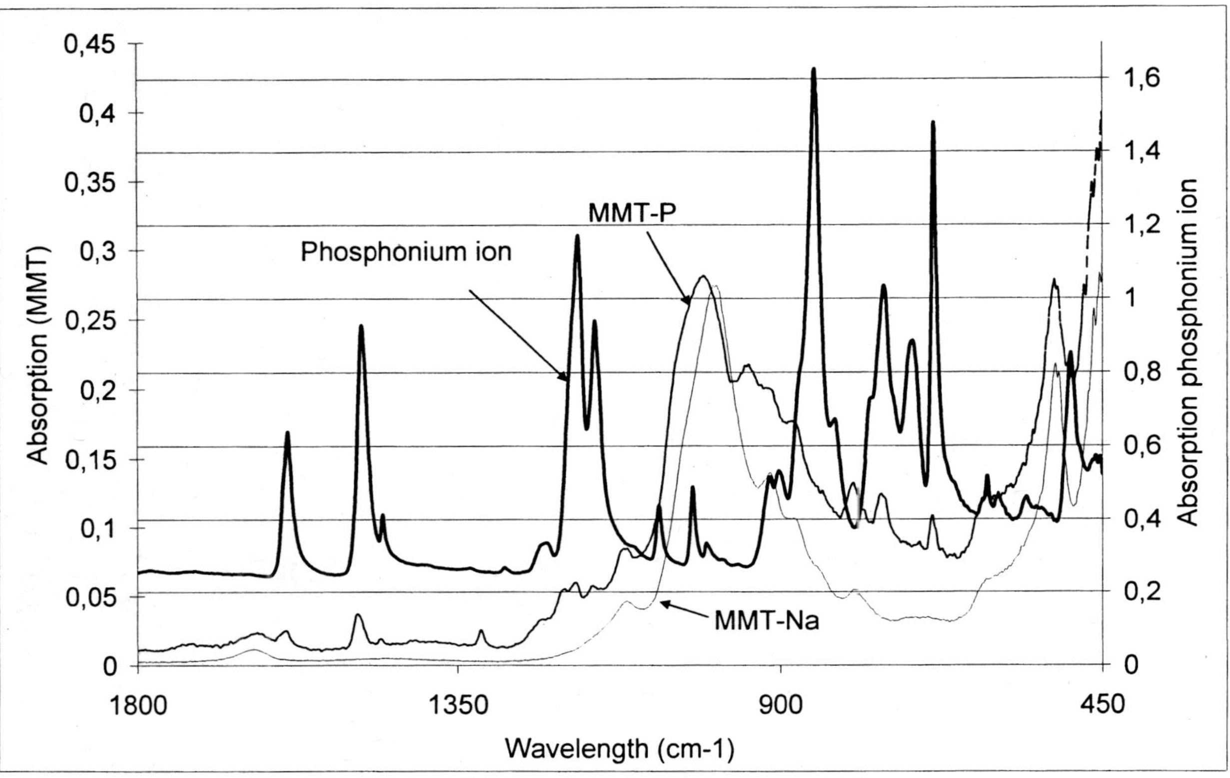


*Figure 1. I.R. Spectra of phosphonium ion, pristine and modified montmorillonite*

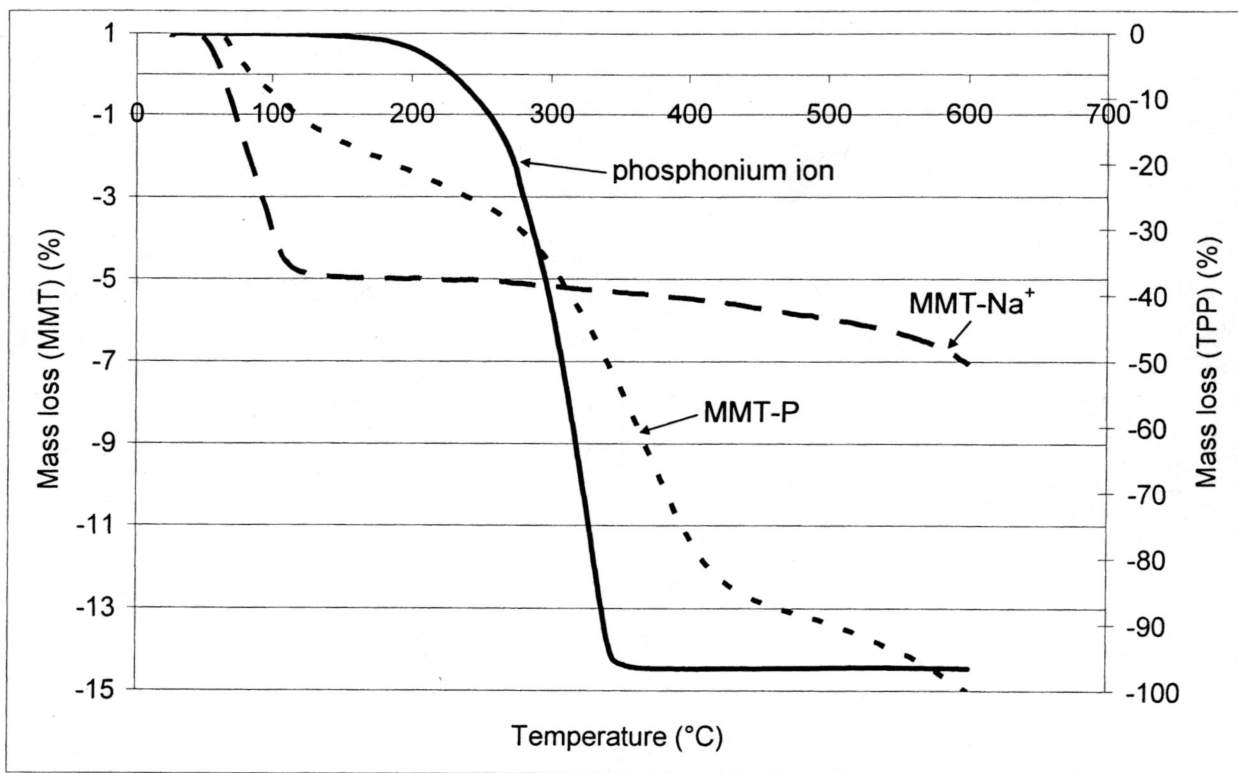


*Figure 2. Thermogravimetric analysis of phosphonium ion, pristine and modified montmorillonite (5°C/min, in air)*

**Table I. Formulations prepared (w./w. fractions). In the case of formulation #9, TPP is added to the melt mixture after 10 minutes and the melt mixing is maintained during 5 minutes**

| *Sample* | *PET* | *PC* | *MMT-$Na^+$* | *MMT-P* | *TPP* | *Catalyst* |
|---|---|---|---|---|---|---|
| 1 | 100 | 0 | 0 | 0 | 0 | 0 |
| 2 | 80 | 20 | 0 | 0 | 0 | 0 |
| 3 | 50 | 50 | 0 | 0 | 0 | 0 |
| 4 | 95 | 0 | 0 | 5 | 0 | 0 |
| 5 | 76 | 19 | 0 | 5 | 0 | 0 |
| 6 | 47.5 | 47.5 | 0 | 5 | 0 | 0 |
| 7 | 76 | 19 | 5 | 0 | 0 | 0 |
| 8 | 76.8 | 19.2 | 0 | 0 | 4 | 0 |
| 9 | 72.8 | 18.2 | 0 | 5 | 4 | 0 |
| 10 | 80 | 20 | 0 | 0 | 0 | $5.10^{-4}$ |

**Characterization techniques**

The dispersion of MMT-P particles in the PET (and PET/PC blends) matrix was studied by X-ray diffraction and dynamic rheological tests at low shear rate at 260 °C using an ARES Rheometrics Scientific apparatus in plate/plate geometry (∅ 25 mm, 1mm gap).

In order to detect the glass transitions of the different polymeric phases, samples of dimensions 4×10×70 $mm^3$ were submitted to Dynamic Mechanical Thermal Analysis (DTMA) in bending conditions (5 Hz, in elastic deformation domain), with a temperature ramp from 20 °C to 180 °C at 3 °C/min. using a Metravib Viscoanalyser.

Static flexural modulus was measured from three point bending tests performed on a Zwick TH010 universal press according to ISO 178 Standard.

The thermo-oxidative degradation of polymer formulations was studied by thermogravimetric analysis (TGA, 5 °C/min, 25-700 °C, samples of typically 25mg, using a Netzsch STA 409 device).

The limiting oxygen index (LOI) was measured using a Stanton Redcroft instrument on barrels (80×10×4 $mm^3$) according to ISO 4589 specifications.

Epiradiateur tests (AFNOR NF P 92-505) were carried out in order to determine the flammability and the self-extinguishability of the different formulations. In this test, a radiator (500 W) is placed above the sample (70×70×4 $mm^3$). When the sample starts burning, the time to first ignition (TTFI) is recorded. After ignition, the radiator is removed and the time of extinction is recorded. As soon as extinction occurs, the radiator (500 W) is replaced above the sample. This procedure is repeated for 5 minutes. The mean inflammation period (MIP) is then calculated. Results presented correspond to mean values obtained from four experiments for each formulation.

Cone calorimeter tests (ISO 5660) were performed on $100\times100\times4$ $mm^3$ samples placed horizontally with an irradiance of 50 $kW/m^2$. Only time-to-ignition (TTI) and peak heat release rate (pHRR) values will be discussed. Results correspond to mean values obtained from three experiments for each formulation, for which a typical variation of 10% was observed.

A vertical UL 94 test was used to determine the self-extinguishability and dripping of the different resins. According to the UL94 standard, samples ($100\times10\times4$ $mm^3$) were exposed to the flame twice during ten second with a break of ten second between the two exposures. For each sample, we measured the time of extinction and the number of liquid drops falling on the cotton wool after each exposition. Each formulation was tested two times. The behavior of the samples allowed the UL94 rating of each formulation (V-0, V-1, V-2 or non rated).

## Results and Discussion

In this section, the results obtained for formulations #1 to #7 will be discussed first. The results obtained for the other formulations are discussed at the end of the section.

### Morphologies

Figure 3 shows the rheological behavior of pure PET (#1) and of PET/MMT-P nanocomposite (#4). The complex viscosity strongly increases at low frequencies, which is typical of a gel-like structure obtained by a fine dispersion of clay platelets. X-ray diffraction analysis supports this conclusion of a strong exfoliation of MMT-P in the PET matrix, since no peak characteristic of clay interlayer distance could be detected.

The behavior and morphology of PET/PC/MMT-P (#5 and #6) formulations are more complex to analyse. The rheological analysis, shown in Figure 3 for formulations #3 and #6, indicates no gel-like behavior at low frequencies.

Actually, in formulations #5 and #6, fractions of the clay particles can be located either in the PET phase and/or in the PC phase and/or at the interface between the two polymers. Given the strong affinity of MMT-P for PET, the fraction of the clay located in that phase is likely to be exfoliated and cannot be detected in X-ray diffraction. Given the incompatibility of PET and PC, it is also likely that MMT-P will not exfoliate or intercalate in the PC phase. Indeed, for formulations #5 and #6, the X-ray diffraction analysis indicates that a fraction of the clay has the same interlayer distance as pure MMT-P: 1.60 nm (Figure 4).

The corresponding peak intensity is also higher for formulation #6. We assume that this fraction of the clay is located in the PC phase.

Finally, the X-ray diffraction of formulation #5 also indicates that another fraction of the clay has an interlayer distance of 3.16 nm, indicating significant intercalation. We assume that this fraction is located at the interface/interphase between PET and PC. This assumption should be confirmed by Transmission Electron Microscopy analysis and will be the subject of future work.

**Blends compatibilizing effect of MMT-P and thermal stability**

The dynamic mechanical analysis results indicate that formulations #2 (shown on Figure 5) and #3 (not shown on Figure 5 for the sake of clarity) present two mechanical relaxations. These are associated with the PET phase (around 100 °C) and the PC phase (around 150 °C), respectively. This behavior is not modified when MMT-$Na^+$ is introduced in the polymer blend, two similar relaxations being present for formulation #7 (not shown on Figure 5 for the sake of clarity). The analysis of formulations containing MMT-P: #5 (shown on Figure 5) and #6 (not shown on Figure 5 for the sake of clarity) indicates the presence of one single broad relaxation around 110 °C. This is the signature of a strong compatibilization of the blend, similar to that we observe on formulation # 10 (shown on Figure 5) in which transesterification is promoted by a catalyst (*2,3*). In the present case, it is not clear whether the compatibilization observed here is due to the promotion of a transesterification reaction by a catalytic effect of the clay (as assumed by other authors in PC/PMMA blends compatibilized by organoclays (*6*)), or mostly to a decrease of the interfacial energy between PET and PC, due to the location of clay platelets at the interface. Recent results of Manias et al. (*5*) on PET/PC blends compatibilized by organomodified MMT could support this second assumption. In our case, if we assume that both compatibilization mechanisms are involved, the observation of an X-ray diffraction peak corresponding to a clay interlayer distance of 3.16 nm could correspond to the intercalation of PET-PC copolymers formed by transesterification.

Figure 6 shows the TGA mass loss curves obtained for formulations #1, 2, 3, 4, 5 and 6. As described in a previous study (*3*), the thermo-oxidative degradation of pure PET (#1) takes place in two steps: a first mass loss takes place between 350 °C and 450 °C followed by a second mass loss at higher temperatures. This second mass loss corresponds to the degradation of a char formed at lower temperatures concurrent to the first mass loss. The introduction of PC into PET (PET/PC blends #2 and #3) tends to reduce the initial thermal stability (the first mass loss starts at lower temperatures), but increases significantly the amount of char formed and its thermal stability (second mass loss is shifted toward a higher temperatures).

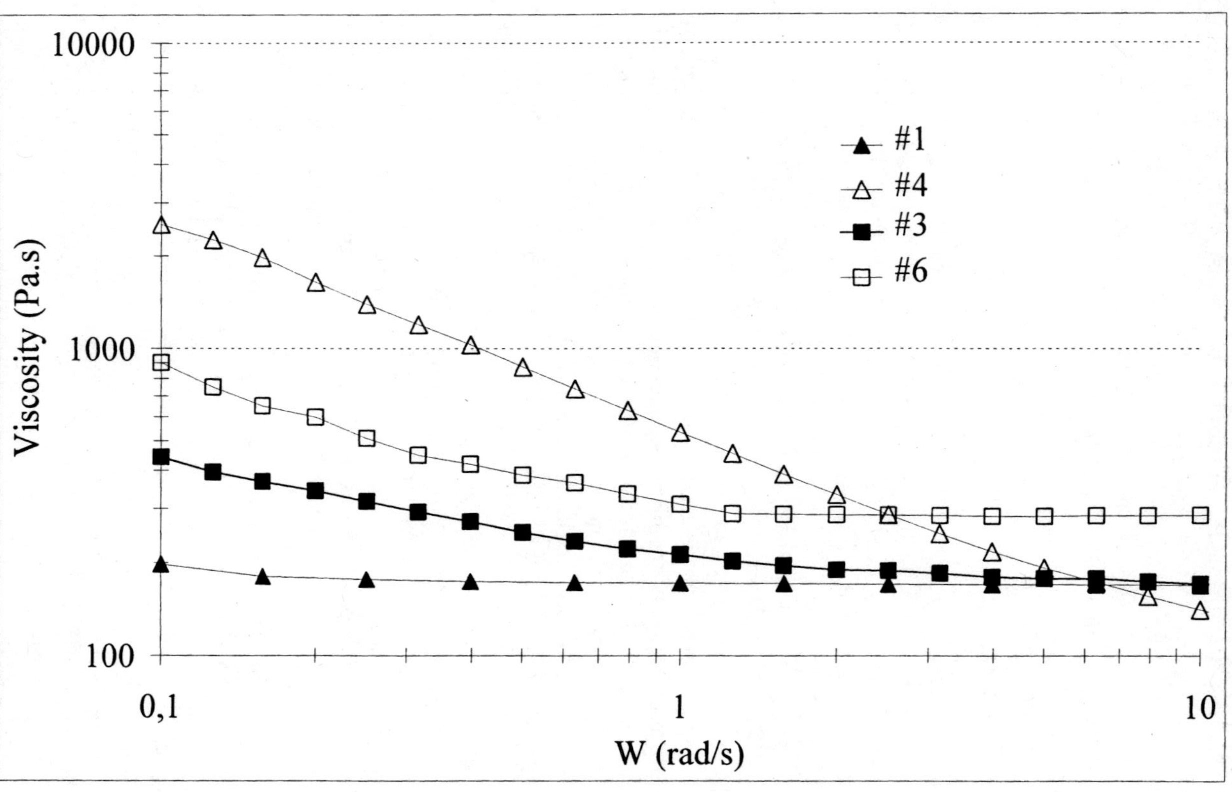


*Figure 3. Rheological behaviour at low frequencies of formulations #1, 3, 4, 6.*

When MMT-P is added to pure PET (#4) or to PET/PC blends (#5 and #6), the initial thermal stability is significantly increased, as well as the amount of char formed and its thermal stability. These results are not in agreement with literature results, since previous authors observed a decrease of initial thermal stability for various organomodified clays in PET (*7*). The increase of thermal stability in our case may be related to a flame retardant action of the phosphorus in MMT-P or to the fact that the recycled PET used has a very low thermal stability.

The fact that the presence of MMT-P tends to increase the initial thermal stability of the blends supports the assumption of a compatibilization mechanism involving a limited transesterification reaction between PET and PC. As observed in previous studies (*3,4*), a strong catalysis of transesterification would have had as a side effect a decrease the average PET chain length, resulting in a significant decrease of initial thermal stability.

### Mechanical properties and reaction to fire

Table II shows the flexural modulus values obtained for the different formulations. It can be seen that blending PET with PC has no significant influence on the modulus, and that the reinforcement induced by MMT-P is much stronger in pure PET than in the PET/PC blends. This is in agreement with the above indications on the morphologies of formulations #1 to 7.

Concerning reaction to fire, both the introduction of PC and/or MMT-P have a positive effect on the LOI which reaches 31.5 % for formulation #6. The influence on cone calorimetry and epiradiateur tests is more complex, as illustrated in Table II and Figures 7 and 8. In both tests, and as described in detail in a previous paper and in agreement with TGA results presented on Figure 6, the introduction of 20% w./w. of PC reduces the time to ignition (TTI and TTFI, respectively), which increases for an higher amount of PC (50% w./w.). At the same time, the pHRR and MIP decrease linearly with PC content due to the increase of charring.

When MMT-P is added to pure PET, the pHRR and the MIP also strongly decrease, while the TTI in cone calorimeter test decreases, the TTFI in epiradiateur tests increases. Actually, this complex behavior has been observed in other systems involving organomodified MMT, and can be attributed to the different fire scenarios involved in the two tests (*8*).

A similar complex behavior is observed when MMT-P is introduced in the PET/PC blends. Nevertheless, for these formulations (#5 and #6), the decrease of pHRR and MIP is less important than when MMT-P is introduced in pure PET. This may be related to the complex morphology of these systems, when a significant fraction of the clay could be located at the interface between PET and PC.

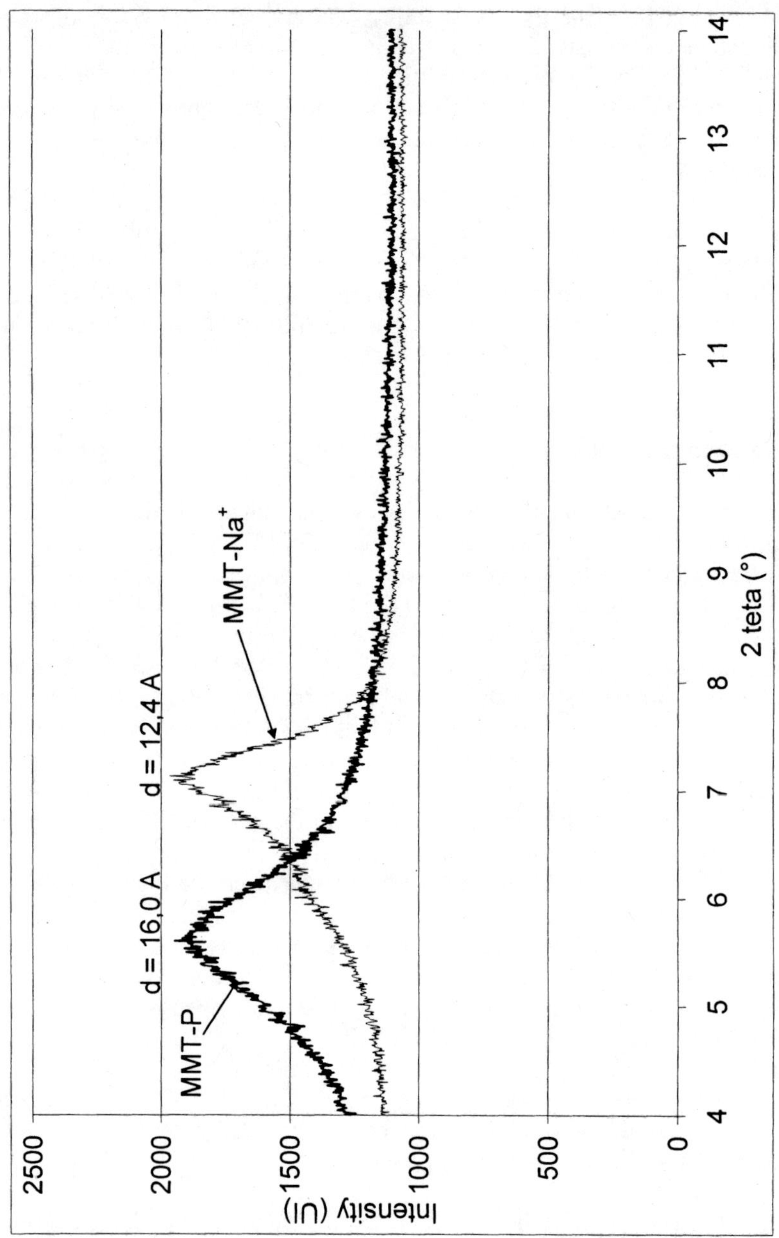

Intensity (UI)
2500
2000
1500
1000
500
0
4
5
6
7
8
9
10
11
12
13
14
2 teta (°)
d = 16,0 A
d = 12,4 A
MMT-P
MMT-Na+

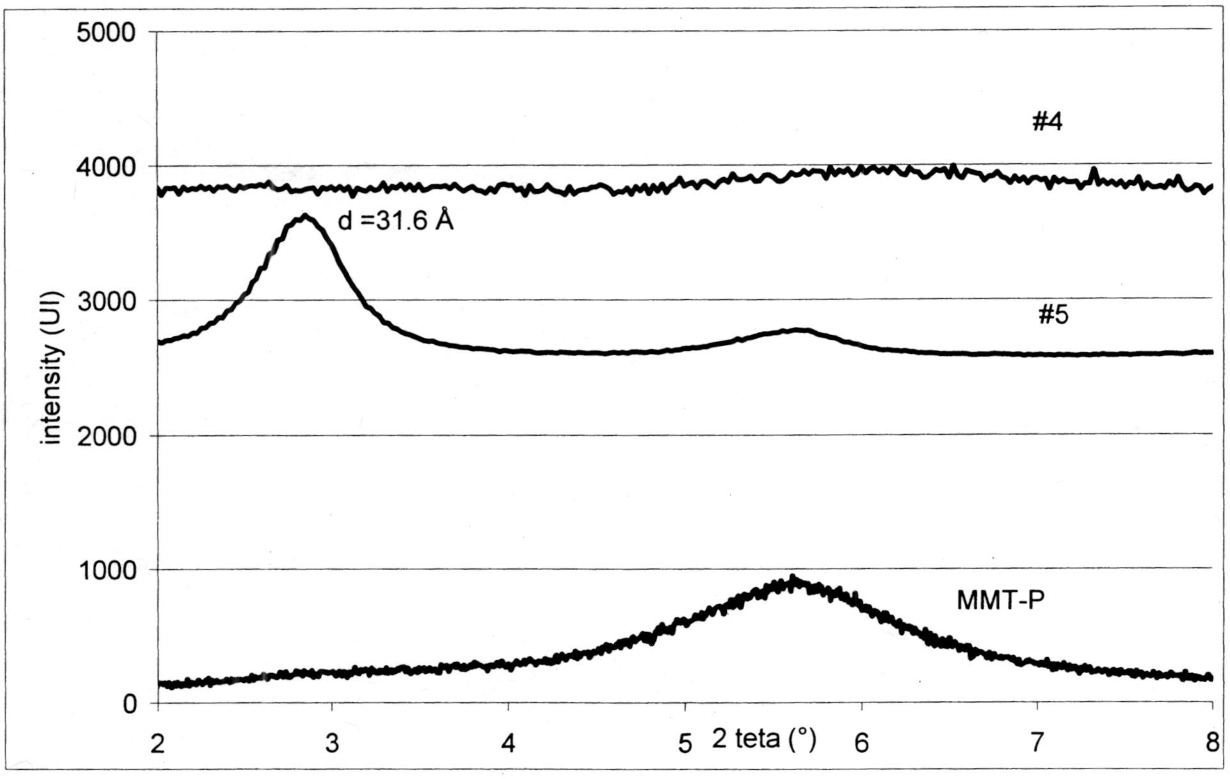


*Figure 4. X-rays diffraction of pristine, organo-modified clay and formulations #4,5.*

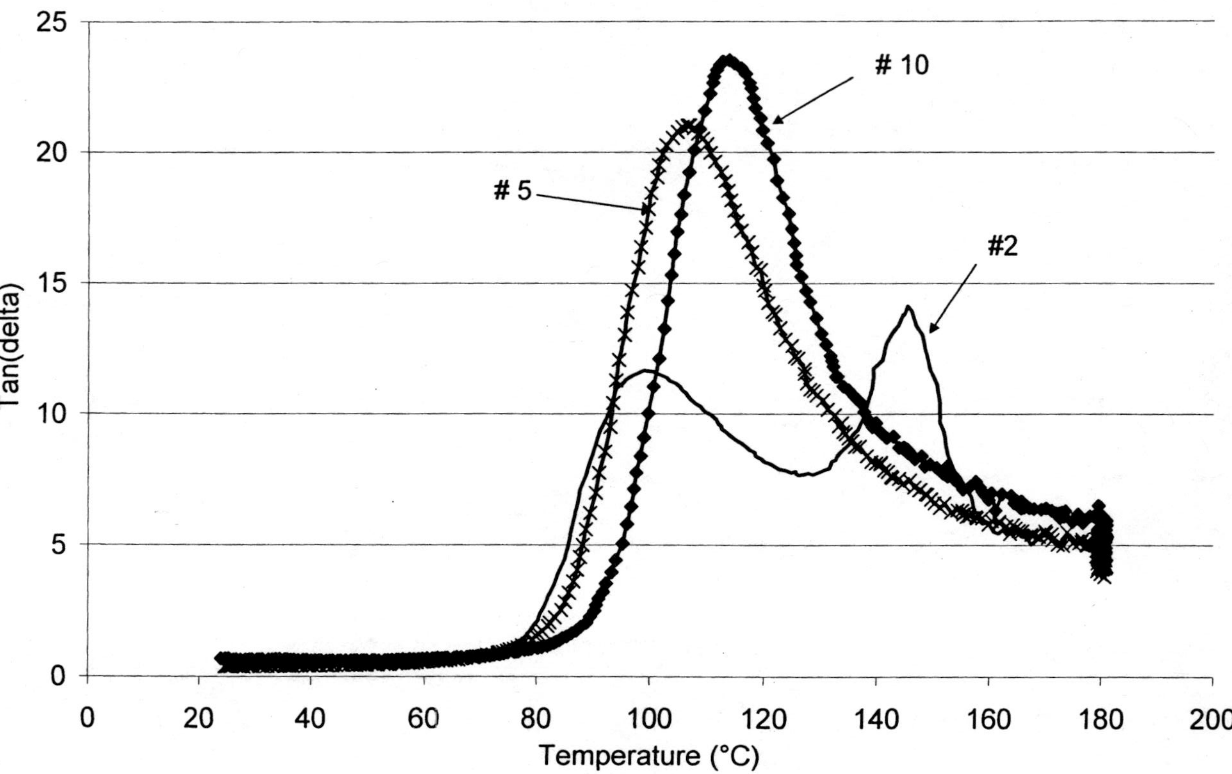


*Figure 5. Dynamic mechanical analysis of formulations #2,5 and 10*

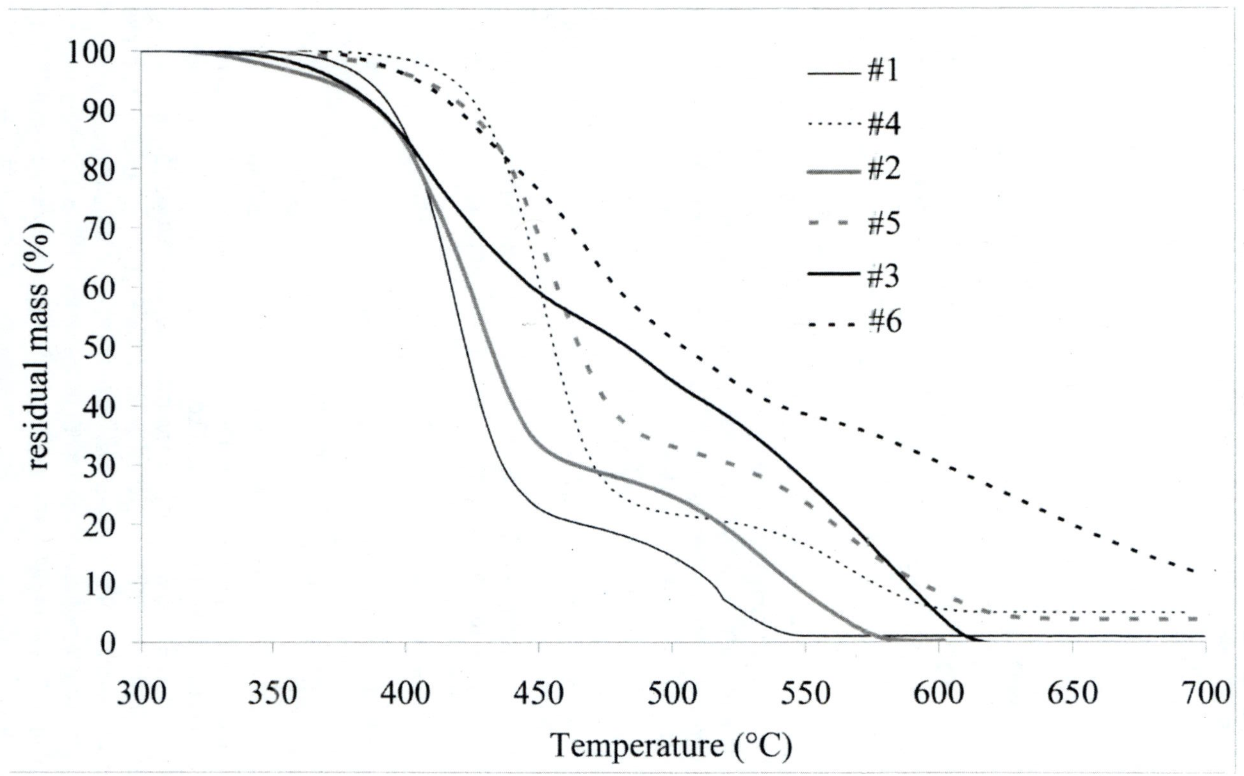


*Figure 6. TGA mass loss curves of formulations #1, 2, 3, 4, 5, 6.*

**Table II. Mechanical properties and reaction to fire results.**

| *Sample* | *modulus (MPa)* | *LOI (%)* | *TTI (s)* | *pHRR (kW/m²)* | *TTFI (s)* | *MIP (s)* | *UL94 rating* |
|---|---|---|---|---|---|---|---|
| 1 | 2017 | 22.5 | 71 | 700 | 97 | 19.0 | V-2 |
| 2 | 2004 | 24.0 | 66 | 550 | 89 | 17.6 | V-2 |
| 3 | 2102 | 27.5 | 95 | 420 | 110 | 10.8 | V-2 |
| 4 | 2915 | 26.0 | 65 | 380 | 103 | 15.5 | V-2 |
| 5 | 2788 | 27.0 | 62 | 400 | 95 | 13.2 | V-2 |
| 6 | 2642 | 31.5 | 82 | 450 | 121 | 10.5 | V-2 |
| 7 | 2664 | 22.9 | 63 | 450 | 67 | 36.5 | V-2 |
| 8 | 2505 | 28.0 | 83 | 449 | 104 | 19.7 | V-2 |
| 9 | 3047 | 29.5 | 88 | 250 | 106 | 9.8 | V-0 |

In this last paragraph, we will discuss the properties of formulations #8 and #9. As stated in the introduction, we observed in previous work that the introduction of triphenylphosphite (TPP), in a PET/PC blend after its compatibilization by a catalysed transesterification reaction, resulted in significantly improved reaction to fire, including V-0 rating in the UL94 test (*4*). The performance obtained here for formulation #9 clearly shows that this formulation strategy can also be applied to optimize the reaction to fire of a PET/PC blends compatibilized by organomodified montmorillonites. The comparison of the results obtained for formulations #2, 5, 8 and 9 on Table 2 and of the HRR cone calorimeter curves shown in Figure 8, shows the fire retardant synergy between MMT-P and TPP, which combines a long time to ignition and a very low pHRR, and to obtain V-0 rating in the UL94 test. In addition, the reinforced blend obtained has a 50% increased flexural modulus compared to initial recycled PET.

## Conclusion

The addition of the combination of a phosphonium modified montmorillonite and triphenylphosphite, to recycled PET/PC blends give good mechanical properties and reaction to fire performance, including V-0 rating in the UL94 test. The organomodified montmorillonites act as flame retardants, mechanical reinforcers and compatibilizers of the PET/PC blend. TPP acts as a flame retardant due to its phosphorus content (9% by weight) and as a chain extender able to promote bridging reactions between PET chain ends. It is also suggests that TPP can regulate the transesterification reaction between PET and PC.

These results illustrate the interest of combining various formulation strategies to obtain engineering polymeric materials, even from recycled plastics

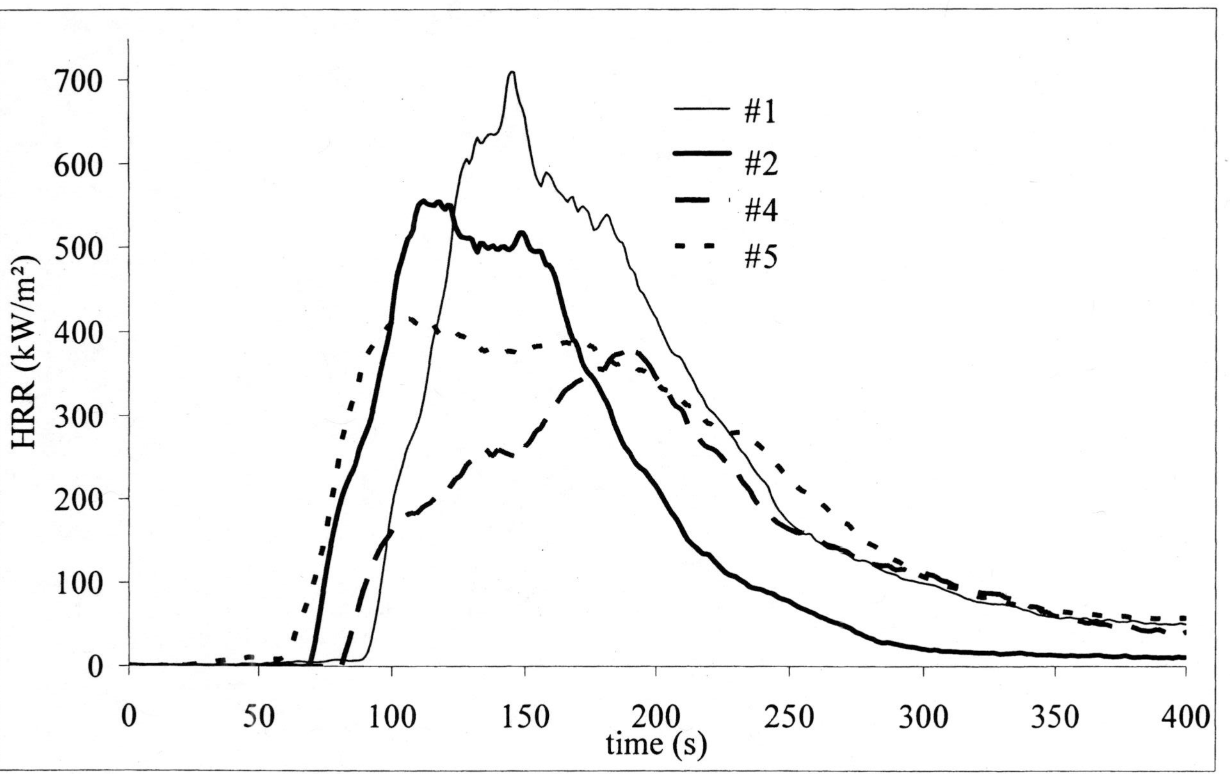


*Figure 7. Heat release rate curves obtained in cone calorimeter tests for formulations #1,2,4,5.*

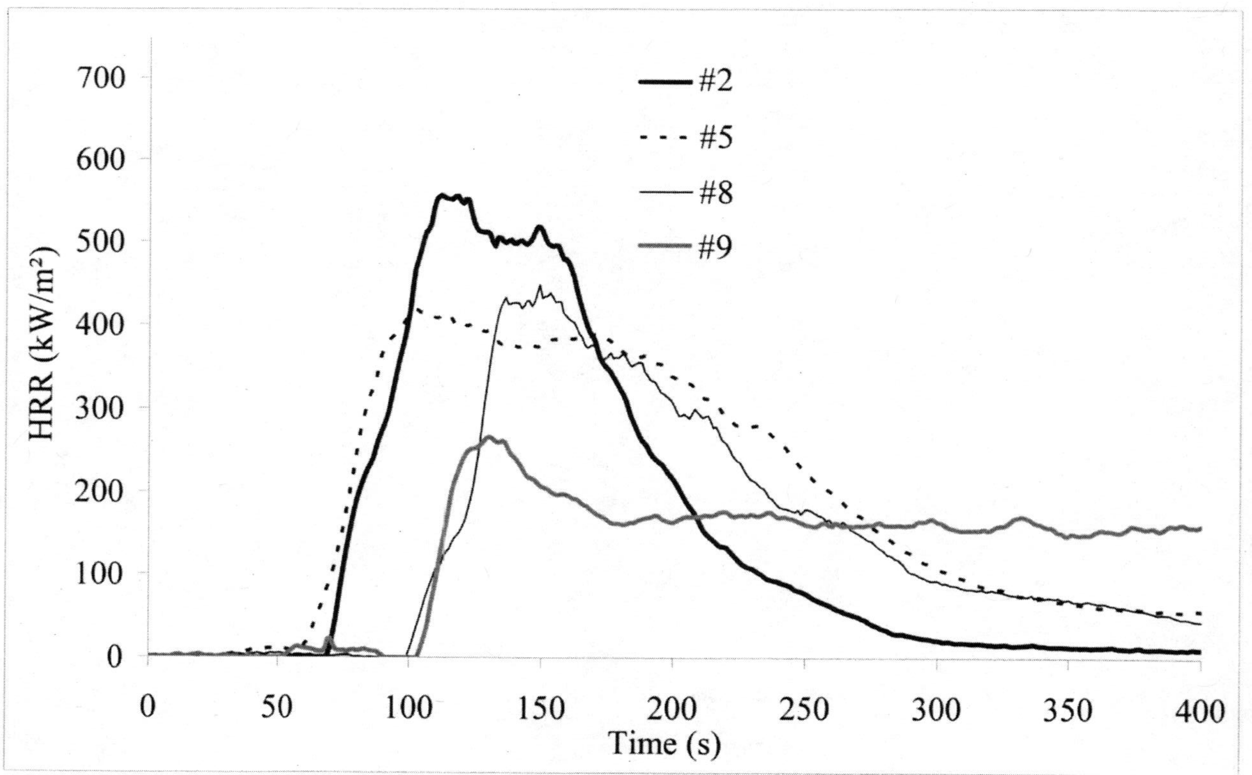


*Figure 8. Heat release rate curves obtained in cone calorimeter tests for formulations #2,5,8,9.*

having intrinsically low performances: polymer blends, nanocomposites and combinations of organoclays with organic fire retardant molecules.

## References

1. Dinger, P.; Proc. of the Identiplast Conference., Brussels, 2001.
2. Torres, N. ; Robin, J.J. ; Boutevin, B. *Eur. Polymer J.* **2000**, *36(10)*, 2075.
3. Swoboda, B.; Buonomo, S.; Leroy, E.; Lopez Cuesta, J.M. *Polym. Deg. Stab.* **2007**, *92(12)*, 2247-2256.
4. Swoboda, B.; Buonomo, S.; Leroy, E; Lopez Cuesta, J.M. *Polym. Deg. Stab.* **2008**, *93(5)*, 910-917.
5. Manias, E.; Heidecker, M.J.; Chung, J.-Y.; Mason J., *ACS Polymer Preprints* **2007**; 138-1054828.
6. Sinha Ray, S.; Bousmina, M.; Maazouz A. *Polym. Eng. Sci.* **2006**, *46(8)*, 1121-1129.
7. Costache, M. C.; Heidecker, M.J.; Manias, E. ; Wilkie, C. A. *Polym. Adv. Tech.* **2006**, *17*, 764-771.
8. Cinausero, N.; Lopez-Cuesta, J. M.; Laoutid, F.; Piechaczyk, A.; Leroy E.. In "Fire Retardancy of Polymers: New Strategies and Mechanisms", Editors, Hull, R.; Kandola, B. K.; The Royal Society of Chemistry, Oxford, 2008.

# Chapter 8

# Development of Polyurea Nanocomposites with Improved Fire Retardancy

**Walid H. Awad[1], Calistor Nyambo[1], Seogjun Kim[1,3], Robert J. Dinan[2], Jeff W. Fisher[2], and Charles A. Wilkie[1,*]**

**[1]Department of Chemistry and Fire Retardant Research Facility, Marquette University, P.O. Box 1881, Milwaukee, WI 53201**
**[2]Air Force Research Laboratory, Tyndall Air Force Base, FL 32403**
**[3]Permanent address: Department of Nano and Chemical Engineering, Kunsan National University, Kunsan, Chonbuk 573 701, South Korea**

The two-component polyurea elastomeric system is a unique technology with a wide range of applications in the coating, lining and caulking industries. Polyureas have shown remarkable properties including, fast cure, abrasion resistance, water repellency, flexibility, exceptional elastomeric qualities, resistance to thermal shock, good adhesion proerties, etc. Despite all these characteristics, one drawback for polyureas is that, like all organic polymers, they have relatively high flammability, which puts restrictions on their broad application. The main objective of this work is to develop a polyurea system with orders of magnitude improvement in fire performance using a nanoadditive approach in combination with conventional flame retardants (FR). The cone calorimeter was used as the tool of evaluation.

Polyurea, as a state of the art coating system, exhibits a variety of remarkable physical and mechanical properties that have a wide range of industrial and commercial applications (1). Pure polyurea is derived from the reaction of an isocyanate and a diamine without a catalyst. The isocyanate can be aromatic or aliphatic in nature.

Polyureas are of interest because they are successfully used in protective systems against blast and ballistic effects. The right combination of tensile strength, stiffness and elongation at rupture are critical in this role. One of the challenges is to develop a suitable fire retardant package while maintaining these qualities. In order to accomplish this, the additive package must be limited.

Over the last 15 years, a new type of technology capable of reinforcing the polymer on the nanoscale level has drawn considerable attention. This developing technology, termed polymer nanocomposites, uses nanoparticles as the reinforcing fillers. By dispersing the nanometer thick, high aspect ratio, particles in a polymer matrix, a wide array of property enhancements can be achieved, e.g., enhanced mechanical properties (2-4), and superior flame retardancy (5,6).

Polymer nanocomposites can be reinforced by iso-dimensional phases, which have three dimensions in the nanometer range such as silica nanoparticles obtained by in-situ sol-gel methods or by polymerization promoted directly from their surface. They can also be reinforced by a phase that has only two dimensions in the nanometer scale, such as carbon nanotubes. A third type of nanocomposite incorporates a reinforcing phase in the form of platelets with only one dimension on the nano-level. Polymer/ layered silicate nanocomposites belong to this class.

The utility of nanoparticles as reinforcing additives and potential flame retardants for the polyurea system will be explored. In previous work (7) we have shown that a significant reduction in the peak heat release rate was obtained by blending the polyurea with phosphorus-based flame retardants such as ammonium polyphosphate (APP), melamine polyphosphate or Fyrol PMP (a proprietary phosphonate). As part of our continuing effort to develop polyurea materials with a reduction in peak heat release rate well above 90%, different classes of nanoparticle additives, such as nanoclays, carbon nanotubes and polyhedral oligomeric silsesquioxane (POSS), have been explored for potential synergistic formulations with conventional flame retardants.

# Experimental

### *Materials*

The materials used throughout this work are: Isocynate (Isonate 143L from Dow Chemical Company), Diamine (Versalink P1000 from Air Products),

Ammonium Polyphosphate APP, (ICL Performance Products Inc), Expandable Graphite (Graftech), Firebrake Zinc Borate (Borax), Cloisite 30B (Southern Clay Products), Carbon nanotubes (Nanocyl) and POSS Poly(vinylsilsesquioxane) (Hybrid Plastics).

### *Sample preparation*

The polyurea samples were prepared in a beaker by mixing the additives in the diamine for several minutes. The dispersion of the additives was achieved using mechanical mixing followed by ultrasonication. Then, the isocyanate was added to the mixture with continuous stirring for one minute, and the contents of the beaker were poured into a mold. The samples were cured at room temperature for 12 hours, and then were placed in a vacuum oven at 70 °C for additional 24 Hours.

### *Instrumentation*

Cone calorimetry was performed using an Atlas Cone 2 instrument according to ASTM E 1354 at an incident flux of 50 $KW/m^2$ using a cone shaped heater. Exhaust flow was set at 24 L/s and the spark was continuous until the sample ignited. Typical results from cone calorimetry are reproducible to within about ± 10%; these uncertainties are based on many runs in which thousands of samples have been combusted.

X-ray diffraction (XRD) data was obtained using a Rigaku MiniflexII diffractometer equipped with Cu-$K_\alpha$ source ($\lambda$ = 1.5404 Å). It was operated at 50 kV and 20 mA and scans were obtained from $2\theta$ = 1 to 10 °, at 0.1 ° step size.

## Results and Discussion

Polyurea, as a state of the art coating system, exhibits extraordinary performance characteristics such as: fast cure (gel time < 5 seconds), abrasion resistance, water repellency, flexibility, exceptional elastomeric quality, resistance to thermal shock and good adhesion properties. The use of nano materials can result in enhanced modulus, strength, elongation at break, toughness, glass transition temperature, reduced thermal shrinkage as well as reduced flammability. The nano-additives-based polyurea materials hold the potential of providing a new generation of materials to withstand blast and ballistic effects.

Two approaches have been employed during the course of this investigation to reduce the flammability of a polyurea system. The first is to use nanoparticles with different dimensionality and aspect ratios, such as organo clay, carbon nanotubes and POSS materials. The second approach is to use conventional flame retardants, such as expandable graphite, ammonium polyphosphate and zinc borate.

### *Polyurea nanocomposite*

X-ray patterns for the pristine organoclay (Cloisite 30B) reveals a broad intense peak around $2\theta = 5°$ corresponding to a basal spacing of 1.85 nm. This peak disappears in a polyurea nanocomposite having an approximate clay concentration of 3 wt.% . The disappearance of the diffraction peak for the clay after compounding with polyurea indicates good dispersion of the clay in the polymer matrix and the formation of a disordered nanostructure.

A graphite-based nanocomposite was obtained by dispersing expandable graphite particles in the polyurea matrix using ultrasonification. Expandable graphite is a sulfuric acid intercalated graphite. XRD for the pure expandable graphite gave a very strong peak at $2\theta = 26.6°$ corresponding to the typical d-spacing between the carbon layers in graphite (0.33 nm). The huge reduction in the intensity of the diffraction peak for the expandable graphite after compounding with polyurea provides evidence for the dispersion of the graphite particles in the polyurea matrix.

## Flammability Measurement

We have characterized the flammability properties of polyurea formulations, under fire like conditions, using the cone calorimeter. A set of fire-relevant properties can be obtained from the cone calorimeter, such as peak heat release rate (PHRR), mass loss rate (MLR), specific extinction area (SEA), ignition time ($t_{ign}$), and specific heat of combustion. These data will give indications about the fire performance of the polyurea materials.

### *Nano-additive formulations*

The effect of nanoclays, carbon nanotubes and POSS on the flammability properties of polyurea have been investigated. Ultrasonication was employed to achieve good dispersion of these nanoparticles in the polyurea matrix. An improvement in the flammability properties of the base polyurea has been obtained with nanoscale additives and these filled systems provide an alternative to conventional flame retardants.

The cone calorimeter data for polyurea-Cloisite 30B nanocomposite showed that the peak heat release rate (PHRR) of polyurea is reduced (33%) upon the addition of the nanoclay. For carbon nanotubes and POSS nanoparticles the results are more significant. A substantial reduction (45%) in the peak heat release rate (PHRR) has been obtained by the dispersion of multiwall carbon nanotubes in the polyurea matrix at 0.5 wt.% loading, as shown in Figure 1. Increasing the loading of carbon nanotubes above this limit has no significant effect on the flammability properties of polyurea. A more significant reduction in PHRR (73%) has been achieved by dispersing POSS nanoparticles in the polyurea matrix (Figure 2). This improvement in flammability properties have been obtained at loading of 5 wt.%. Again, increasing the loading of POSS above this limit has little effect on PHRR value.

Two mechanisms have been reported in the literature to describe the improvement in flame retardancy of nanocomposites. The first mechanism proposed by Gilman et. al (8) states that during the burning process, a carbonaceous silicate multilayer structure is formed. This layer acts as a barrier to protect the underlying polymer from further degradation. Recently, a nanoconfinement mechanism has been proposed to describe the improvement in the flame retardancy of nanocomposites (9). According to this mechanism, the radicals produced from the polymer degradation are nanoconfined permitting a variety of bimolecular reactions to occur.

On the other hand, Kashiwagi et. al (10) have indicated that the enhancement in the flammability properties of nanocomposites is due to the formation of a network structure protective layer, and that the aspect ratio of the nanoparticles has an impact on the flammability properties. The particles with higher aspect ratio tend to form a more protective network structure that is critical to the reduction of the heat release rate of the nanocomposites (11).

*Conventional flame retardants formulations*

As a flame retardant, the important property of expandable graphite is the onset of expansion temperature and the expansion volume. Therefore, using expandable graphite, the temperature should be controlled to avoid its expansion during the processing. The d-spacing in expandable graphite increases from 0.335 nm to approximately 0.789 nm, and upon heating one obtains expanded graphite, in which an increase in the d-spacing of several hundreds times along the c axis is observed.

Our results indicated that expandable graphite (EG) is effective in reducing the flammability of polyurea. Figure 3 compares the heat release rate curves for different loadings of expandable graphite. The peak heat release rate (PHRR) was reduced by up to 85% compared to the pure resin at 3% loading content. Increasing the loading of expandable graphite made a little improvement in PHRR value. As shown in Figure 3, the reduction obtained in PHRR values were 86% and 88% at 5 wt.% and 10 wt. % respectively.

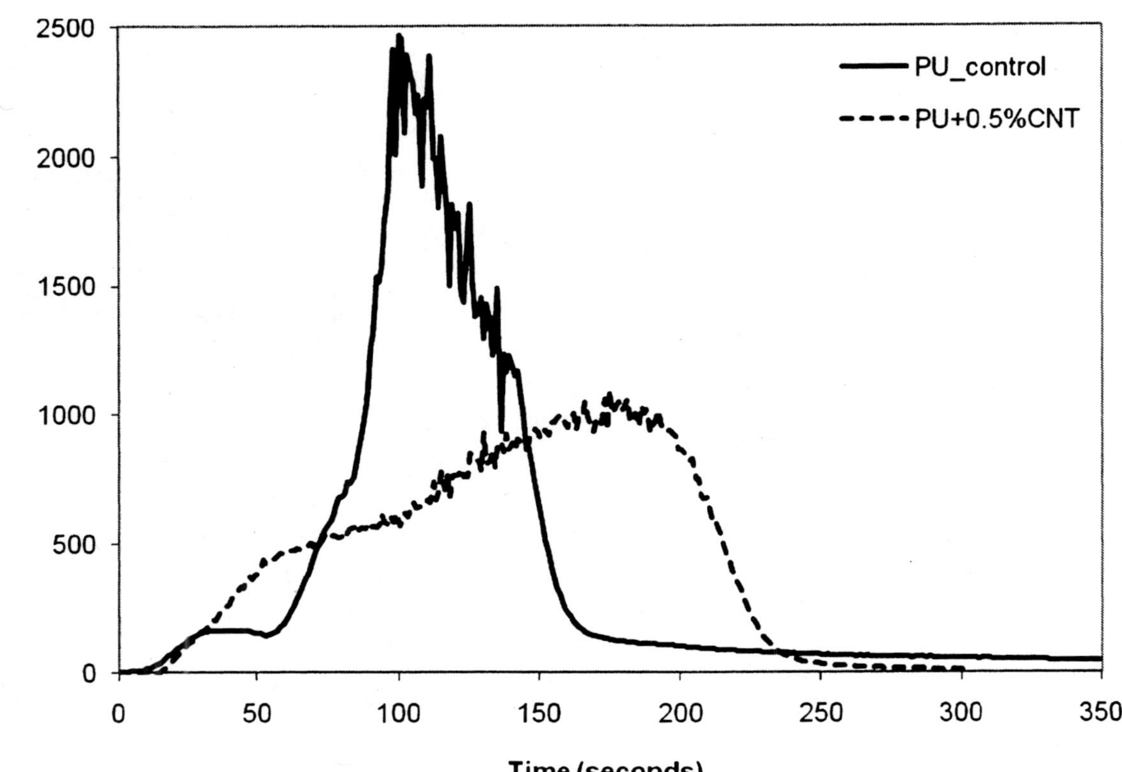


*Figure 1. heat release rate curves for polyurea/CNT nanocomposite.*

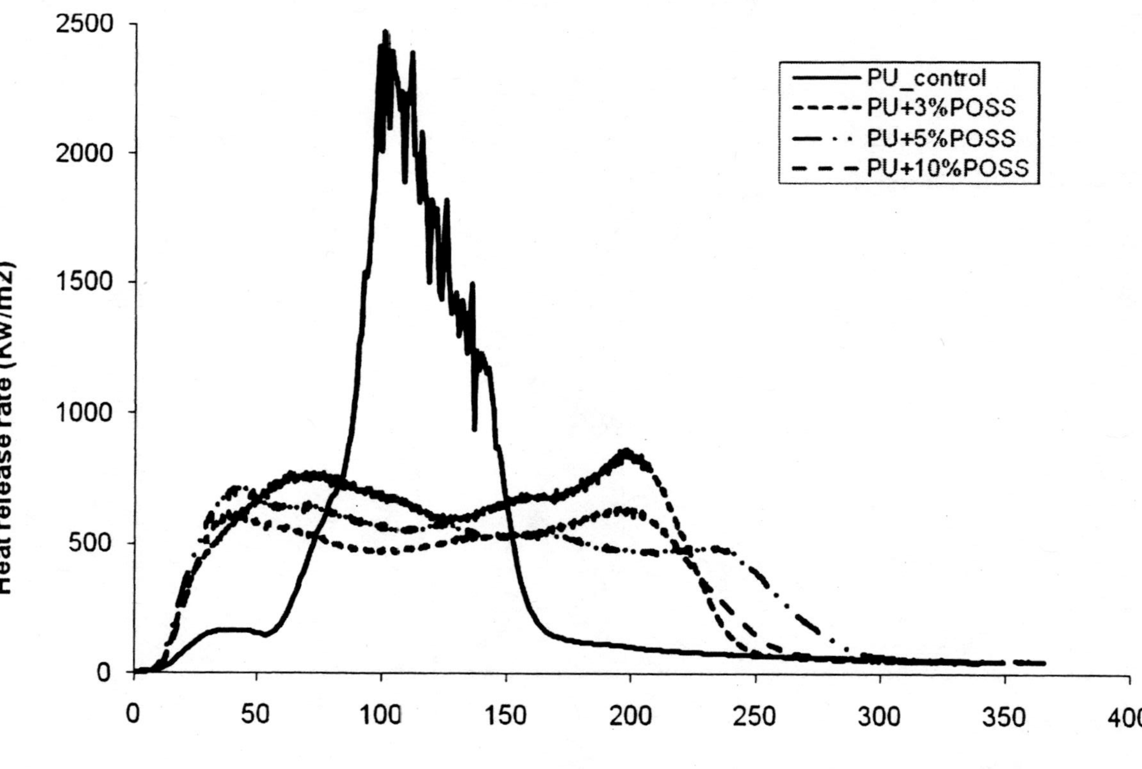


*Figure 2. heat release rate curves for polyurea/POSS nanocomposites.*

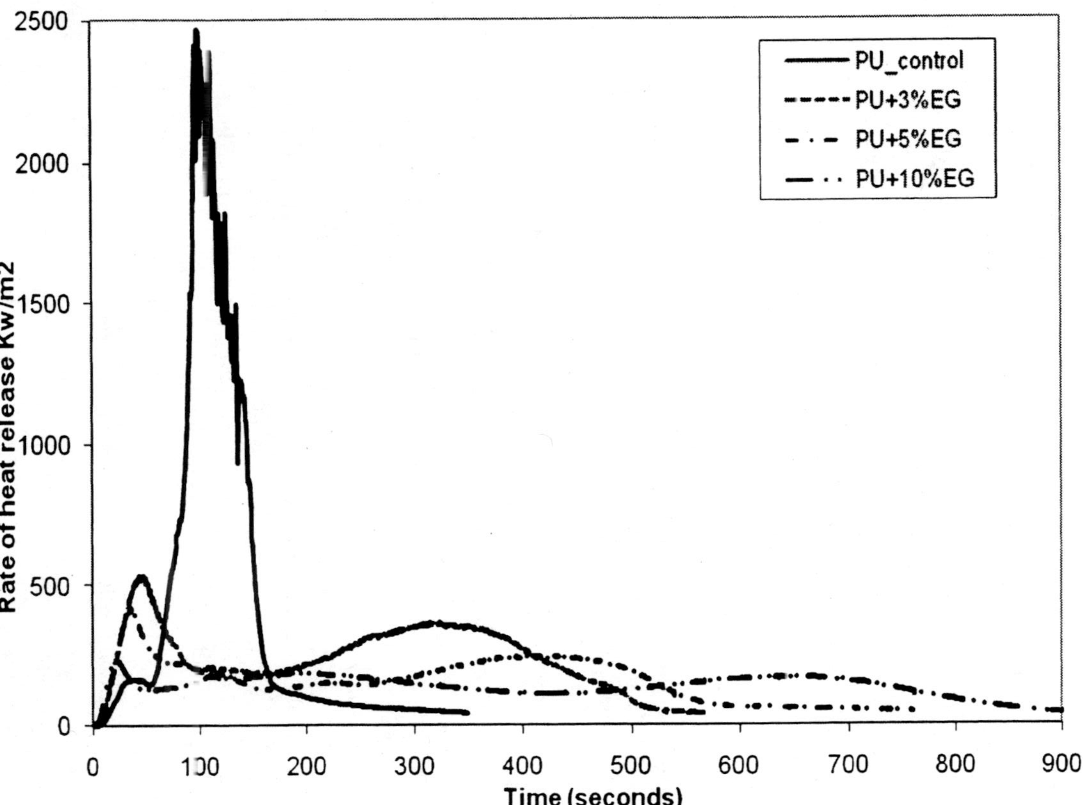


*Figure 3. Heat release curves for polyurea with different loading contents of expandable graphite*

Expandable graphite acts as a blowing agent as well as a carbonization agent. The intumescent shield formed after burning is composed of expanded graphite 'worms' embedded in the degraded matrix of PU. Its structure consists of a carbonaceous and polyaromatic structure. This expanded carbon layer works as an insulating layer to reduce the heat transfer (12-14).

The influence of ammonium polyphosphate (APP) on the flammability of polyurea has also been examined. The chain length (n) of this polymeric compound is both variable and branched, and can be greater than 1000. Short and linear chain APPs ($n < 100$) are more water sensitive (hydrolysis) and less thermally stable than longer chain APPs ($n > 1000$), which show a very low water solubility.

Cone calorimetry at an external heat flux of 50 $KW/m^2$ for polyurea compounded with varying amounts of APP reveals significant reduction in PHRR values. There is a 67% reduction in PHRR value at 3 wt. % loading content. Increasing the APP content to 5 wt.% , 10 wt.% and 15 wt. %, the reduction in PHRR values were 68%, 75% and 79% respectively (plots are not shown).

Ammonium polyphosphate is known to have an intumescent effect, when exposed to heat. When exposed to fire, APP starts to decompose, commonly into polymeric phosphoric acid and ammonia. The polyphosphoric acid then converts into a non-stable phosphate ester. The dehydration of the phosphate ester follows and carbon foam is formed on the surface of the sample preventing further decomposition of the polymer (15,16).

Zinc borate (ZB) also proved to be effective in imparting flame retardancy to the polyurea. A 36 % reduction in PHRR was obtained at 3 wt.% loading content (not shown). Increasing the zinc borate content to 5 wt.% only gave a slight improvement in the flammability properties of polyurea (41% reduction in PHRR value). The principle mechanism of zinc borate is the formation of a glassy char in the condensed phase. This acts as a fire resistant coating that protects unburned polymer from contact with heat and further combustion. Approximately 14% of the zinc borate crystal is composed of water, which contributes to flame retardation by reducing the energy in a burning system when it is released at 290 °C (an endothermic reaction).

*Conventional flame retardants with nano-additives*

The question of synergy between the traditional flame retardants and nanoparticles has been addressed during this investigation. Adding carbon nanotubes at 0.5 wt% to 3 wt% expandable graphite yieded a slight decrease in PHRR value from 85% to 88%. Adding zinc borate to the previous formulation had no impact on the PHRR value. The cone calorimeter residue indicated that both the carbon nanotubes and zinc borate enhanced the integrity of the char structure.

Figure 4 shows that combining the intumescent additives, expandable graphite and ammonium polyphosphate in polyurea at 10 wt.% loading from each, gave a 86% reduction in the PHRR value. Adding 0.2 wt% carbon nanotubes to this intumescent formulation delivered 91% reduction in PHRR value .

Another remarkable improvement in the flammability properties of polyurea was achieved by adding either zinc borate or POSS at 3 wt.% to the following formulation: PU/10% EG/10% APP. The PHRR was reduced 92% and 93%, respectively (Figure 5). The char residue indicates that zinc borate imparts integrity to the char structure while the POSS nanoparticles made the char structure more brittle.

### *Mechanical properties*

Polyureas are characterized by high elongation, good tear strength, and a moderate combination of tensile strength and modulus of elasticity. The extensive intermolecular hydrogen bonding of polyureas leads to "tough" mechanical properties. Polyureas are used as energy-absorbing material in systems that provide protection against ballistic and blast effects (17, 18). Such applications motivate the mechanical response study of the material under high strain rates (200-800 in/in per second). The PU molecular mechanism during blast and ballistic response is not entirely understood. A contributing factor is the substantial energy dissipation within due to a strain-induced transition from the rubbery to the glassy state (19). The glass transition zone of polymers is the region of greatest energy dissipation and this transition can be induced by fast deformation rates. Our goal is to develop polyurea nanocomposites that are flame retardant yet provide the needed mechanical resistance.

Figures 6-8 show the engineering tensile stress-strain curves (340/s) for polyurea with different additives (APP, Cloisite 30B, and EG). The data indicated that the addition of 5% APP does not have a significant effect on the mechanical behavior but 10% APP does cause a signficant reduction in tensile strength, effective modulus of elasticity and elongation at rupture.(Figure 6). Likewise, an organically-modified montmorillonite at 1% loading does not have as great effect as 3% clay (Figure 7). Finally, expandable graphite does cause a significant loss in all three key properties at either 5% or 10% (Figure 8).

Based upon this testing, the total additive package must be kept as low as possible to achieve fire retardancy with minimal effect on the mechanical material properties.

## Conclusions

The main objective was to develop a polyurea system with orders of magnitude improvement in fire performance. The impact on tensile strength,

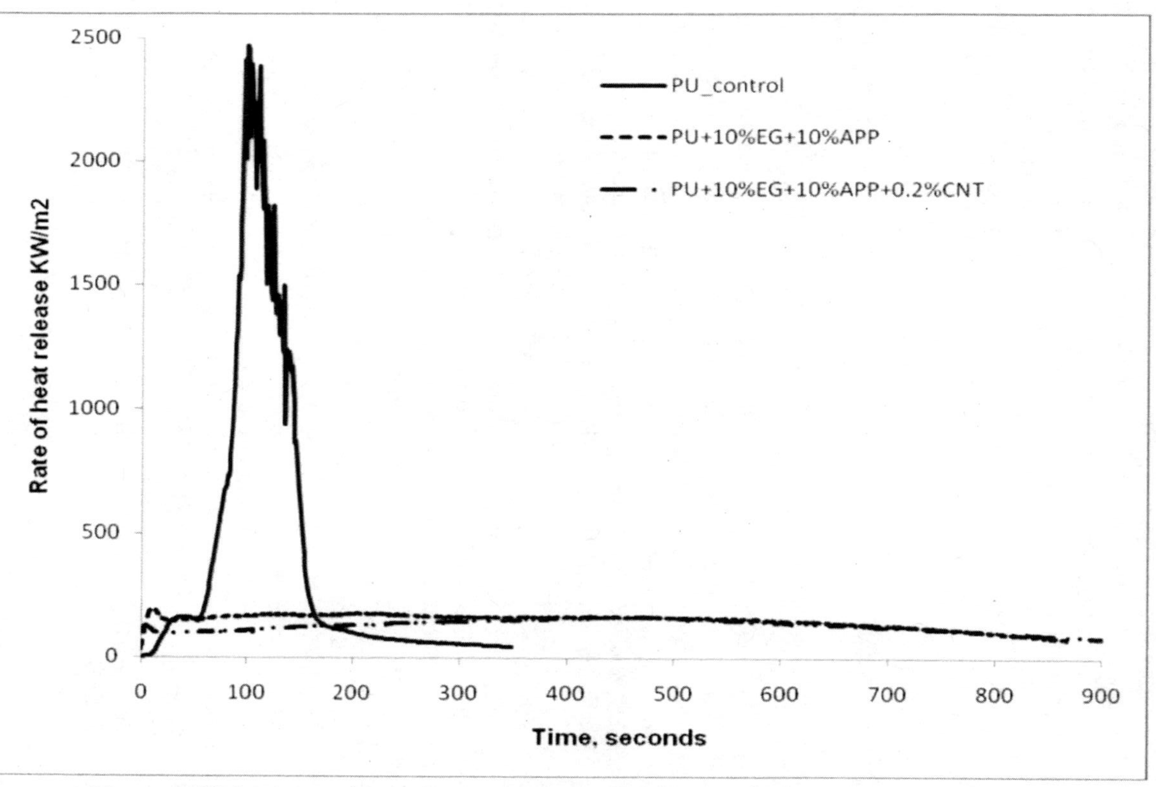


*Figure 4. HRR curves of formulations of polyurea with expandable graphite, ammonium polyphosphate and carbon nanotubes.*

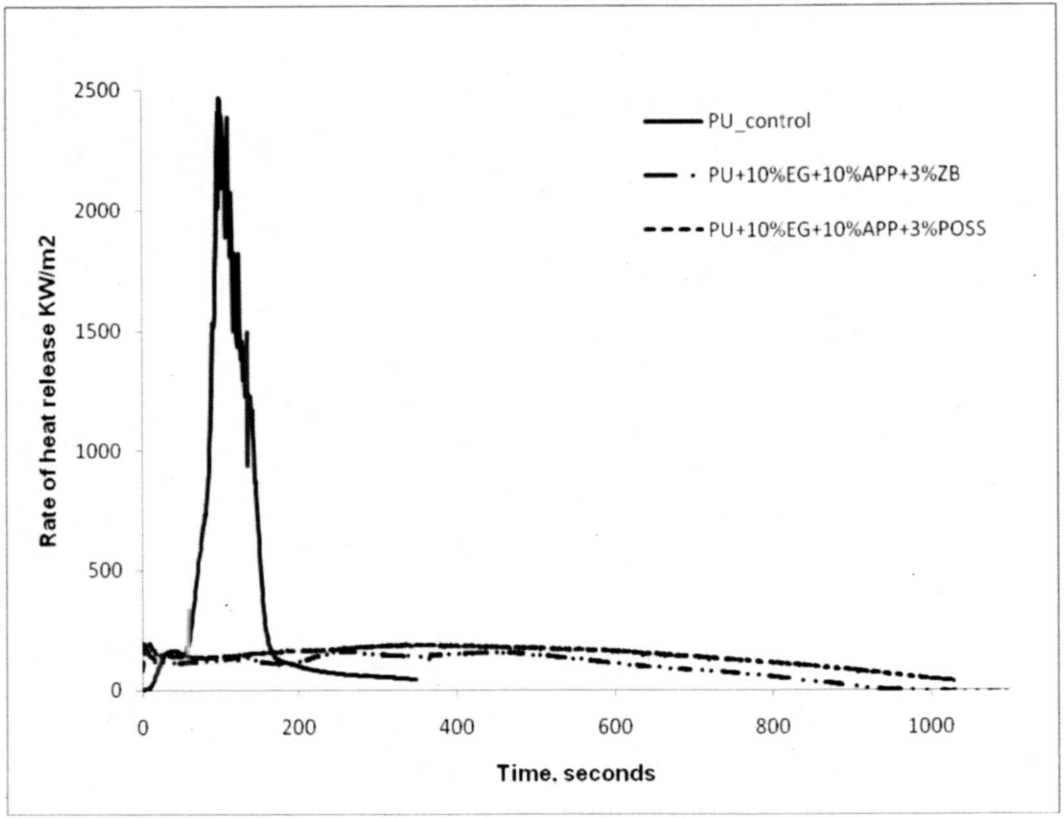

*Figure 5. HRR curves of polyurea formulations with EG, APP, ZB and POSS.*

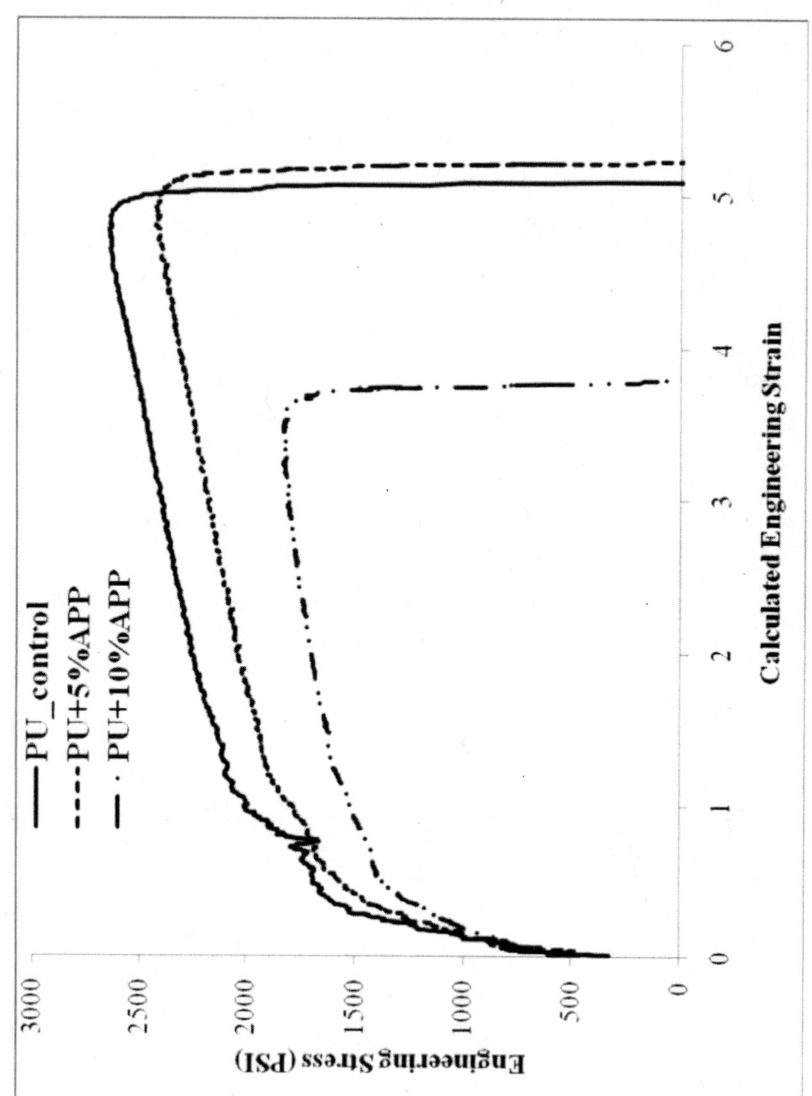


*Figure 6. High strain rate (340/s) of polyurea with ammonium polyphosphate*

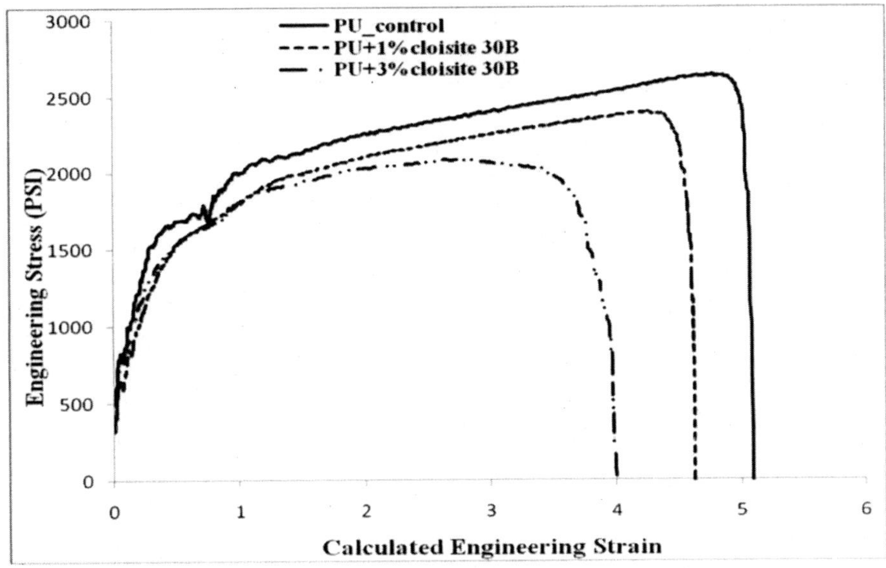


*Figure 7. High strain rate(340/s) measurement of polyurea with cloisite 30B*

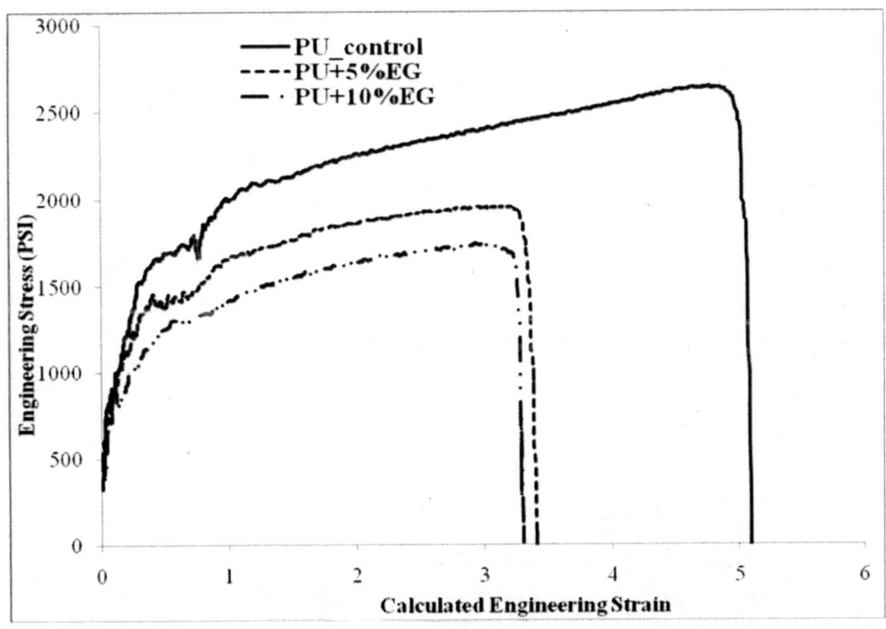


*Figure 8. High strain rate(340/s) of polyurea with expandable graphite*

effective modulus of elasticity and elongation at rupture was evaluated during the process. Several combinations of nanoadditives with conventional flame retardants were evaluated for this purpose. The following can be concluded:

- The nano particles, clay, carbon nanotubes and POSS, showed a substantial improvement in flammability of polyurea.
- Expandable graphite and ammonium polyphosphate as intumescent materials proved to be very effective in improving the flammability of polyurea.
- Combining the nano-additives with conventional flame retardants provided slight improvement in the flammability performance of polyurea.

## Acknowledgment

Work carried out at Marquette University has been supported by the US Air Force under award number FA8650-07-1-5901 and this support is gratefully acknowledged.

## References

1. http://www.pda-online.org/
2. Wang, Z; Pinnavaia, TJ. *Chem Mater* 1998; **10**(12): 3769.
3. Fornes, TD; Yoon, PJ; Keskkula, H; Paul, DR. *Polymer* 2001; **42**(25): 9929.
4. Fornes, TD; Yoon, PJ; Hunter, DL; Keskkula, H; Paul, DR. *Polymer* 2002; **43**(22): 5915.
5. Zhu J; Uhl FM; Morgan AB; Wilkie CA. *Chem. Mater.* 2001; **13**: 4649.
6. Zhu J; Morgan AB; Lamelas FJ; Wilkie CA. *Chem. Mater.* 2001; **13**: 3774.
7. Costache, M.C.; Kanugh, E.M.; Sorathia, U.; Wilkie, C.A. *J. Fire Sci.*, 2006, **24**, 433.
8. Gilman, J.W.; Jackson, C.L.; Morgan, A.B.; Harris, R. Jr., Manias, E.; Giannelis, E.P. Wuthenow, M.; Hilton, D. and Phillips, S.H. Chem Mater, 2000, **12**, 1866.
9. Chen, K.; Wilkie, C.A. and Vyazovkin, S. *J. Phys. Chem B,* 2007, **111**, 12685.
10. Kashiwagi, T; Du, F; Douglas, J.F.; Winey, K.I.; Harris, R.H; Shields, J.R. *Nat Mater* 2005, **4**, 928.
11. Cipiriano, B. H.; Kashiwagi, T.; Raghavan, S. R.; Yang, Y.; Grulke, E. A.; Yamamoto, K.; Shields, J. R.; Douglas, J. F. *Polymer,* 2007, **48**, 6086
12. Uhl, F.M.; Yao, Q.; Wilkie, C.A. *Polym. Adv. Tech.* 2005, **16**, 533

13. Uhl, F.M.; Yao, Q; Nakajima, H; Manias, E. Wilkie, C.A. *Polym. Degrad. Stab.* 2002, **76**, 111.
14. Duquesne,S.; Le Bras, M.; Bourbigot, S.; Delobel, R.; Vezin, H.; Camino, G.; Berend, E.; Lindsay, C. and Roels, T. *Fire Mater.* 2003; **27**:103.
15. Duquesne, S.; Le Bras, M.; Bourbigot, S.; Delobel, R.; Camino, G.; Eling, B.; Lindsay, C.; Roels, T.; Vezin, H. *J. Appl. Polym. Sci,* 2001, **82**, 3262.
16. http://www.specialchem4polymers.com/
17. Davidson, J.S.; Porter, J.R.; Dinan, R.J.; Hammons, M.I. and Connell, J.D. *J. Perform. Constr. Facil.,* 2004, **18(2)**, 100.
18. Davidson, J.S.; Fisher, J.W.; Hammons, M.I.; Porter, J.R. and Dinan, R.J. *J. Struct. Eng.* 2005, **131**, 1194.
19. Roland, C.M. ; Twigg, J.N. ; Vu, Y. ; Mott, P.N. *Polymer*, 2007, **48**, 574.

# Chapter 9

# Fire-Retardant Mechanism of Acrylonitrile Copolymers Containing Nanoclay

T. Richard Hull, Anna A. Stec, and Adam May

Centre for Fire and Hazards Science, University of Central Lancashire, Preston PR1 2HE, United Kingdom

The thermal decomposition and burning behaviour of acrylonitrile, containing nanoscopic sodium cloisite, and co-polymerised with diethyl(acryloyloxy-1-ethyl) phosphonate (DEAEP), and 2-acrylamido-2-methylpropane sulphonic acid (AMPS) has been investigated. The presence of these copolymers which decompose to form nucleophiles improves the thermal stability and reduces the flammability. This occurs through the formation of a more stable char, containing more nitrogen, during the early stages of decomposition, which is less dependent on the presence of oxygen during its formation, and so continues to form when engulfed by a flame, in essentially anaerobic conditions.

 

## Introduction

Polacrylonitrile (PAN) is a significant bulk polymer, extensively employed in the textile industry in fibre form, which is significantly more flammable than its natural counterparts, such as wool or cotton. It is also used extensively as a copolymer, for example in acrylonitrile-butadiene-styrene (ABS) formulations. Depending on conditions, polyacrylonitrile either decomposes by cyclisation, or chain scission resulting in volatile formation, or chain stripping with the release of hydrogen cyanide. The resulting structure may either decompose further producing volatiles, such as ethyne by random chain scission, and rearrangements such as cyclisation, resulting in ethyne, ethane, benzene and higher aromatics, or, under different (usually slower) conditions undergo rearrangements such that the polyene may cross-link, ultimately resulting in the formation of a protective char. This may shield the remaining polymer from radiative heating, while acting as a barrier to fuel and oxygen. Indeed, under controlled conditions of slow heating this process is used to manufacture carbon fibre from polyacrylonitrile fibre, and has been studied extensively (*1*).

## PAN decomposition

Horrocks (*2*) identified 3 pairs of competing volatilisation and cyclisation/carbonisation decomposition reactions in pure PAN which depended on atmosphere and heating rate. At the slowest heating rates in air, cyclisation (~350°C) was followed by the first carbonisation processes, around 450°C, followed by volatilisation at around 550°C. At higher heating rates, chain scission leading to volatilisation, at around 350°C competes with cyclisation or carbonisation processes. The residue then undergoes the first carbonisation stage, and the final volatilisation stage. It is this volatilisation stage which is responsible for the high flammability of acrylonitrile polymers in general.

In the production of carbon fibres from PAN, the initial stabilisation stage is conducted under carefully controlled conditions in the presence of oxygen, where a strongly exothermic process occurs around 200°C. In DSC studies, a sharp, narrow exothermic peak is observed under inert conditions in contrast to a broad exotherm, some 50°C higher, associated with oxidising atmospheres (*3,4*). The atmosphere is then changed for nitrogen in the subsequent thermal decomposition stages, in order not to oxidise the carbon fibres.

During the stabilisation stage, the following processes, shown in Figure 1 are believed to occur (*5*). Polyacrylonitrile (A), particularly as a result of nucleophilic attack (X), undergoes cyclisation of nitrile groups leading to the formation of hydronaphthiridine rings (*6*) (B). The hydronaphthiridine rings may undergo oxidative dehydrogenation leading to acridone and other structures(*7*) (C) increasing the aromaticity (*8*).

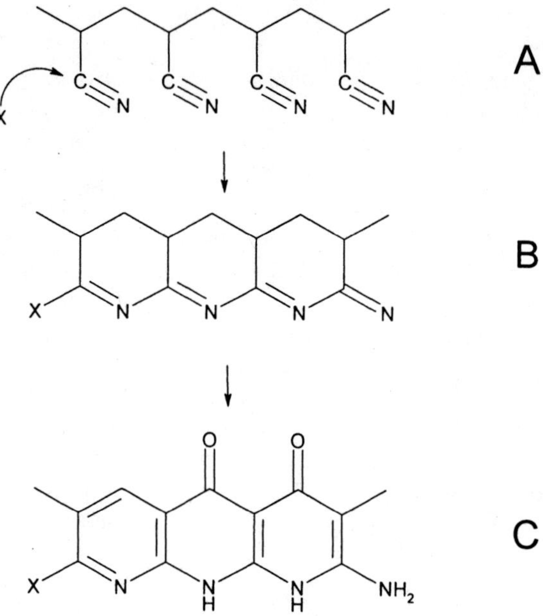


*Figure 1. Stages in the decomposition of PAN*

The result is a ladder polymer containing a mixture of acridone, pyridine, hydronaphiridine and other structures. In the early stages of decomposition this ladder polymer alternates with unchanged PAN (*9*). Figure 2 shows typical ratios of these different components obtained under certain conditions, which started to form the ladder polymer at 180°C, as shown by FTIR (*5*).

Memetea et al (*5*) showed from FTIR of residues that heating PAN for 15 minutes in air resulted in a large increase in the amount of >C=N relative to -C≡N, a large increase in the >C=O relative to $>CH_2$ and a slight decrease in the ->CH relative to -C≡N. In order to investigate the process of aromatisation, they established that the production of hydroperoxides and hydronaphthiridine rings are strongly associated, suggesting that cyclisation of PAN could be entirely initiated by hydroperoxide decomposition. Heating PAN in air at 180°C results in the formation of hydronaphthiridine and acridone rings together with significant concentrations of hydroperoxides, as identified by chemiluminesence and FTIR of residues.

Liggat (*9*) described the intense exotherm in air around 300°C, which results from the cyclisation of PAN leaving the ladder polymer alternating with short sequences of unchanged PAN. The cyclisation process involved free radical

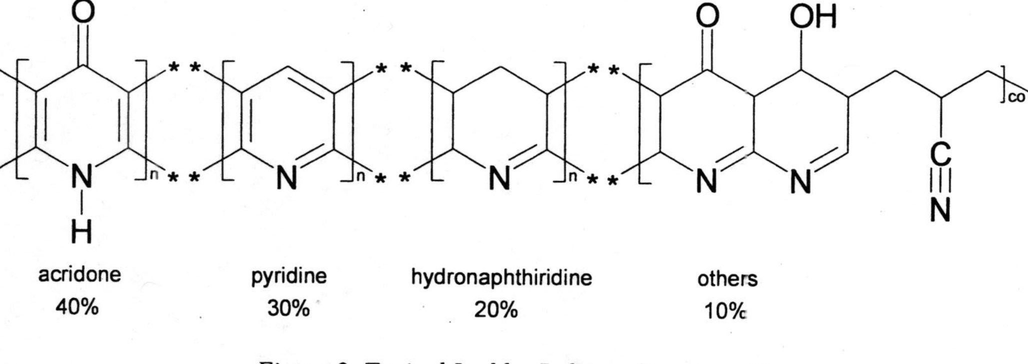


*Figure 2. Typical Ladder Polymer Structure (5)*

initiated intermolecular cross-linking, which could occur through azomethine. Although ammonia ($NH_3$) is a widely observed decomposition product, the mechanism for its formation remains the subject of speculation. Xue (*10*) et al proposed a mechanism involving tautomerisation to the more stable polyenamine, from which terminal amines could be lost as ammonia. However, rapid cyclisation and insufficient dissipation of heat can lead to sudden volatile loss which fragments the chains. In carbon fibre production this can damage the fibre structure, and in extreme cases, fusion or burning of the fibres has been known to occur. Elimination of HCN or $NH_3$ as the primary volatile or dehydrogenation products at lower temperatures would create uncyclised gaps in the polymer structure which retards the process of cyclisation. The presence of nucleophiles such as sulphonates or phosphonates has been shown to promote cyclisation (*11, 12*).

## Fire Retardants for PAN

The most frequently reported studies of fire retardants (FRs) for polyacrylonitrile include decabromodiphenyloxide, which acts by gas phase inhibition, and ammonium polyphosphate (*13*), which combines a phosphate barrier layer with acid catalysed char formation, and the release of a blowing agent to swell the char ($NH_3$). Horrocks observed differences when adding APP and sodium hydrogen phosphate to PAN, both of which decompose to form a phosphate barrier layer, but only APP shows other chemical effects (*11*). Environmental pressures are driving the decline in use of halogenated FRs, whilst the difficulties in ensuring a fire retardant effect after washing have limited the use of the latter. An alternative to additive fire retardants, is to use reactive fire retardants, by incorporation of a fire retardant comonomer into the polymer chain. This traps the fire retarding moiety within the polymer, allowing washability and meeting the current European demands of the Reach directive, which requires all "chemicals", such as non-polymeric materials to undergo comprehensive hazard assessment before they may be used commercially), and maintains the integrity of the polymer for spinning into a textile fibre.

This work describes the investigation of two reactive fire retardant comonomers diethyl(acryloyloxy-1-ethyl) phosphonate (DEAEP), and 2-acrylamido-2-methylpropane sulphonic acid (AMPS) in the presence or absence of nanoscopic sodium cloisite (SC) on the thermal decomposition and burning behaviour of acrylonitrile (AN) comonomers. The structure of these two comonomers is shown in Figure 3. Clearly the fire retardant goal of this work is to increase the rate of cross-linking, and hence char formation, while reducing the random chain scission and formation of flammable volatile species. The preparation and characterization of these copolymers has been reported separately (*14*).

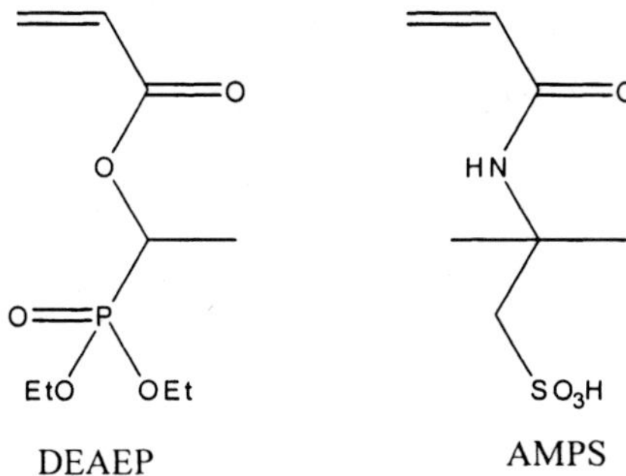


*Figure 3. Diethyl(acryloyloxy-1-ethyl) phosphonate (DEAEP) and 2-acrylamido-2-methylpropane sulphonic acid (AMPS) monomers*

## Experimental

**Materials** The clay containing PAN polymer and the AN copolymers were prepared by collaborators at Sheffield, as reported previously (*12, 15*). Nanoscopic sodium cloisite is an unmodified clay (1nm x 70-150nm x 70-150nm). This synthesis was repeated on a larger scale in order to produce sufficient material for flammability testing. The composition of the materials is shown in Table I. Unfortunately, the scale on which the synthesis was conducted made it impractical to make more than single plaques for cone calorimetry.

**Table I. Composition of Samples**

| *Sample* | *Nominal ratios in feed* |
|---|---|
| PAN | 100 |
| PAN_SC | 97 : 3 |
| PAN_AMPS | 97 : 3 |
| PAN_AMPS_SC | 94 : 3 : 3 |
| PAN_AMPS_DEAEP_SC | 91 : 3 : 3 :3 |

**Sample Preparation** The samples were prepared by cold pressing the dried, precipitated polymer using the method described by Horrocks et al (*16*) directly into plaques 100 mm x 100 mm, and approximately 2 mm thick.

**Cone Calorimetry** The samples were tested using a Fire Testing Technology cone calorimeter, using an applied heat flux of 35 kW $m^{-2}$. However, the limited availability of sample material meant that only single plaques could be tested, rather than the normal triplicate sets.

## Simultaneous Thermal Analysis with Fourier Transform Infrared analysis of evolved gases (STA-FTIR)

**Thermogravimetric Analysis (TGA)** Analysis was carried out using a TA Instruments 2960 SDT simultaneous thermal analyzer. Samples (approximately 10 mg) were decomposed at a heating rate of 10°C $min^{-1}$ in air, and also in nitrogen, from ambient temperature to 900°C. This produced a record of the mass changes (TGA) and sample temperature changes (differential thermal analysis DTA) during the furnace heating programme.

**Fourier Transform Infrared Analysis of Volatiles (FTIR)** Evolved volatile species were analysed by direct connection of a heated sampling line (at 180°C) to a 20 cm path length infrared cell (also at 180°C) using a Thermo Electron TGA interface located in a Nicolet Magna 550 FTIR spectrometer.

**Pyrolysis GC-MS** A CDS 5200 Pyrolyser was used to inject volatiles onto an HP 5890 Series II gas chromatograph interfaced to a Trio 2000 mass spectrometer. Approximately 2-3 mg samples were loaded into 25mm quartz tubes and pyrolysed in helium. The sample were heated then held at either 250°C and evolved gases were vented to waste. Then the sample was pyrolysed quickly to 300°C over a period of 20mS and the evolved products were trapped on a Tenax sorption tube. Following capture the flow was reversed though the trap and the collected sample was desorbed at 280°C for 5 minutes onto the GC with a fused silica capillary column (CP-Sil 5CB). Once transferred onto the column the GC begins a temperature gradient program heating at a steady rate of 10°C $min^{-1}$ to a maximum of 280°C and a Trio 2000 mass spectrometer, running in positive ion electron ionization mode, is used as method of detection.

## Results

### Cone calorimetry

***Heat Release Rate*** The heat release curves are shown in Figure 4. These show noticeable increases in the ignition delay time of two of the samples, PAN_SC (the sample of PAN containing only sodium cloisite) and PAN_AMPS (the comonomer containing the sulphonic acid group). It is interesting to note that PAN_AMPS_SC (the sample containing both) does not show an increased ignition delay time. The other clearly observable feature is that only PAN and PAN_SC show a very minor shoulder of lower heat release rate, rising to a single maximum value, in a manner typical of a thermally thin material (*17*).

After ignition, the AN_AMPS and AN_AMPS_SC show similar behaviour, with AN_AMPS showing a slightly larger slightly later peak of heat release rate (pHRR) after ignition. The AN_AMPS_DEAEP_SC shows a significantly lower steady and pHRR and overall a lower total heat release (THR). This indicates that the combination of DEAEP, AMPS and SC is more effective than either AMPS or SC alone or in combination. Unfortunately no cone plaques of AN_DEAEP or AN_DEAEP_SC were available.

**Table II. Effective Heat of Combustion and Char Yield**

| | PAN | PAN_SC | AN_AMPS | AN_AMPS_SC | AN_AMPS_ _DEAEP_SC |
|---|---|---|---|---|---|
| EHC* | 34.3 | 35.3 | 29.7 | 31.3 | 27.7 |
| Char Yield %** | 27.6 | 23.9 | 42.9 | 39.3 | 50.4 |

(* EHC from tti to tti + 260 s; **char yield at 525s )

Table II shows the effective heat of combustion (EHC) for each of the AN samples together with the char yield. The slightly higher EHC for PAN_SC and the lower char yield, compared to AN may be due to a catalytic enhancement of char oxidation. The presence of AMPS and DEAEP comonomers clearly results in a strong fire retardant effect, linked to the formation of more stable char.

The lower HRR for the AN_AMPS samples is probably due char formation, as carbonisation effectively removes the most calorific component (carbon) from the vapour phase, so the effective heat of combustion decreases as char yield increases.

Samples of PAN_AMPS and particularly PAN_AMPS_DEAEP_SC showed considerable swelling, leaving well-developed chars approximately 5 and 15 time the thickness of the original sample, respectively. In addition, the char from the burning sample containing DEAEP shows the most coherent structure, with less cracks and voids (Figure 5).

While the shapes of the $CO_2$ production rate curves more or less mirror the heat release curves, as expected, each CO production rate curve (Figure 6) shows a minima as the heat release rate starts to decrease, followed by a prolonged char oxidation stage. Most notably, the AN_AMPS_DEAEP_SC sample shows a significantly higher CO production rate, and a 10% lower $CO_2$ production rate than would be expected form the HRR curves. Both the higher CO yield and the lower $CO_2$ for AN_AMPS_DEAEP_SC indicate that a gas phase inhibition mechanism is also operating.

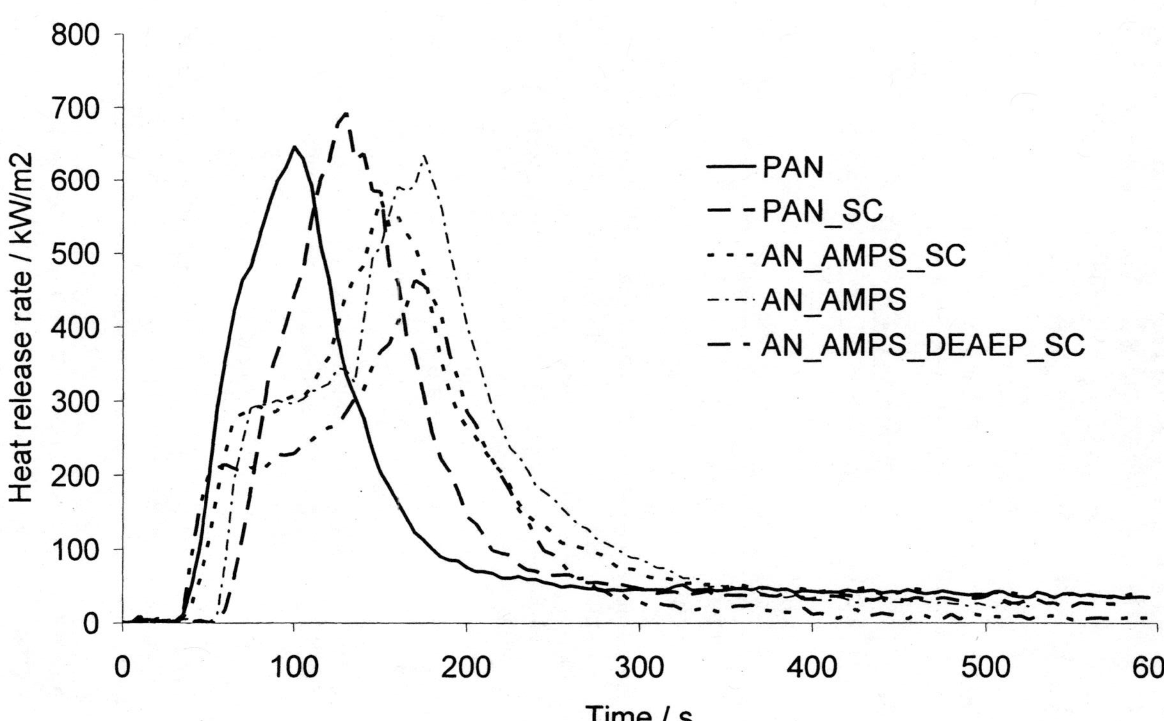


*Figure 4. Heat release rate of acrylonitrile based polymers at 35 $kWm^{-2}$ measured in the cone calorimeter*

*Figure 5. Cone Residues a) PAN_AMPS and b) PAN_AMPS_DEAEP_SC*

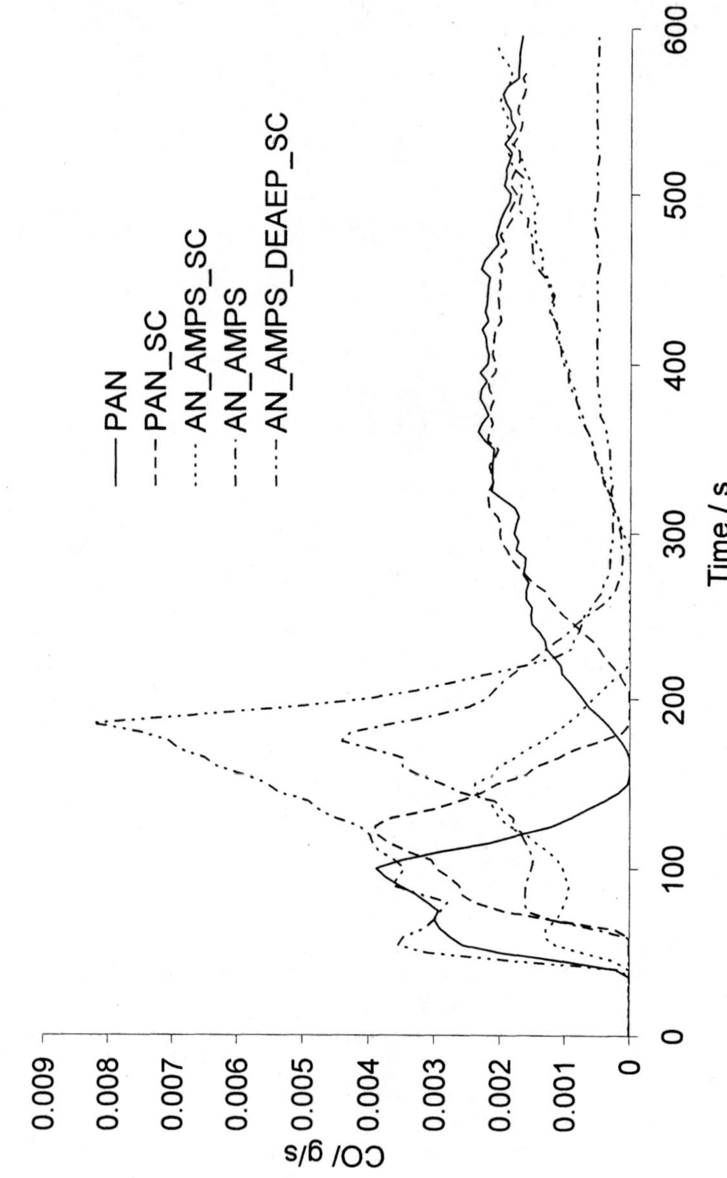


*Figure 6. CO production rate*

## TGA/DTA

**Thermogravimetric Analysis** Thermograms are presented for decomposition in air (corresponding to decomposition prior to ignition) and in nitrogen (representative of the atmosphere during flaming combustion). Figure 7 and Figure 9 show the decomposition in air and nitrogen of the 5 samples. The kink, in the PAN and PAN_SC thermograms around 300°C, and corresponding inclination of the DTA peaks occurs when the intense exotherm associated with cyclisation causes the TGA furnace to stop heating. Use of smaller samples eliminated this effect, but lead to reduced sensitivity in FTIR gas analysis.

PAN and PAN_SC samples show very rapid early mass loss around 300 °C, and less stable char formation, with 50% mass loss around 550 °C. The AN_AMPS and AN_AMPS_SC samples show the slowest decomposition, until 600 °C, with ~50% mass loss, but ultimately, the char is less stable than the AN_AMPS_DEAEP_SC which shows more rapid early mass loss but radically improved char stability.

The differential thermal analysis (DTA) curves, shown in Figure 8 in air and Figure 10 in nitrogen, demonstrate the characteristically intense isotherm for PAN and PAN_SC around 300 °C in air, with much smaller exotherms for the copolymers. In nitrogen, all materials show a smaller, broader exotherm between 270 °C and 300 °C. In air, smaller exotherms corresponding to the competing volatilisation/carbonisation processes described by Horrocks (*2*) are evident around 450 °C and a larger exotherm of char oxidation around 550 to 700 °C whereas in nitrogen the pyrolysis of char appears slightly endothermic.

The TGA in nitrogen shows very stable residues up to 900 °C. Again, PAN loses about 25% of its mass rapidly at around 300 °C, while PAN_SC loses about 15%, more rapidly that any of the copolymer samples. The AN_AMPS and AN_AMPS_SC show similar mass loss rates, with greater material retention in the condensed phase than PAN. The AN_AMPS_DEAEP_SC shows earlier mass loss and the most stable residue.

The stabilisation effect of oxygen in PAN and PAN_SC is clear ($T_{10\%}$ is around 290 °C in nitrogen, and $T_{10\%}$ is around 300 °C in air). In the case of the copolymers, this stabilisation effect is minimised, suggestion that AN copolymers are less dependent on air to form a stable char than PAN. This has been observed elsewhere, for AN copolymers with methyl acrylate in nitrogen and air (*2*).

## Evolved gas analysis by TGA-FTIR

The TGA and DTA data show considerable differences in terms of char stability for the different AN materials. However, as discussed in the introduction, the stability of the char is a function of its formation, about which relatively little information is available. Coupling TGA with evolved gas analysis, using FTIR, provides additional insight into the crucial char formation process.

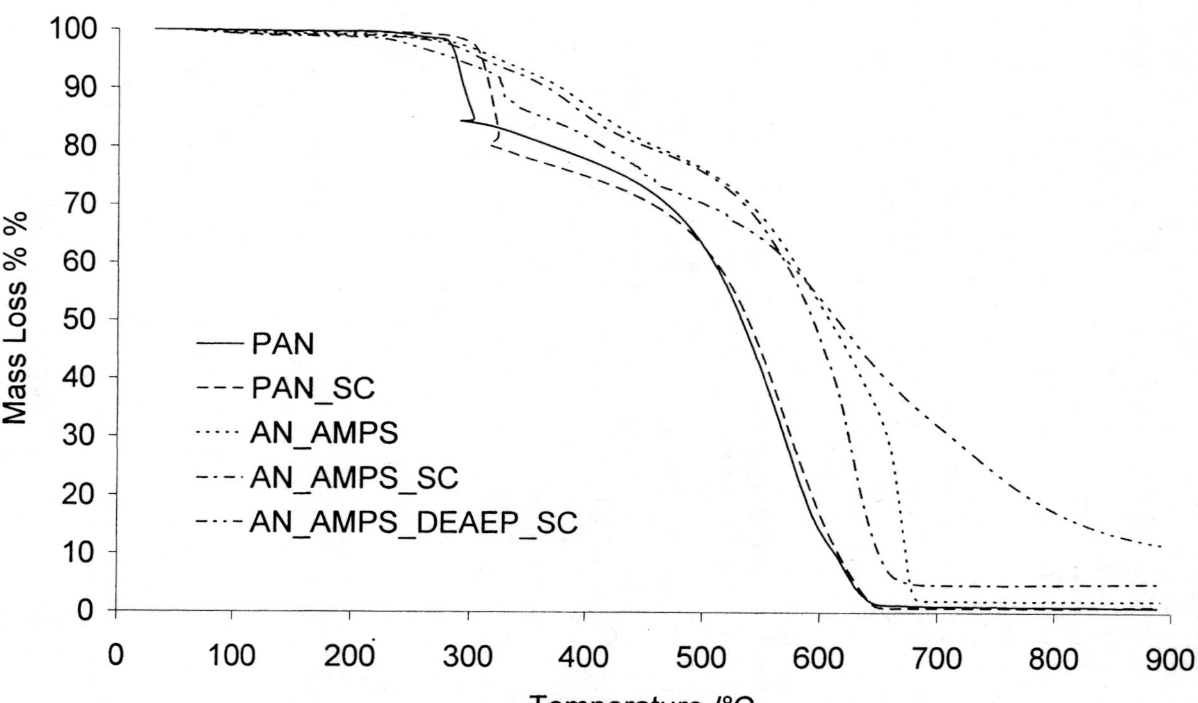


*Figure 7. Thermogravimetric analysis in air of acrylonitrile based materials*

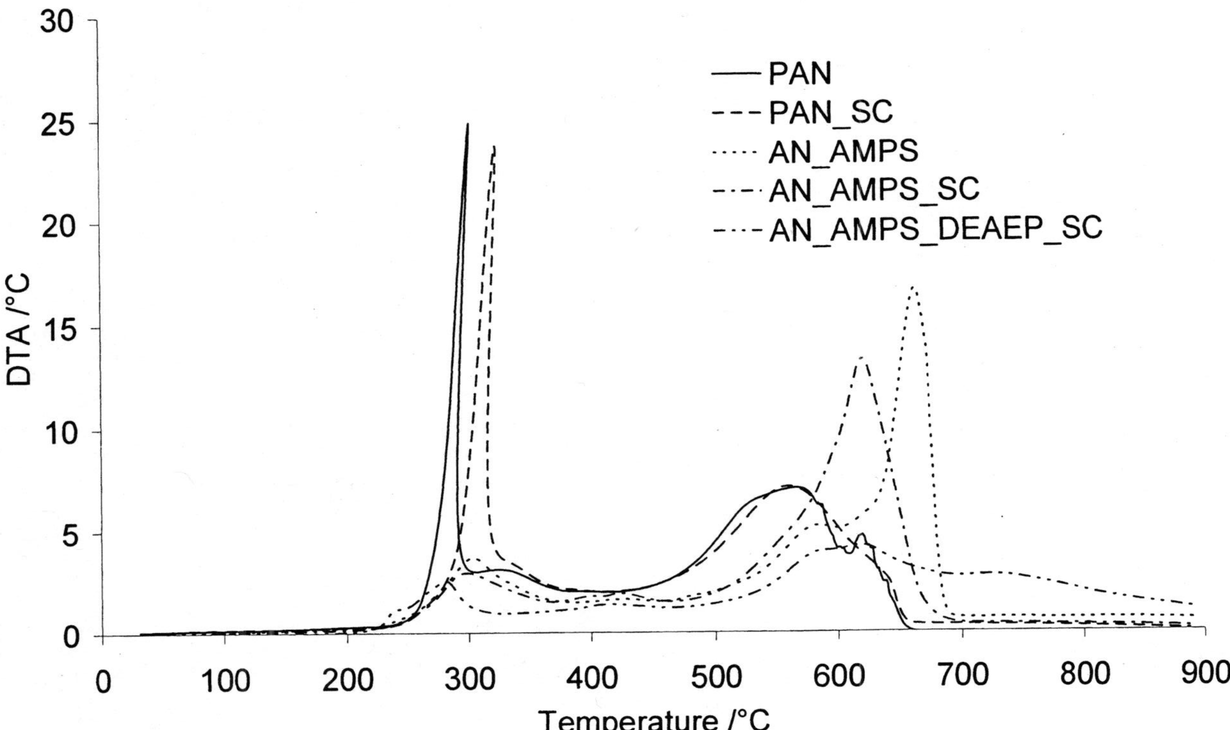


*Figure 8. Differential thermal analysis in air of acrylonitrile based materials*

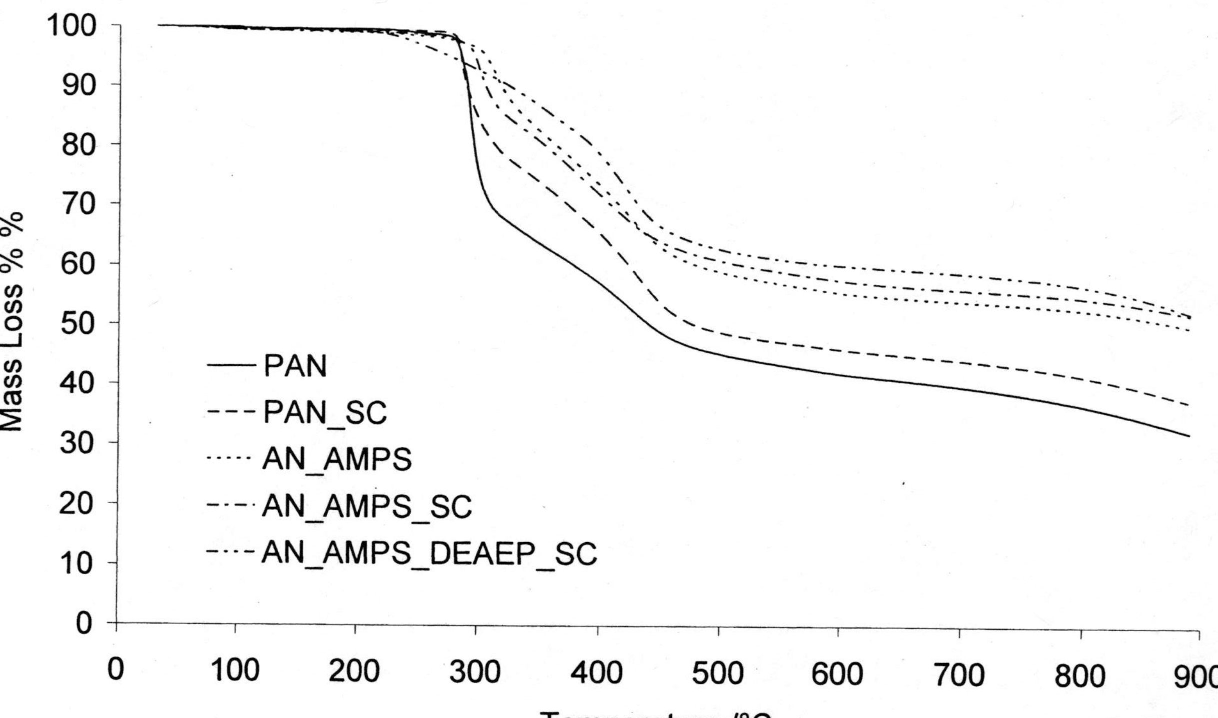


*Figure 9. Thermogravimetric analysis in nitrogen of acrylonitrile based materials*

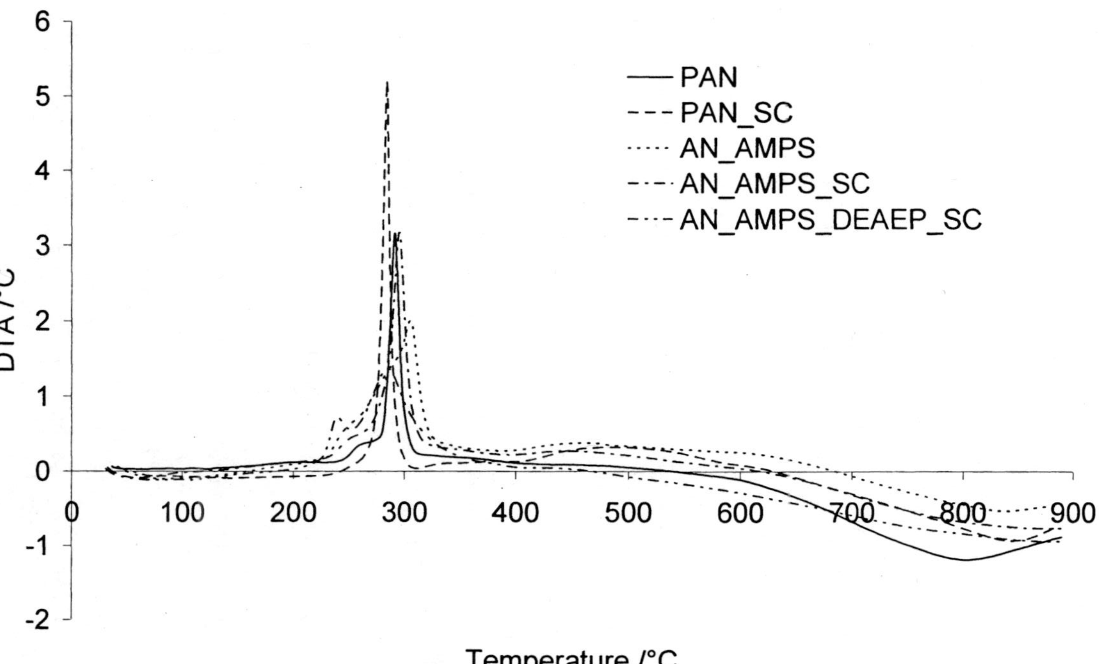


*Figure 10. Differential thermal analysis in nitrogen of acrylonitrile based material*

Figure 11 shows the variation in absorption corresponding to the hydrogen cyanide (HCN) peak (715 $cm^{-1}$), which is approximately proportional to the HCN concentration, for the TGA decomposition in air. This shows clearly a significant evolution of HCN corresponding to the intense exotherm in air for the PAN and PAN_SC samples only. It also shows that there is no other sharp or significant loss of HCN during the remainder of the decomposition. Similar, but noisier data exist for acrylonitrile, also showing a sharp peak of evolution, around 300°C, which is strongest for PAN, but also evident for PAN_SC. This indicates a chain stripping mechanism, by the loss of HCN, presumably leading to char formation from the resultant unsaturated polyene, competing with the formation of volatiles by chain scission, with the loss of AN.

In nitrogen, in addition to small losses of ammonia for PAN and PAN_SC, all the samples loose significant quantities of HCN around 400°C, and again, between 600-700°C. In the early stages, around 300°C, PAN also shows a sharp peak of ammonia evolution, while PAN_SC shows sharp peaks for carbon monoxide and carbon dioxide (presumably because of the catalytic effect of clay in oxidising organic materials, as reported elsewhere (*18*)). In contrast, two broad peaks appear between 1000 – 1100 $cm^{-1}$ and 2800 – 3000 $cm^{-1}$, which seem to correspond to the butan-1-ol spectrum, but require further investigation, between 250-300 °C are observed for AN_AMPS_DEAEP_SC. In the later stages, in air, where the cyclised structure starts to volatilise, methane is evolved for all the AN copolymer samples between 450-650 °C, but not for the PAN and PAN_SC samples. In nitrogen, all samples show a broad peak of methane production between 400 and 650 °C, with the exception of PAN, they also show a peak in ammonia evolution between 600-700 °C. This suggests that the char formed by PAN alone had a lower nitrogen content.

### Evolved gas analysis by pyrolysis-GC MS

Preliminary results from a pyrolysis-GCMS investigation using a CDS pyroprobe, shown for AN_AMPS_DEAEP_SC in Figure 12, and collecting the decomposition products in helium between 250 and 300°C suggests the presence of 1,3-butanediol, at 4.01 minutes (possibly corresponding to the butan-1-ol seen in the infrared) and shows the presence of diethylphosphite at 10.73 minutes from DEAEP in the gas phase, where it would be able to inhibit flaming combustion. While the main results from this investigation are outside the scope of this work, it provides useful supporting evidence.

## Discussion

The essential information about the burning behaviour of these novel materials is derived from the cone calorimetry data. In summary, this shows

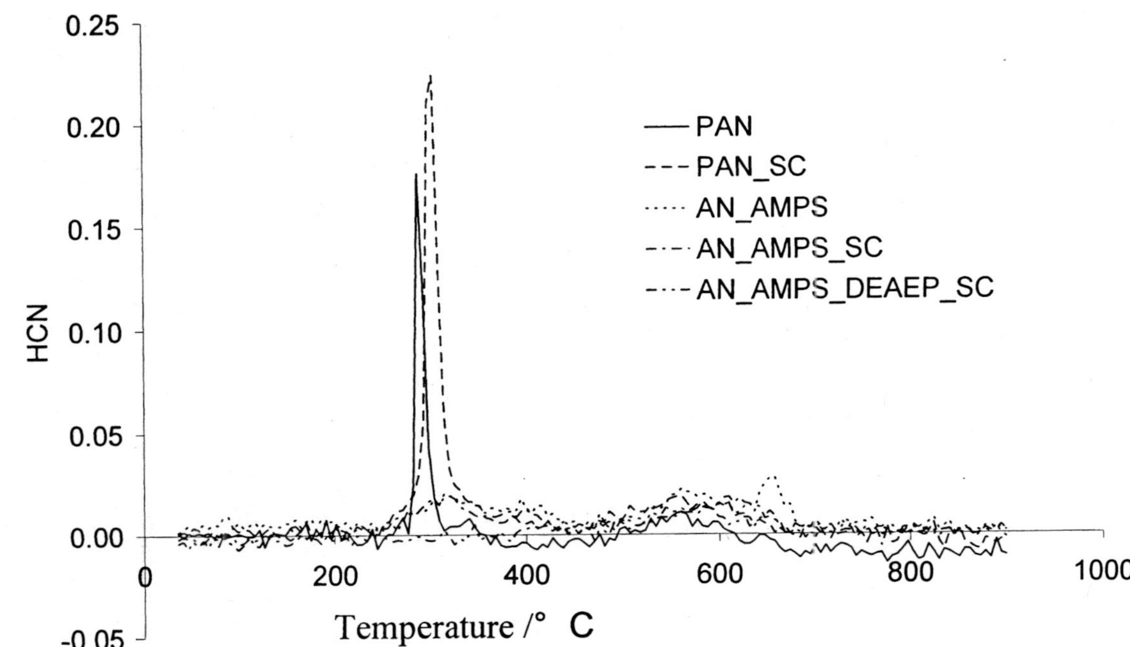


*Figure 11. Variation of HCN absorbance a function of temperature for 5 AN materials during TGA in air.*

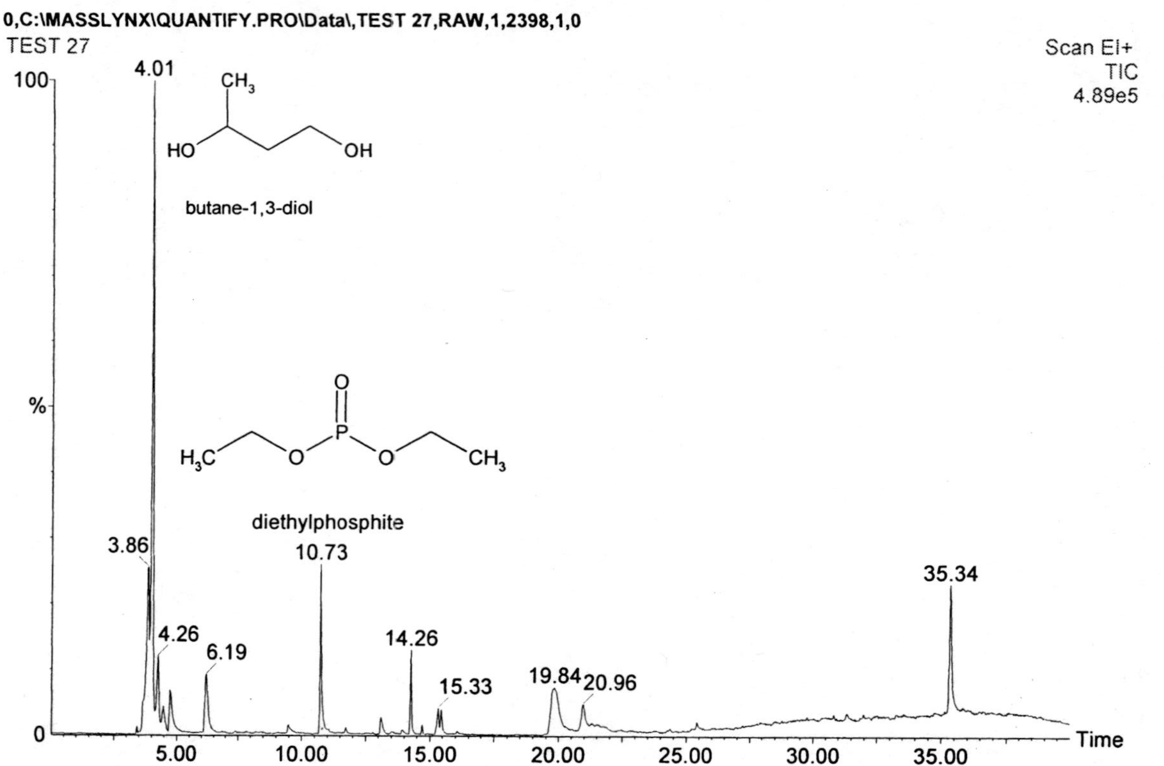


*Figure 12. py_GCMS trace for AN_AMPS_DEAEP_SC in air with volatile collection between 250-300°C*

interesting differences in time to ignition, and also differences between PAN and PAN_SC, which almost burnt as thermally thin materials, and the copolymers, which showed steady burning ending with a peak and thermally thick behaviour. This demonstrates the importance of the char as the main fire retardant mechanism. Only in the case of AN_AMPS_DEAEP_SC is there any evidence of gas phase action, from lower carbon dioxide, higher carbon monoxide, lower EHC and the presence of diethylphosphite in the gas phase from pyrolysis GC-MS.

The relationship between the char yield formed in the TGA in nitrogen, and in the cone calorimeter is shown in Figure 13. The TGA data in nitrogen is used since the atmosphere above the surface of a sample burning in the cone calorimeter is practically oxygen free (17). This shows a positive correlation, with slightly higher char yields in the TGA. It is interesting to see the apparent catalytic char oxidation effect of the sodium cloisite, lowering the char yields in the cone calorimeter, despite the 3% non-volatile clay component.

In order to demonstrate the importance of char formation on the burning behaviour, the relationship between the steady heat release rate in the cone calorimeter and the char formation in the TGA in nitrogen was investigated. In the TGA, the char forms uniformly throughout the whole of the ~15mg sample. In the cone calorimeter, the char forms on the surface, and slows down the rate of pyrolysis of the material beneath it. The steady heat release rate was plotted against the char yield at 600 and 700°C. This is shown in Figure 14, and shows that at both 600°C, a surface temperature corresponding to early or underdeveloped flaming, or at 700°C corresponding to more fully developed flaming, controls the steady heat release rate.

This agrees with an empirical observation (*11*), which demonstrated a relationship linking the limiting oxygen index of a series of PAN polymers and AN-acrylate and AN-acetate compolymers containing additive fire retardants to the char yield at 500°C. It has been reported that the additive fire retardant ammonium dihydrogen phosphate had no systematic effect on the intense exotherm associated with cyclisation of PAN during carbon fibre production (*11*). However, incorporation of a comonomer of methyl acrylate or vinyl acetate reduced the enthalpy and intensity of this process. This suggests that the presence of a non-cyclisable comonomer interrupts the highly exothermic process as it progresses along the polymer chain, and has benefits in terms of more stable char formation and hence flammability reduction.

## Reaction Schemes

In order to explain the differences in burning behaviour of the different AN materials, which seem to correspond to the differences in their decomposition behaviour when heated at 10°C $min^{-1}$ in TGA, outline reaction schemes have been proposed based on the evidence from the evolved gas analysis.

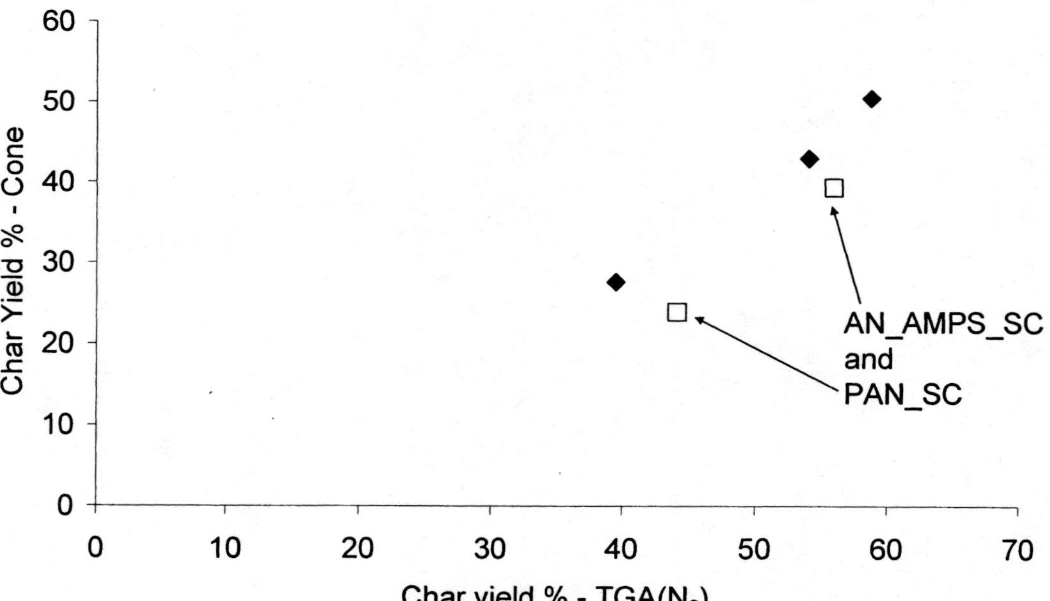


*Figure 13. Char yield comparison between TGA and cone calorimeter*

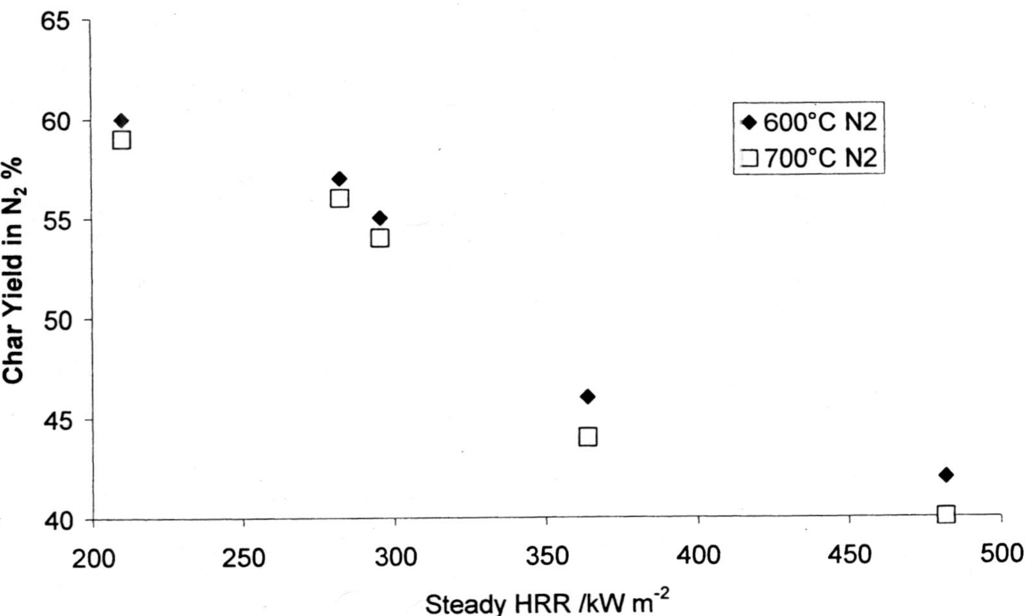


*Figure 14. Relationship between steady state heat release and char yield in nitrogen*

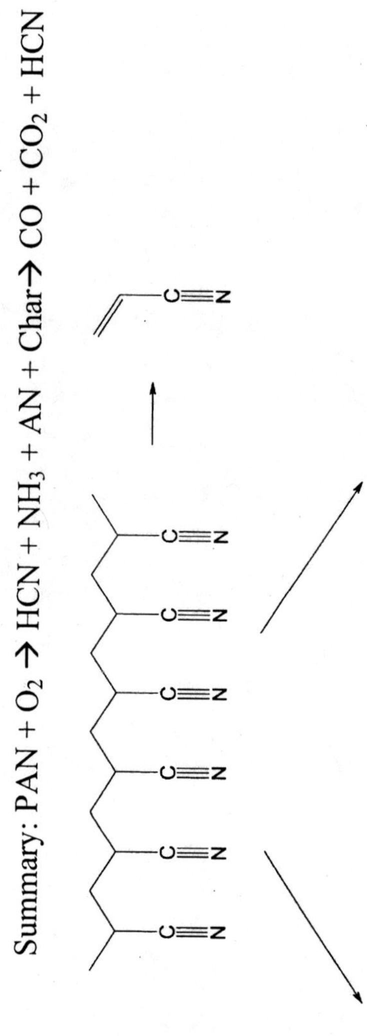
Summary: PAN + $O_2$ → HCN + $NH_3$ + AN + Char → CO + $CO_2$ + HCN
C≡N

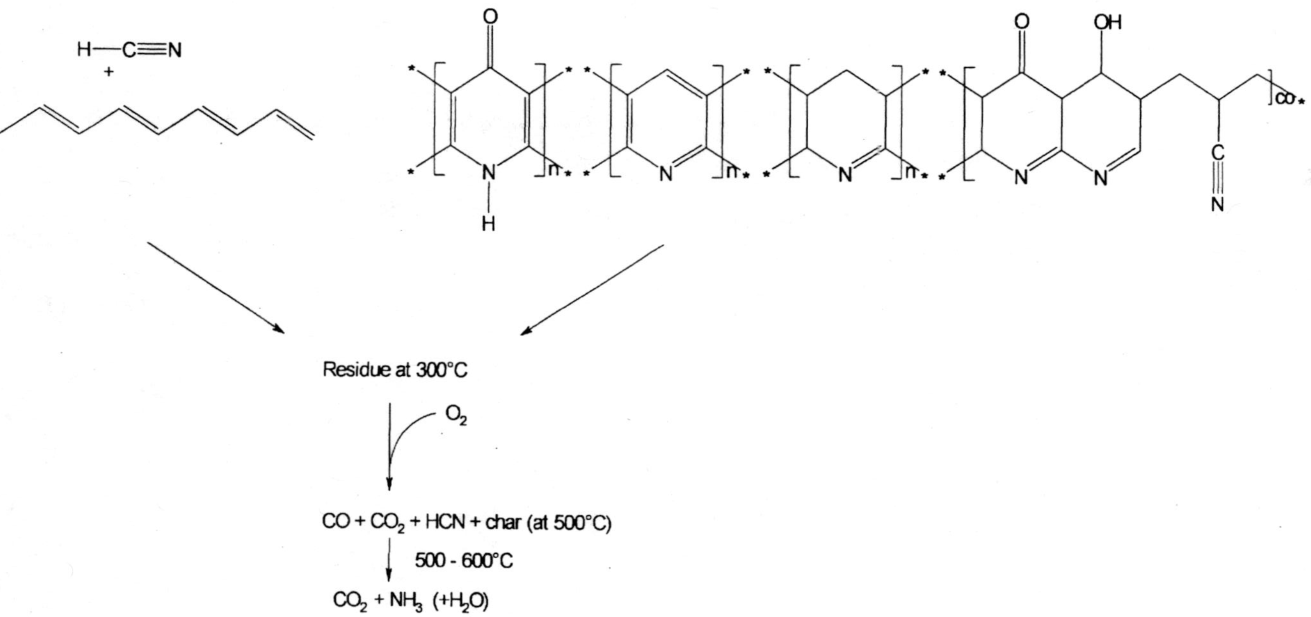


*Scheme 1. The major steps in the decomposition of PAN*

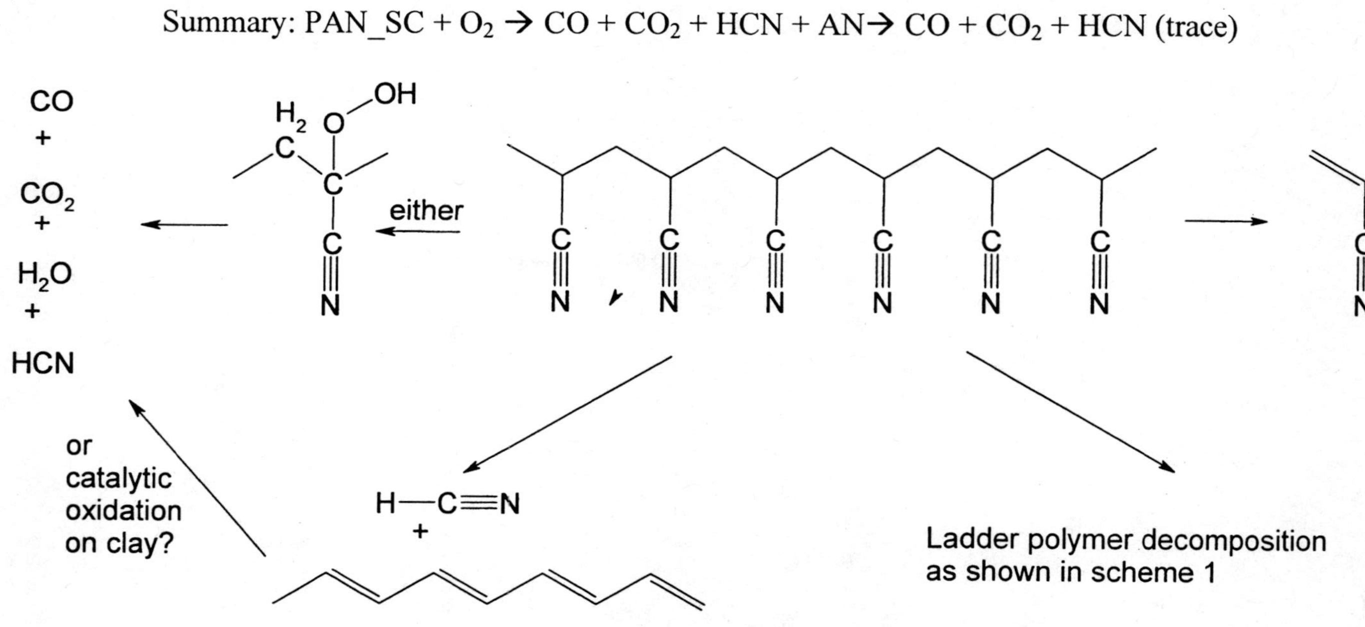


*Scheme 2. The major steps in the decomposition of PAN with sodium cloisite.*

**Scheme 1** shows the three competing routes in PAN decomposition which determine the amount and stability of the char. The lower stability of the char, relative to the copolymers, may be due to the rapid char formation, during the intense exotherm. Generally, chars formed slowly are more stable than those formed quickly. Alternatively, it may also be due to a lack of nitrogen in the char, suggested by the correspondingly intense peak of HCN during the intense exotherm.

**Scheme 2** shows the early evolution of CO and $CO_2$ from the PAN containing sodium cloisite, possibly via the hydroperoxide structure observed elsewhere[5], or possibly through oxidation on the clay surface, which leads to a fourth decomposition route.

**Scheme 3** shows the two influences of a nucleophilic comonomer on the decomposition of the samples containing AMPS. The nucleophile will promote cyclisation and effectively eliminate chain stripping and chain scission. In addition, the AMPS residue on the polymer chain will block the cyclisation process, leading to slower char formation, and apparently greater stability. No sulphur products were observed in FTIR spectra.

The AN_AMPS_DEAEP_SC sample shows the following processes:

$$\text{AN\_AMPS\_DEAEP\_SC} + O_2 \rightarrow \text{butanol} + \text{hydrocarbons} \rightarrow CO_2 + CO$$

This material leads to the most stable char. There is more than 10% remaining at 900°C. The breakdown chemistry of DEAEP requires further investigation, but it seems to follow a pathway that results in the release of a butanol type structure, which is not readily explained. This sample showed the smallest exotherm around 300°C and had the greatest concentration of nucleophilic agents and the greatest number of comonomers blocking rapid exothermic cyclisation. The later loss of ammonia suggests its incorporation into the char structure, only being released late in the decomposition, while the earlier loss of methane leaves an unsaturated structure better able to cross-link to form a more stable structure.

## Conclusions

This work shows that AMPS and DEAEP comonomers influence the decomposition processes by improving the thermal stability of the char structure. The loss of $NH_3$ and HCN leads to a less stable char, while the loss of $CH_4$ leads to a more stable char. As a fire retardant, DEAEP acts in both the condensed phase and in the gas phase.

In flaming combustion, the lack of oxygen at the surface prevents the formation of a stabilising char. However, the additive FRs trigger the formation

Summary: AN_AMPS(+_SC) + $O_2$ → HCN(trace) → $CH_4$ → CO + $CO_2$ + $NH_3$

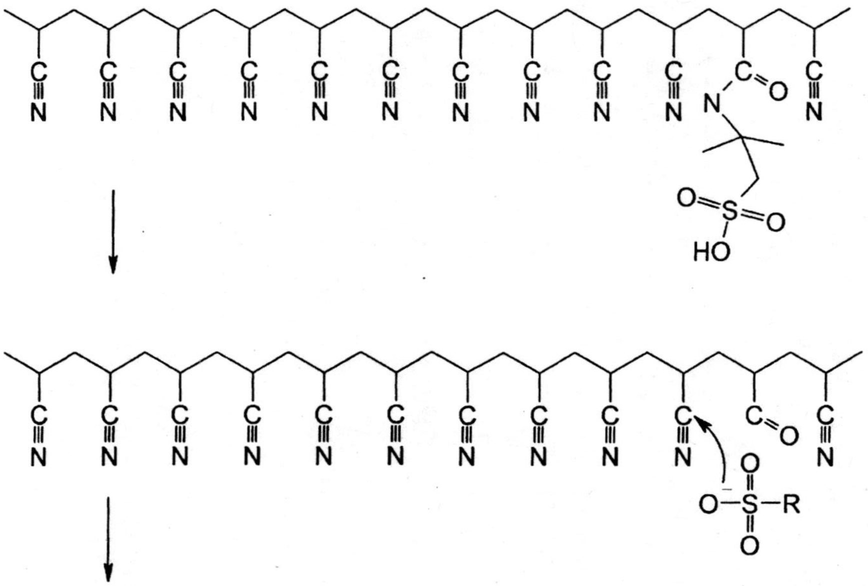

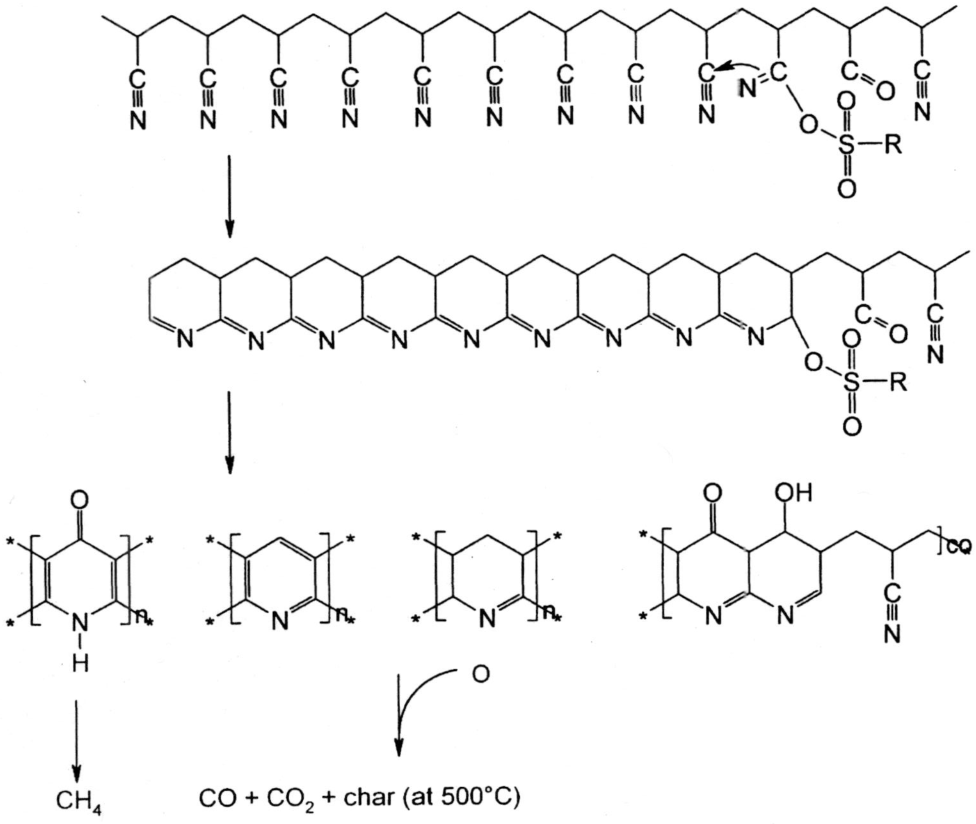

$CH_4$ $CO + CO_2$ + char (at 500°C)

*Scheme 3. The major steps in the decomposition of AN_AMPS and AN_AMPS_SC (and AN_AMPS_DEAEP_SC)*

of a stable char through nucleophilic attack, promoting cyclisation, while providing breaking points in the polymer chain to prevent a runaway exotherm.

The incorporation of sodium cloisite had a small effect in promoting char oxidation, but provided no major flammability improvements. However, the enhancement of physical properties associated with the incorporation of nanoclay were not investigated, because only precipitated samples were available.

## Acknowledgements

Grateful thanks are due to colleagues, particularly John Ebdon and Barry Hunt and their team (University of Sheffield, UK) and Paul Joseph, (now of University of Ulster, UK) for selection and careful preparation and characterisation of the samples, and for invaluable inputs of knowledge and wisdom. Thanks are also due to Jon Taylor (formerly of Acordis), Baljinder Kandola, Dick Horrocks and Dennis Price (University of Bolton) for their support over the years.

This work would not have been possible without financial support from the Engineering and Physical Sciences Research Council of the UK (EPSRC) and the UK Ministry of Defence (MoD) for the award of funding (Grant No. GR/S24350/01). Acordis, UK, and Rhodia Consumer Specialities, Ltd., for access to facilities and expertise.

## References

1. Bashir, Z.; *Carbon*, **1991**, *29*, 1081.
2. Horrocks, A.R.; Zhang, J.; and Hall, M. E.; *Polym Int.*, **1994**, *33*, 303-314.
3. Grassie, N.; In: Grassie N, editor. Developments in polymer degradation, vol. 1. London: Applied Science, **1977**. p. 137 and references therein.
4. Fitzer, E.; Muller, D.; *Carbon*, **1975**, 13:63.
5. Memetea, L.T.; Billingham, N.C.; and Then, E.T.H.; *Polym Degrad. Stab.*, **1995**, *47*, 189-201.
6. Grassie, N.; Hay, J.N.; and McNeill, I.C.; *J Polym. Sci.*, **1958**, *31*, 205.
7. Brandrup, J.; and Peebles, L.H.; *Macromolecules,* **1968**, *1*, 64.
8. Geiderikh, M.A.; et al, *J Polym. Sci.*, **1961**, *54*, 621.
9. Martin, S.C.; Liggat, J.J.; and Snape, C.E.; *Polym. Degrad. Stab.*, **2001**, *74*, 407-412.
10. Xue, T.J.; McKinney, M. A.; and Wilkie, C.A.; *Polym. Degrad. Stab.*, **1997**, *58*, 193-202.
11. Hall, M.E.; Horrocks, A.R.; and Zhang, J.; *Polym. Degrad. Stab.*, **1994**, *44*, 379-386.

12. Wyman, P.; Crook, V.; Ebdon, J.R.; Hunt, B.J.; and Joseph, P.; *Polym Intl.* **2006**, 55: 764.
13. Zhang, J.; Silcock, G.W.H.; and Shields, T.J.; *J Fire Sci.,* **1995**, 13, 141.
14. Ebdon, J.R.; Cook, A.G.; Hunt, B.J.; and Joseph, P.; ACS Preprint for this meeting (Fire and Polymers, PMSE, **2008**)
15. Ebdon, J.R.; Cook, A.G.; Hunt, B.J.; and Joseph, P.; Flame Retarding Polyacrylonitrile (PAN) Through Use of Comonomers and Nanofillers ACS Preprint 238th National Meeting New Orleans **2008** (in press).
16. Zhang, J.; Hall, M. E.; and Horrocks, A.R.; *J. Fire Sci.*, **1993**, 11, 442.
17. Schartel, B.; and Hull, T.R.; *Fire Mater.*, **2007**, 31, 327.
18. Hull, T.R.; Price, D.; Liu, Y.; Wills C.L.; and Brady J.; *Polym. Degrad. Stab.*, **2003**, 82, 365–371.

# Chapter 10

# Fire and Engineering Properties of Polyimide-Aerogel Hybrid Foam Composites for Advanced Applications

**Trent M. Smith[1], Martha K. Williams[1], James E. Fesmire[2], Jared P. Sass[2], and Erik S. Weiser[3]**

**[1]John F. Kennedy Space Center, Polymer Science and Technology Laboratory, NASA, Kennedy Space Center, FL 32899**
**[2]John F. Kennedy Space Center, Cryogenics Test Laboratory, NASA, Kennedy Space Center, FL 32899**
**[3]Langley Research Center, Advanced Materials and Processing Branch, NASA, Hampton, VA 23861**

NASA has had a growing need for high-performance materials for cryogenic insulation, fireproofing, energy absorption, and other advanced applications. New monolithic aerogels are good candidates except for their inherently fragile nature. Modern commercially available aerogels are tougher, but still lack adequate structure and can become a contamination concern in certain applications. Polyimide foam materials have been used for demanding applications that require extreme thermal, chemical, and weathering stability. The combination of aerogel with a high performance polyimide foam fabricated into a hybrid material provides structure to the aerogel, reduces heat transfer, improves vibration attenuation, and retaines the excellent fire properties of the polyimide foam. Incorporation of aerogel material into TEEK-H polyimide foam provides three favorable effects: thermal conductivity is reduced by the aerogel content in the hybrid foams, Peak Heat Release remains low, and the time to Peak Heat Release increases with increasing aerogel content.

 

# Introduction

Aerogel materials have been known for many years as excellent thermal and acoustic insulators (*1-6*), but have suffered limited use due to their inherently fragile nature. There have been two main commercialization thrusts of aerogel. In the late 1940's Monsanto Chemical commercialized silica aerogel under the trade name Santocel®. Santocel® products, in a number of hydrophilic grades with relatively low surface areas, were used in paints, thermal insulation, and a diverse number of applications (7-13) through the 1960's. Production was stopped due to the high cost of maintenance and labor, and in conjunction with the sale of Monsanto's silica business (*14*). Today's commercial aerogel materials, characterized by their high surface areas and low densities, were developed along two different lines, loose fill and composite blanket type, in the late 1990's. These modern aerogels are fully hydrophobic and have been demonstrated to have the lowest thermal conductivity, at ambient pressure 101.3 kPa (760 torr), of any material in the world (15,16). Modern commercial aerogel materials are much more resilient than monolithic forms of aerogel, however many of these new forms of aerogel still lack sufficient structural integrity to be considered for use in many applications. Commercially available aerogel beads/granules and aerogel composite blanket materials are excellent thermal insulators. For example, these aerogel materials are finding their way into space vehicle applications solving specialized problems (17-19). Aerogel materials though can cause contamination concerns with aerogel dust which becomes airborne during normal installation procedures and can be liberated during system operations which include mechanical loads or vibrations. Specialized containment systems must be employed to mitigate dust contamination concerns. The dust is considered low health risk to humans because of the amorphous nature of the silica, but can be problematic in electrical systems, optical systems, and in general use. Silica based aerogel materials are expected to be inherently fire retardant, but commercially available silica aerogel materials from Cabot Corp. and Aspen Aerogels Inc. are treated to be super-hydrophobic. The treatment is normally a branched carbon-based silane such as a trimethylsilane moiety which covalently bonds with the silica. Fire performance of surface functionalized aerogel materials is expected to be good, but limited data are available and no data are available under forced combustion conditions such as cone calorimeter.

New crosslinked aerogel (x-link aerogel) materials are under development at Glenn Research Center, but the process currently limits production to small sizes. The crosslinked aerogel materials are non-dusting and can be made structural (*20-22*). Fire properties of x-link aerogel is currently unknown.

For many years, polyimide foams have been used in demanding applications that require extreme structural stability in both hot and cryogenic environments, excellent chemical stability, and good long term weathering and aging characteristics *(23-31)*, but a further reduction in thermal conductivity would be preferred.

In general, state-of-the-art polyimide foam systems have a thermal conductivity around 10 mW/m-K higher than the current insulation systems used on the Space Shuttle External Tank. The Ares launch vehicle currently being designed by NASA may require an insulation system with a higher use temperature than the current polyurethane and polyisocyanurate foams used on the Space Shuttle External Tank while maintaining the low thermal conductivity inherent in these systems.

The combination of aerogel materials with a robust polyimide foam such as TEEK developed at Langley Research Center would be desirable for such demanding applications. Incorporating modern day commercial aerogel materials in a polyimide foam matrix mitigates contamination concerns of aerogel and gives structural integrity to the aerogel while the aerogel in the polyimide foam reduces the thermal conductivity. Fire properties of commercially available aerogels, TEEK polyimide foam, and TEEK-aerogel hybrid foam composites were measured to understand fire safety risks of using these materials. Mechanical and thermal conductivity data were also produced to provide selection criteria of particular advanced applications.

# Experimental

## Materials

Polyimide-aerogel hybrid foams were made from 4,4′-oxydiphthalic anhydride / 3,4′-oxydianiline (TEEK-H) friable balloons from Unitika Ltd., with combinations of Spaceloft® and Cryogel® aerogel blanket (AB) from Aspen Aerogels Inc., and/or Nanogel® aerogel beads from Cabot Corp. Cryogel aerogel blanket is produced by a solution-gelation (sol-gel) process and is a composite of hydrophobic aerogel within a polymeric fiber matrix; the drying step to remove the solvent is a supercritical process. Nanogel® beads, nominally 1-mm in diameter, are also produced by a sol-gel process but with an ambient pressure drying step. Friable balloons are partially cured TEEK-H balloons or powder that contain some blowing agent. When processed, these balloons form polyimide foams of various open/closed cell content and densities. Open/closed cell content and densities can be targeted by varying the processing parameters.

## Fabrication of Composites

TEEK aerogel bead composites (TEEK XX% aero) were prepared by mixing the appropriate amount of Nanogel® beads with TEEK friable balloons followed by placing the mixture into a mold and curing in a convection oven at 200°C for 2 hours. The aerogel beads are uniformly distributed in the resulting foam composite. The TEEK aerogel blanket (TEEK AB X layer) composites were made in a similar manner: a) Half of the TEEK friable balloons were disbursed evenly across the mold surface, b) The aerogel blanket was placed in the center of the mold, c) The remaining TEEK friable balloons were poured over the blanket and finally, d) The foam composite was cured at 200°C for 2 hours. The foam composite with the pocket of aerogel beads was made in a similar manner as the aerogel blanket composites by substituting a bed of aerogel beads in place of the blanket. The combined TEEK aerogel bead and blanket composites were made by substituting a balloons/beads mixture in place of the pure balloons. The diagonal strips designation refers to the aerogel blanket being cut into strips and placed diagonally across the mold instead of a single blanket.

## Test Methods

Thermal conductivity was measured using a Netzsch Lambda 2300 heat flow meter per test method ASTM C518. This instrument is a comparative heat flow meter that uses a standard reference material for calibration. Our tests were performed at ambient pressure with boundary temperatures of 34°C (warm) and 13°C (cold).

Fire performance was measured at an irradiance of 50 kW/m$^2$ on a CSI cone calorimeter per a modified ASTM E1354 (non-standard thickness and limited number of specimens). All specimens were 4″ (0.102 m) wide and 4″ (0.102 m) long. The virgin polyimide foam specimens and polyimide-aerogel hybrid foam specimens were 0.50″ (0.013 m) thick, the Pyrogel® blanket was 0.24″ (0.006 m)thick, and the Cryogel® blanket was 0.39″ (0.010 m) thick rather than the standard 1″ (0.025 m) and only two specimens of each material were tested. Cryogel blanket was two strips placed together in an aluminum foil boat. Aerogel beads and granules were 1″ (0.025 m) thick and were poured into an aluminum foil boat. All aerogel materials were tested unrestrained and all aerogel foam hybrid composites were tested in a restrained configuration. TEEK polyimide foam was tested both restrained and unrestrained. Samples were restrained with a metal grid that fits inside a sample housing which fits over the normal sample holder. The grid surface area of 0.00088 m$^2$ was substracted from the sample surface area.

Vibration attenuation measurements were accomplished by sandwiching 1.25″ (0.032 m) x 6″ (0.152 m) x 6″ (0.152 m) foam specimens between aluminum plates and striking with an instrumented impact hammer. Accelerometers strategically placed on the opposite aluminum plate recorded the vibration that was transmitted through the foam.

Mechanical properties were measured in an Instron 4507 tensile test machine by compressing 1″ (0.025 m) x 1″ (0.025 m) x 1″ (0.025 m) foam specimens to greater than 75% strain at a rate of 0.25″/min (6.35 mm/min).

## Results

### Cone Calorimeter Results

Cone calorimeter results for aerogel beads manufactured by Cabot Corp. showed a peak heat release of 60 kW/m$^2$ under an irradiance of 50 kW/m$^2$ as shown in Figure 1.

Translucent aerogel granules had a peak heat release rate of 55 kW/m$^2$ under an irradiance of 50 kW/m$^2$ as shown in Figure 2. The aerogel beads and granules both quickly ignited under the irradiance of 50 kW/m$^2$. The aerogel materials burned with a low flame. The peak heat release for both beads and granules occurred at approximately 18 seconds. The combustion is a result of trimethylsilane and other organosilane surface modification agents vaporizing and igniting. Aerogel beads and granules can be calcined which removes the surface modification agents and once calcined aerogel will combust as well as sand.

Aerogel blanket materials manufactured by Aspen Aerogels Inc. were also tested in the cone calorimeter. Cryogel aerogel blanket 0.39″ (0.010 m) thick had a peak heat release of 96 kW/m$^2$ under an irradiance of 50 kW/m$^2$ as shown in Figure 3.

Pyrogel aerogel blanket manufactured by Aspen Aerogels Inc. is an insulating material designed for higher temperature applications and uses carbon and glass fibers instead of polymeric fibers for the aerogel particles. Pyrogel aerogel blanket 0.24″ (0.006 m) thick had a peak heat release rate of 66 kW/m$^2$ under an irradiance of 50 kW/m$^2$ as shown in Figure 4.

Both aerogel blanket materials quickly ignited under the irradiance of 50 kW/m$^2$. The peak heat release for both blankets occurred at approximately 12 seconds. Combustion of the Cryogel is a result of surface modification agents and polymer fiber. Combustion of Pyrogel is a result of organosilane surface modification agents.

Cone calorimeter results for virgin TEEK polyimide foam made from friable balloons showed a low heat release rate. The first TEEK sample was run unrestrained and stress relieved under the thermal load causing increased surface

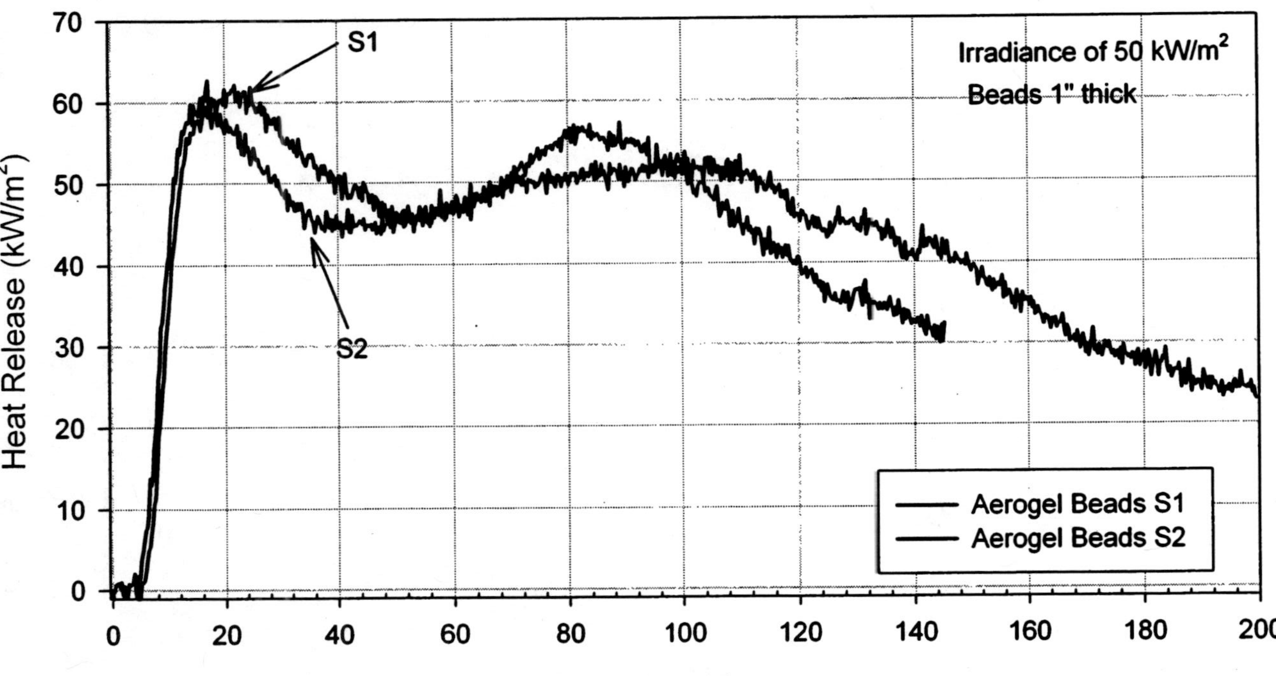


*Figure 1. Heat release data of aerogel bead, Nanogel®, manufactured by Cabot Corp. under 50 kW/m² irradiance.*

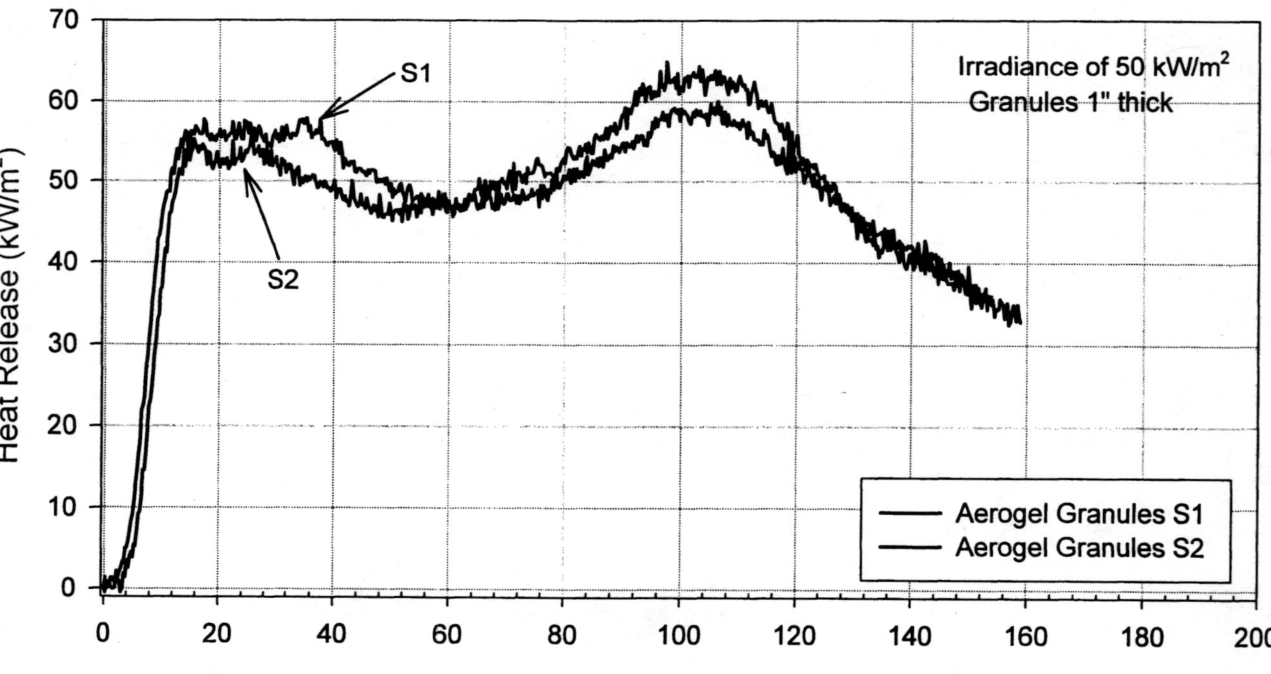


*Figure 2. Heat release data of aerogel granules, Nanogel®, manufactured by Cabot Corp. under 50 $kW/m^2$ irradiance.*

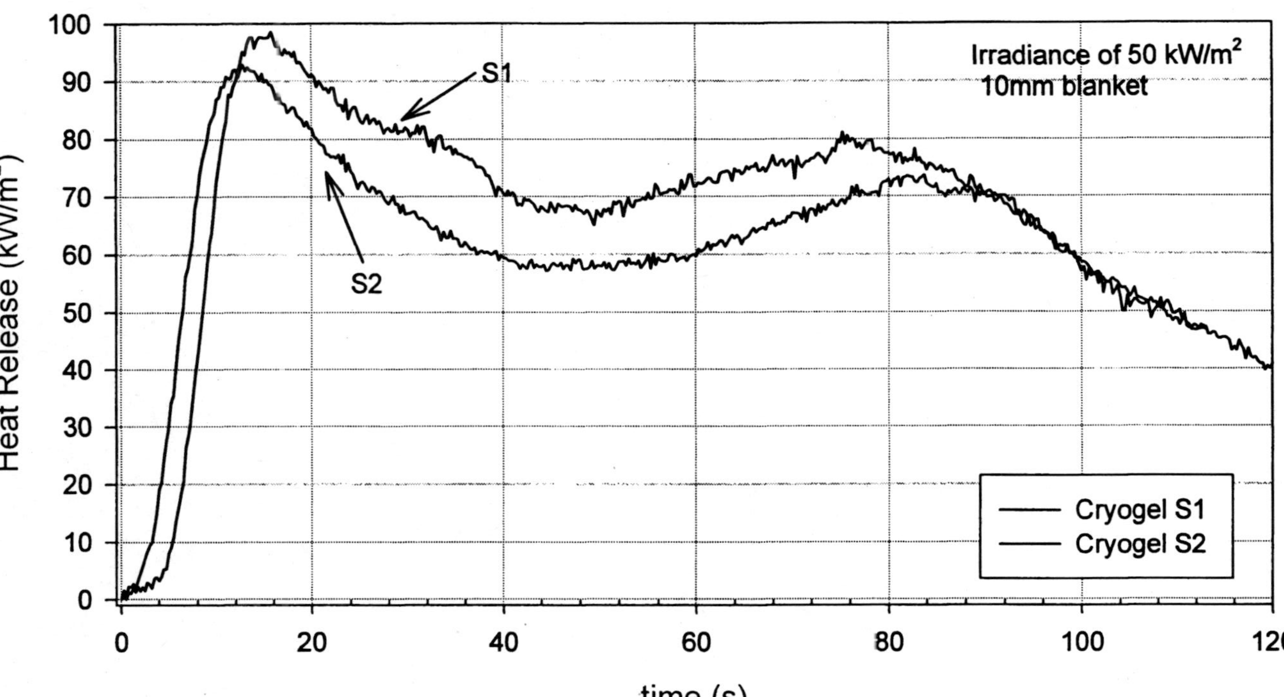


*Figure 3. Heat release data of 0.39" (0.010 m) thick aerogel blanket, Cryogel®, manufactured by Aspen Aerogels Inc. under 50 kW/m² irradiance.*

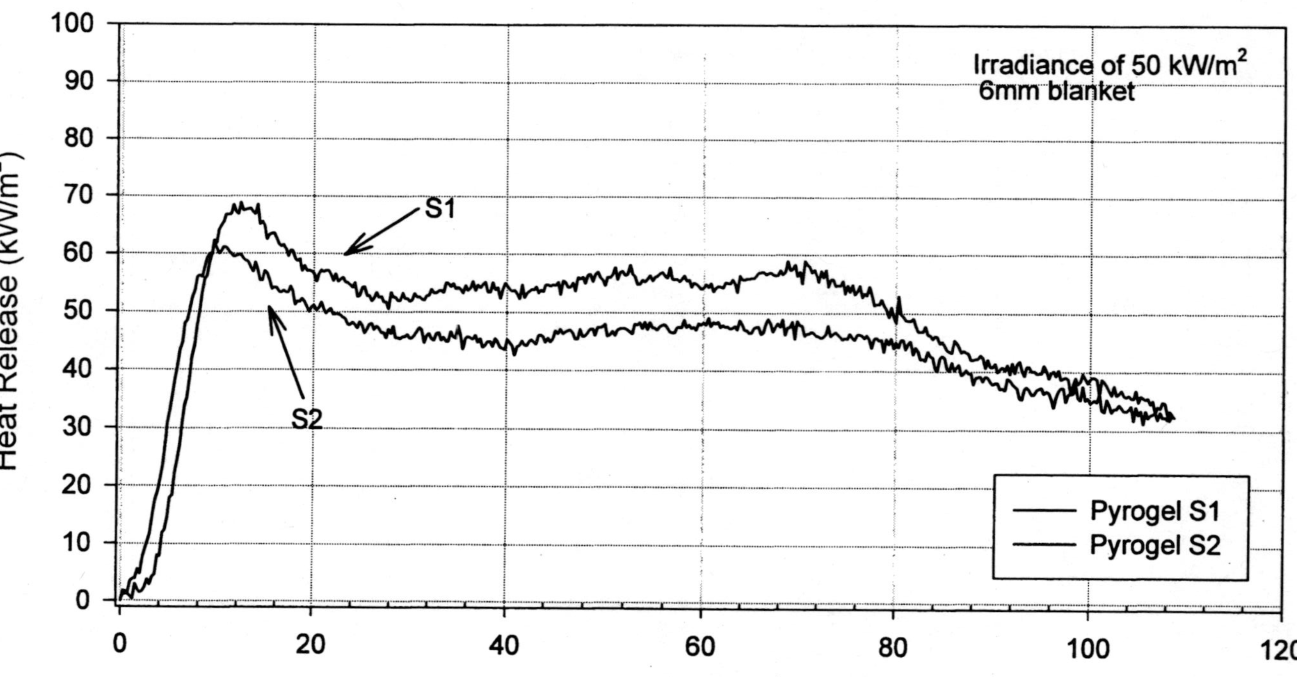


*Figure 4. Heat release data of 0.24" (0.006 m) thick aerogel blanket, Pyrogel®, manufactured by Aspen Aerogels Inc. under 50 $kW/m^2$ irradiance.*

area combustion. This observed behavior correlated to the peak heat release of 86 kW/m$^2$ under an irradiance of 50 kW/m$^2$ for the S1 sample shown in Figure 5. Subsequent foam samples and foam aerogel hybrid composites were tested restrained. The second virgin TEEK sample, S2, did not have an observable flame until 162 seconds and it burned with a low flame. The peaks for S2 before 162 seconds were cracks which formed allowing more offgassing products. The peak heat release for S2 was 25 kW/m$^2$ under an irradiance of 50 kW/m$^2$ around 170 seconds as shown in Figure 5.

TEEK polyimide foam with 10% aerogel beads took more than two minutes to ignite under 50 kW/m$^2$ irradiance. Upon ignition the foam composite burned with a low flickering flame. Both tests were terminated early because it appeared the flame was out. S2 had a peak heat release of about 30 kW/m$^2$ at around 170 seconds then low level combustion around 25 kW/m$^2$as shown in Figure 6. S1 had an initial peak heat release similar to S2, but a crack formed in the sample which resulted in a peak heat release of 35 kW/m$^2$ just over 200 seconds as shown in Figure 6.

TEEK polyimide foam with 20% aerogel beads took more than two minutes to ignite under 50 kW/m$^2$ irradiance. Upon ignition the foam composite burned with a low flickering flame. The first test was terminated early because it appeared the flame was out. S1 showed a peak heat release of about 25 kW/m$^2$ at around 180 seconds then low level combustion around 25 kW/m$^2$ as shown in Figure 7. S2 showed an initial peak heat release of about 30 kW/m$^2$ around 160 seconds. A crack opened in the specimen which pushed the peak heat release to 37 kW/m$^2$ at 280 second followed by low level combustion around 30 kW/m$^2$ as shown in Figure 7. The peaks after 200 seconds are the foam deforming/ shrinking and exposing new surface area.

TEEK polyimide foam with 25% beads took more than two minutes to ignite under 50 kW/m$^2$ irradiance. Upon ignition the foam composite burned with a low flickering flame. Sample S1 had a peak heat release of about 30 kW/m$^2$ at around 200 seconds then low level combustion around 25 kW/m$^2$ as shown in Figure 8. S2 had an initial peak heat release of 30 kW/m$^2$ around 180 seconds, but then S2 deformed some while constrained pushing the peak heat release to 36 kW/m$^2$ just after 300 seconds as shown in Figure 8.

After fire testing aerogel beads on the surface appeared similar to before testing, but below the surface soot appeared to be deposited as shown in Figure 9, photo A. Translucent aerogel granules appeared white on the surface and appeared to have soot deposited under the surface after fire testing as shown in Figure 9, photo B. Cryogel aerogel blanket before testing was white and after fire testing appeared to have some soot deposited as shown in Figure 9, photo C. Pyrogel aerogel blanket is black and did not appear different after testing, though the color could mask soot as shown in Figure 9, photo D. No char was evident or expected.

TEEK polyimide foam and TEEK-aerogel hybrid foam composites did have good char formation and aerogel beads were evident in and on the char. The

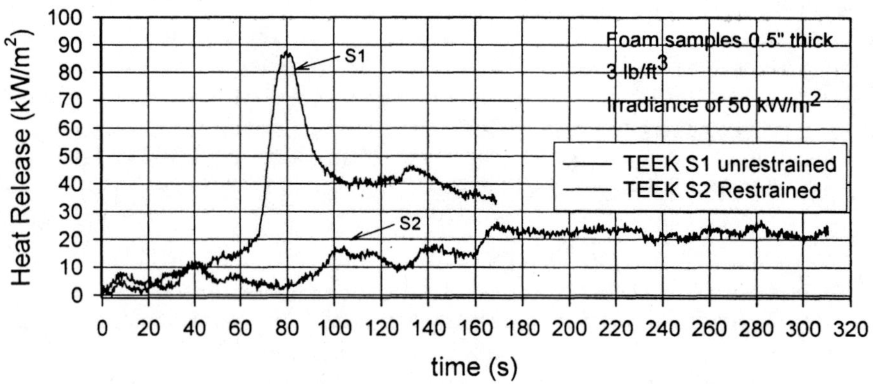


*Figure 5. Heat release data of 0.5" (0.013 m) thick, 3 lb/ft$^3$ TEEK polyimide foam under 50 kW/m$^2$ irradiance. S1 was tested unrestrained and test was ended early. S2 was tested restrained.*

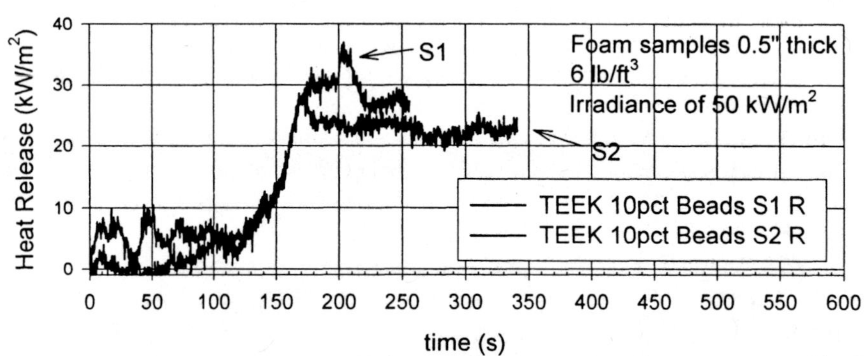


*Figure 6. Heat release data of 0.5" (0.013 m) thick, 6 lb/ft$^3$ TEEK polyimide foam with 10% aerogel beads w/w under 50 kW/m$^2$ irradiance.*

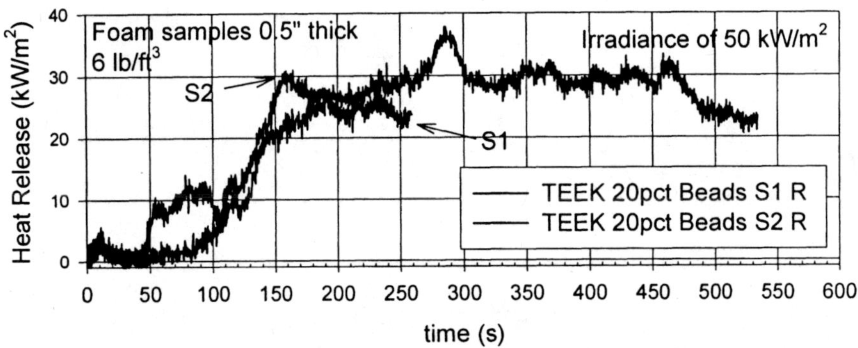


*Figure 7. Heat release data of 0.5" (0.013 m) thick, 6 lb/ft$^3$ TEEK polyimide foam with 20% aerogel beads w/w under 50 kW/m$^2$ irradiance.*

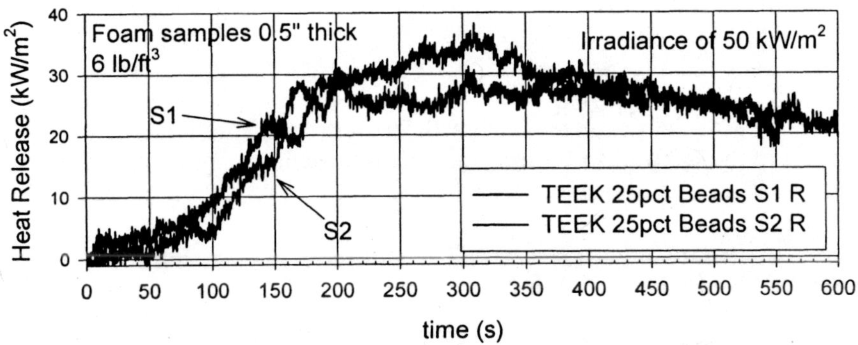


*Figure 8. Heat release data of 0.5" (0.013 m) thick, 6 lb/ft$^3$ TEEK polyimide foam with 25% aerogel beads w/w under 50 kW/m$^2$ irradiance.*

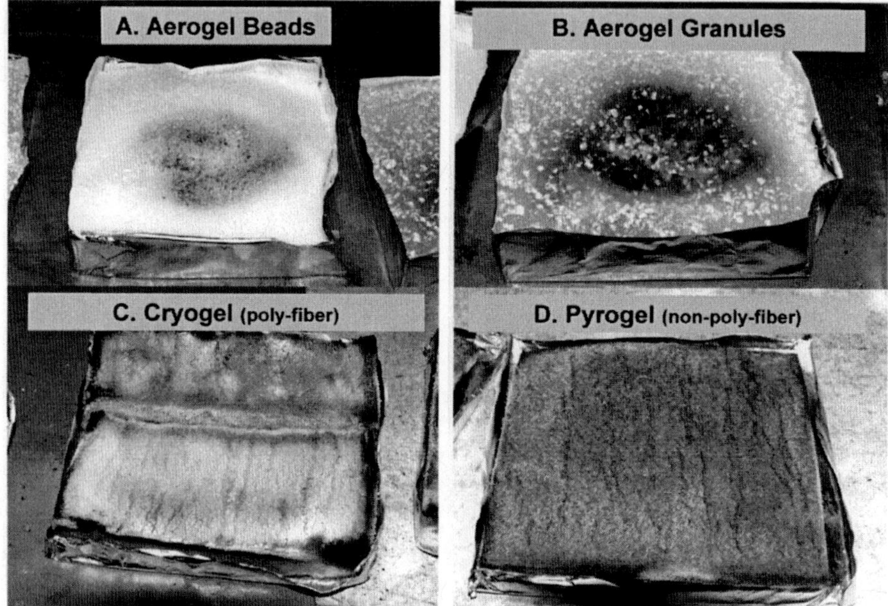


*Figure 9. Photographs of aerogel materials after fire testing in the cone calorimeter. Aerogel beads are shown in the top left corner (A), aerogel granules in the top right corner (B), aerogel blanket with polymer fiber in the bottom left (C), and aerogel blanket with non-polymer fiber in the bottom right corner (D).*

square shaped pattern was where the metal restraint grid was placed. The surface area of the grid was accounted for in the fire results. Higher aerogel loading in the hybrid composites was clearly evident as more aerogel was left on/in the char as shown in Figure 10.

## Thermal Conductivity Results

Thermal conductivity results for aerogel beads, aerogel granules, aerogel blankets, TEEK polyimide foam, and TEEK-aerogel hybrid foam composites show that all of these materials perform well as thermal insulators. The amount of aerogel loading in TEEK-aerogel hybrid foam composites was the main factor in thermal performance rather than density. Although foam density is normally the main factor which effects thermal conductivity, this dependence is only true for similar materials with comparable pore sizes or cellular structures. Figure 11 shows thermal conductivity (k-value) and density for TEEK foam samples and various TEEK-aerogel hybrid foam composites. Different aerogel architectures

*Figure 10. Photographs of TEEK polyimide foam and TEEK-aerogel hybrid foam composites after fire testing in the cone calorimeter. TEEK polyimide foam is shown in the top left corner (A), TEEK polyimide foam with 10% aerogel beads w/w is shown in the top right corner (B), TEEK polyimide foam with 20% aerogel beads w/w is shown in the bottom left corner (C), and TEEK polyimide foam with 25% aerogel beads w/w is shown in the bottom right corner (D).*

within the foam hybrid composites were produced and tested. Clearly higher aerogel loading decreased thermal conductivity and even mitigated density effects on thermal conductivity. Spikes in thermal conductivity coincide with spikes in density for TEEK and less so for TEEK 10% aero w/w. There was no increase of thermal conductivity with increased density for TEEK 20% aero w/w and TEEK 25% aero w/w as shown in Figure 11. The best performer with respect to thermal conductivity for the TEEK-aerogel hybrid foam composites was the sample with two layers of aerogel blanket placed in the center of the interior portion. The sample with one aerogel blanket also performed well. The sample which had a pocket of aerogel beads in the interior performed comparable to the single aerogel blanket sample. Surprisingly, the sample with nine diagonal strips of aerogel blanket in the interior showed a much higher thermal conductivity. A dissection will be performed to ascertain what may have happened.

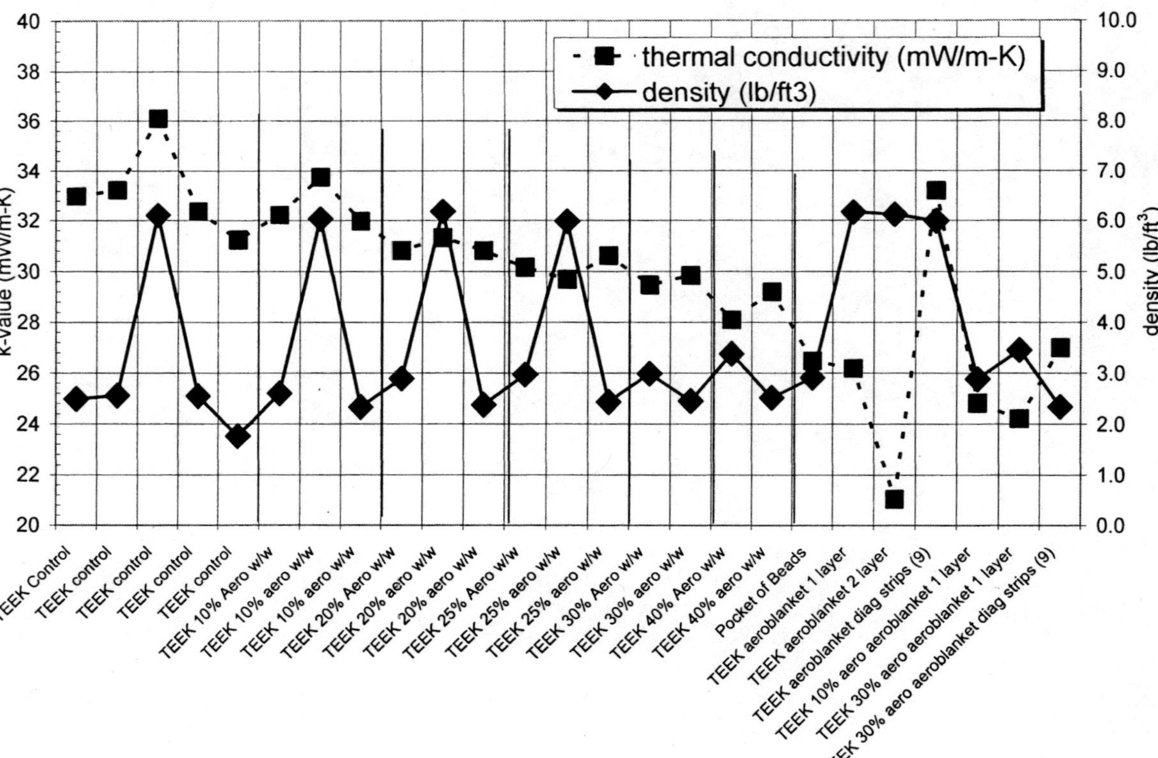


*Figure 11. Thermal conductivity and density data for TEEK polyimide foam and TEEK-aerogel hybrid foam composites. TEEK with 2 layers of aerogel blanket had the lowest thermal conductivity.*

Thermal conductivity values for 6 lb/ft$^3$ aerogel bead foam composites are shown graphically in Figure 12. Once again, the results show that thermal conductivity decreases with increased loading of aerogel. Finally plotting thermal conductivity with respect to density clearly shows that aerogel loading was the main factor for thermal performance and not density as shown in Figure 13.

Thermal conductivity (k-value) of TEEK polyimide foam like most foams is strongly dependent on density, which increases solid heat conduction through the material. However, thermal conductivity of TEEK with 10% aerogel beads has less dependence on density. Thermal conductivity TEEK with 20% aerogel beads was nearly indifferent to density. When the aerogel loading was greater than 25% the k-value decreased with respect to density. This result shows that the aerogel is doing more and more of the work of preventing the heat transfer, both solid conduction and gaseous convection, through the material.

## Compression Test Results

TEEK polyimide-aerogel hybrid foam composites were tested in compression to understand how incorporating aerogel beads into the polyimide foam affected mechanical properties. Foam samples tested in compression were from the same samples burned in the cone calorimeter. Foam samples were 6 lb/ft$^3$. TEEK polyimide foam sample at a density of 6 lb/ft$^3$ could not be located for fire and compression testing. However the aerogel loading affect was still evident in the mechanical data as shown in Tables I through III. Compression data is graphically shown in Figure 14 for samples S14, S24, and S34.

**Table I. Compression data for TEEK polyimide-aerogel hybrid foam composites with 10% w/w aerogel beads homogeneously distributed throughout the material.**

| *TEEK 10% Beads 6.03 lb/ft$^3$* | | | |
|---|---|---|---|
| | *yield (psi)* | *yield (%)* | *stress@50% (psi)* |
| S11 | 64.5 | 6.0 | 156.3 |
| S12 | 60.5 | 5.5 | 166.4 |
| S13 | 64.5 | 6.0 | 175.0 |
| S14 | 61.9 | 5.5 | 164.1 |
| S15 | 64.5 | 5.0 | 171.6 |
| S16 | 64.1 | 4.5 | 180.6 |
| Average | 63.3±1.7 | 5.4±0.6 | 169.0±8.6 |

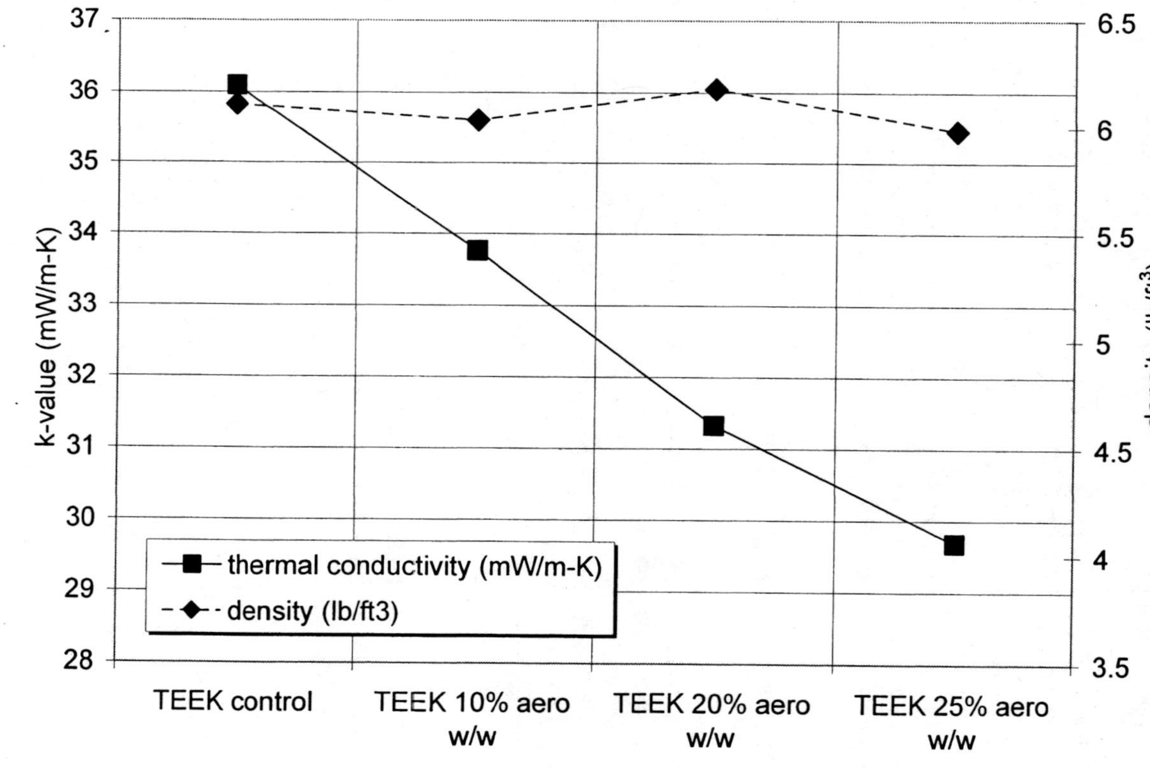


*Figure 12. Density and thermal conductivity measurements for rigid TEEK foam and TEEK-aerogel hybrid foam composites. Thermal conductivity clearly decreases with increased loading of aerogel beads while foam density stays relatively constant.*

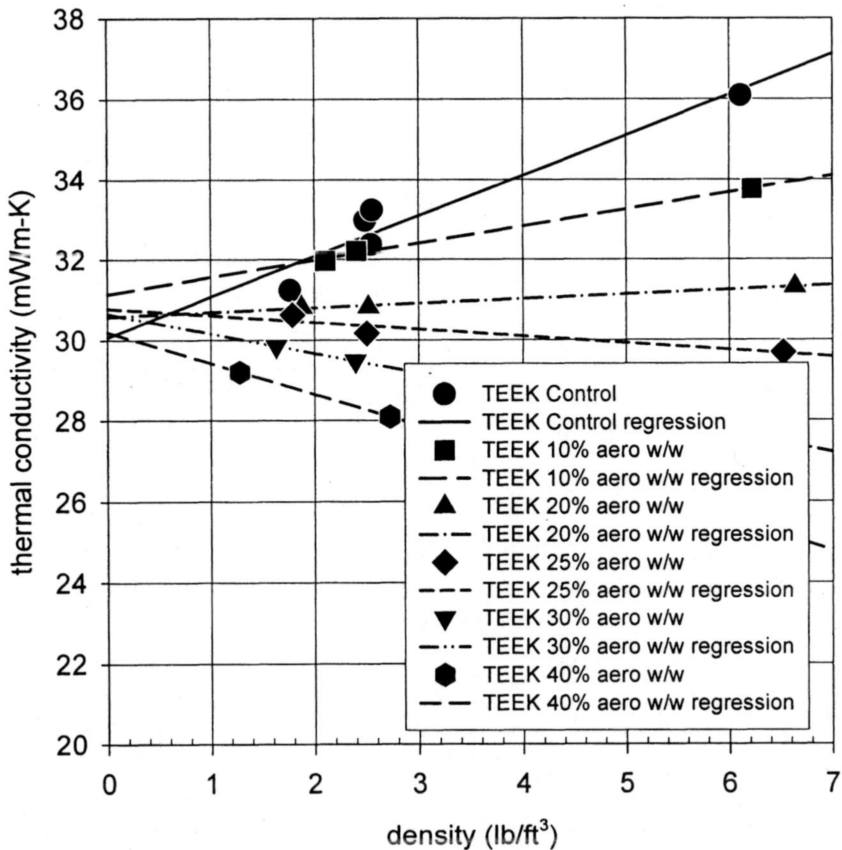


*Figure 13. Thermal conductivity data for TEEK control and TEEK-aerogel hybrid foam composites with respect to density. The data clearly indicates that aerogel loading is the main factor effecting thermal conductivity instead of density.*

**Table II. Compression data for TEEK polyimide-aerogel hybrid foam composites with 20% w/w aerogel beads homogeneously distributed throughout the material.**

| *TEEK 20% Beads 6.18 lb/ft$^3$* | | | |
|---|---|---|---|
| | *yield (psi)* | *yield (%)* | *stress@50% (psi)* |
| S21 | 36.6 | 4.0 | 135.2 |
| S22 | 41.8 | 6.0 | 162.9 |
| S23 | 44.3 | 6.0 | 165.7 |
| S24 | 47.8 | 8.0 | 153.6 |
| S25 | 45.1 | 6.0 | 147.2 |
| S26 | 48.6 | 6.0 | 143.3 |
| Average | 44.0±4.4 | 6.0±1.3 | 151.3±11.7 |

**Table III. Compression data for TEEK polyimide-aerogel hybrid foam composites with 25% w/w aerogel beads homogeneously distributed throughout the material.**

| *TEEK 25% Beads 5.98 lb/ft$^3$* | | | |
|---|---|---|---|
| | *yield (psi)* | *yield (%)* | *stress@50% (psi)* |
| S31 | 28.5 | 6.0 | 114.0 |
| S32 | 26.0 | 5.0 | 125.1 |
| S33 | 31.3 | 6.0 | 120.8 |
| S34 | 34.1 | 7.0 | 136.8 |
| S35 | 34.0 | 6.0 | 145.3 |
| S36 | 28.1 | 7.0 | 110.5 |
| Average | 30.3±3.3 | 6.2±0.8 | 125.4±13.4 |

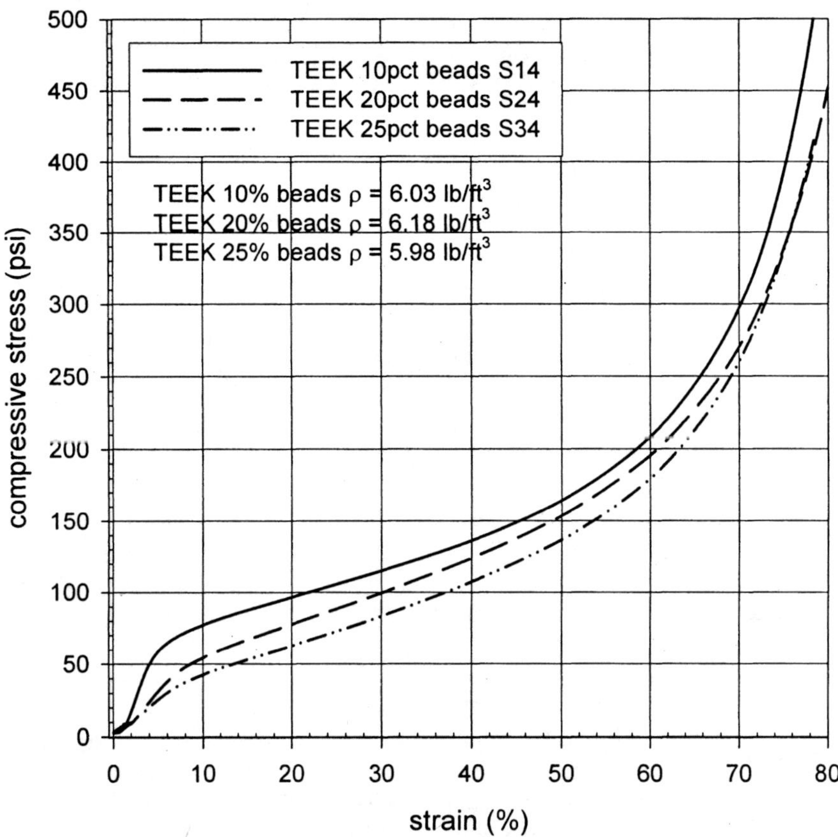


*Figure 14. Compression test data for TEEK polyimide-aerogel hybrid foam composites with 10%, 20%, and 25% aerogel beads w/w.*

Incorporating aerogel beads into the TEEK polyimide foam decreased mechanical properties with respect to compressive yield strength. Increasing loading from 10% to 25% aerogel decreased yield strength by half and decreased stress at 50% compression by about a quarter. However, the strain to yield was increased and the modulus decreased indicating that TEEK-aerogel hybrid foam composites are more elastic in compression.

### Vibration Attenuation Results

TEEK-aerogel hybrid foam composites made with aerogel blankets were tested for vibration attenuation properties. TEEK-aerogel composite with one blanket performed slightly better than the TEEK polyimide foam as shown in Figure 15. TEEK-aerogel composite with two aerogel blankets performed much better than the TEEK polyimide and foam and single blanket composite as shown in Figure 15.

The reduced acceleration over time clearly shows that TEEK foam and the TEEK-aerogel composite foams are good attenuation materials at 5000 hertz with the two blanket composite dramatically improved over the control specimen.

## Discussion

Incorporating commercially available loose fill aerogel and aerogel blanket into polyimide foam improved thermal performance by decreasing heat transfer through the foam and did not compromise the excellent fire performance of the polyimide foam. The addition of aerogel blanket to TEEK polyimide foam also improved vibration attenuation. Although some loss of compression strength did occur, the foam composites maintained yield strengths greater than 25 psi and can be considered for some structural applications. However, this same mechanical testing did show that strain to yield was improved. The fire performance of the commercially available aerogel materials was good with a low heat release rate (HRR), but with a relatively fast time to PHR. The aerogel materials immediately ignited then burned with a low flame under an irradiance of 50 $kW/m^2$. TEEK foam control took over 120 seconds to ignite and then burned with a low flame under the same irradiance. The addition of aerogel into TEEK caused the time to PHR to extend which was not expected based on the fire data, rather it was expected based on the decreased heat transfer as measured by the heat flow meter method. The aerogel materials are not very effective at blocking infrared and thus the hydrophobic organic groups that make commercial aerogels super-hydrophobic can quickly be excited and vaporized thus contributing fuel to the combustion process.

In comparison to the parent foam, the TEEK-aerogel foam composites were found to be more effective as vibration attenuation materials and had significantly lower thermal conductivity. Finally, commercially available aerogel materials, TEEK foam, and TEEK-aerogel foam composites were all found to have low peak heat release under forced combustion conditions and can probably be considered flame retardant. Cone calorimeter tests at higher heat fluxes; 75 $kW/m^2$ and 100 $kW/m^2$ should be performed to verify these fire test results, since the flame would go out while the sample continued to offgas under the thermal load.

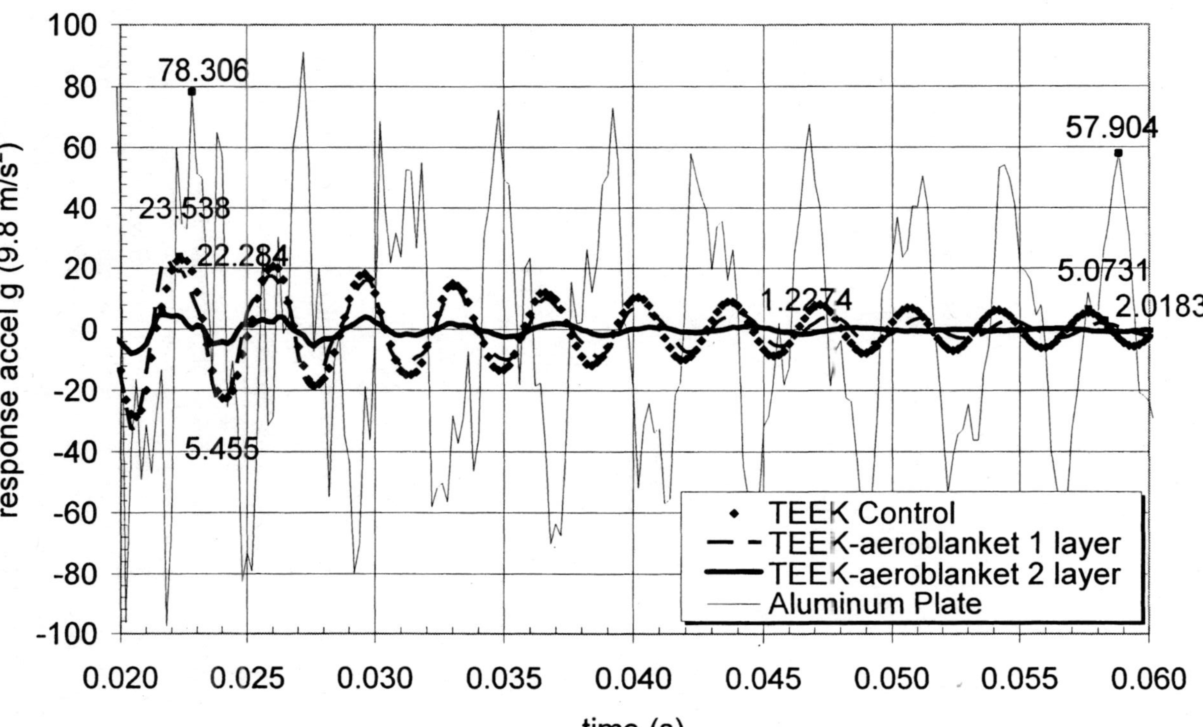


*Figure 15. Vibration attenuation data for aluminum plate, TEEK control, TEEK aerogel blanket 1 layer, and TEEK aerogel blanket 2 layer. Vibration frequency was 5000 hertz.*

## Conclusions

Incorporating commercially available loose fill aerogel and aerogel blanket into polyimide foams improved thermal performance by decreasing heat transfer through the foam, maintained good fire properties of TEEK polyimide, and only moderately reduced the overall mechanical properties. Table IV summarizes the density, mechanical data, thermal data, and fire data.

Increased loading of aerogel beads resulted in decreased heat transfer. A greater reduction in heat transfer occurred when a pocket of aerogel beads was placed in the center of the foam in the form of a thin pocket. The most effective configuration for minimizing heat transfer employed the use of aerogel blanket materials placed in the center of the foam. Keeping density constant and comparing k-values for TEEK control and the two-layer aerogel blanket foam hybrid (TEEK AB 2 layer) demonstrated a 42% decrease in heat transfer as measured by ASTM C518. The compression strength was degraded somewhat by integrating aerogel into the system, but strain to yield was improved. Vibration attenuation measurements were collected for a rigid TEEK foam sample and two aerogel blanket foam composites: 6 lb/ft$^3$ TEEK control, TEEK AB 1 layer, and TEEK AB 2. The two aerogel blanket composite foam exhibited the best vibration attenuation. Most importantly, the outstanding fire retardant nature of TEEK-H polyimide foam was not compromised by incorporating in aerogel. All aerogel materials, TEEK foam, and TEEK-aerogel hybrid foam composites exhibited excellent fire properties. Tests at higher heat fluxes should be performed.

## Acknowledgments

The authors wish to thank Rudy Werlink and Dr. LaNetra Tate for vibration attenuation measurements, Wayne Heckle for thermal conductivity measurements, David Smith for assistance in performing cone calorimeter fire tests, and Courtney Hornsby for sample fabrication.

This document was prepared under the sponsorship of the National Aeronautics and Space Administration. Neither the United States Government nor any person acting on behalf of the United States Government assumes any liability resulting from the use of the information contained in this document, nor warrants that such use will be free from privately owned rights. The citation of manufacturers' names, trademarks, or other product identification in this document does not constitute an endorsement or approval of the use of such commercial products.

**Table IV. Fire, mechanical, and thermal properties of aerogel materials, TEEK polyimide foam, and TEEK polyimide-aerogel hybrid foam composites. PHR for TEEK control was one test of a restrained specimen. PHR for TEEK XX% aero composites is the average initial PHR of two specimens from one 6 lb/ft$^3$ sample. Initial PHR discounts sample deformation under thermal load.**

| *Material* | *Density lb/ft$^3$* | *Thermal Conductivity mW/m-K* | *Peak Heat Release kW/m$^2$* | *Compressive Strength @ 50% strain psi* |
|---|---|---|---|---|
| Nanogel Beads | 5.0 | 18.0* | 60.3±3.5 | --- |
| Nanogel Granules | 5.9* | 18.0* | 56.6±1.8 | --- |
| Cryogel Blanket | 8.0** | 15.0** | 95.8±4.0 | --- |
| Pyrogel Blanket | 10.7** | 15.5** | 65.6±4.5 | --- |
| TEEK control | 2.3±0.3 | 32.46±0.69 | 25.5 | --- |
| TEEK control | 6.1 | 36.10 | --- | --- |
| TEEK 10% aero | 2.5±0.2 | 32.09±0.17 | --- | --- |
| TEEK 10% aero | 6.03 | 33.76 | 30.6±3.2 | 169.0 ± 8.6 |
| TEEK 20% aero | 2.6±0.4 | 30.82±0.00 | --- | --- |
| TEEK 20% aero | 6.18 | 31.32 | 29.6±3.3 | 151.3 ± 11.7 |
| TEEK 25% aero | 2.7±0.4 | 30.40±0.32 | --- | --- |
| TEEK 25% aero | 5.98 | 29.69 | 31.5±1.3 | 125.4 ± 13.4 |
| TEEK 30% aero | 2.72 | 29.66 | --- | --- |
| TEEK 40% aero | 2.95 | 28.65 | --- | --- |
| TEEK pocket of aero | 2.91 | 26.48 | --- | --- |
| TEEK AB 1 layer | 6.17 | 26.18 | --- | --- |
| TEEK AB 2 layer | 6.12 | 21.04 | --- | --- |
| TEEK AB diag strips (9) | 5.99 | 33.20 | --- | --- |
| TEEK 10% aero + AB 1 layer | 2.88 | 24.83 | --- | --- |
| TEEK 30% aero + AB 1 layer | 3.45 | 24.22 | --- | --- |

## References


1. Kistler, S. S. "Coherent expanded aerogels and jellies", Nature **1931**, *127*, 741.
2. Kistler, S. S. "Coherent expanded aerogels", J. Phys. Chem. **1932**, *36*, 52-64.
3. Kistler, S. S.; Caldwell, A. G. "Thermal conductivity of silica aerogel", Ind. Eng. Chem. **1934**, *26*, 658-662.
4. Kistler, S. S. "The relation between heat conductivity and structure in silica aerogel", J. Phys. Chem. **1935**, *39*, 79-85.
5. Kistler, S. S. "The calculation of the surface area of microporous solids from measurements of heat conductivity", J. Phys. Chem. **1942**, *46*, 19-31.
6. Kistler, S. S.; Fischer, E. A.; Freeman, I. R. "Sorption and surface area in SiO2 aerogel", Journal of the American Chemical Society **1943**, *65*, 1909-1919.
7. Jones, C. L. "Santocel-a new raw material for the coating industry", Paint Industry Magazine **1947**, 62, 390-391.
8. Shand, L. D. "Santocel in coatings", Official Digest, Federation of Paint and Varnish Production Clubs **1948**, *No. 277*, 175-184.
9. Church, H. F.; Scott, J. R. "Properties of ebonite. XXXVI. Silica as a loading material", J. Rubber Res. **1949**, *18*, 13-27.
10. Duchacek, C. F., Stabilized paraffin wax. US2827387, 1958.
11. Hughes, E. C.; Milberger, E. C., Thickened lubricants. US2820763, 1958.
12. Marotta, R., Hydrophobic siliceous insecticidal compositions. US3159536, 1964.
13. Rao, K. S.; Das, B. "Sorptive properties of fibrous silica gel (Santocel C) activated at different temperatures", Current Science **1968**, *37*, 599-602.
14. Smith, D. International Aerogels Conferences, "A short history of aerogel commercialization and their use in thermal insulation", Boston, December 2007.
15. Begag, R.; Fesmire, J. E. In *Thermal Conductivity*; DEStech Publications: Lancaster, PA, 2008, p 323-333.
16. Fesmire, J. E.; Rouanet, S.; Ryu, J. In *Advances in Cryogenic Engineering*; Plenum Press: New York, 1998, p 219-226.
17. Fesmire, J. E.; Augustynowicz, S. D.; Rouanet, S. "Aerogel Beads as Cryogenic Thermal Insulation System", Adv. Cryogenic Eng. **2002**, *47*, 1541-1548.
18. Fesmire, J. E. "Aerogel insulation systems for space launch applications", Cryogenics **2006**, *46*, 111-117.
19. Fesmire, J. E.; Sass, J. P. "Aerogel Insulation Applications for Liquid Hydrogen Launch Vehicle Tanks", Cryogenics **2008**, *48*, 223-231.

20. Capadona, L. A.; Meador, M. A. B.; Alunni, A.; Fabrizio, E. F.; Vassilaras, P.; Leventis, N. "Flexible, low-density polymer crosslinked silica aerogels", Polymer **2006**, *47*, 5754-5761.
21. Katti, A.; Shimpi, N.; Roy, S.; Lu, H.; Fabrizio, E. F.; Dass, A.; Capadona, L. A.; Leventis, N. "Chemical, physical, and mechanical characterization of isocyanate crosslinked amine-modified silica aerogels", Chem. Mater. **2006**, *18*, 285-296.
22. Meador, M. A. B.; Capadona, L. A.; McCorkle, L.; Papadopoulos, D. S.; Leventis, N. "Structure-Property Relationships in Porous 3D Nanostructures as a Function of Preparation Conditions: Isocyanate Cross-Linked Silica Aerogels", Chemistry of Materials **2007**, *19*, 2247-2260.
23. Weiser, E. S.; Baillif, F. F.; Grimsley, B. W.; Marchello, J. M. "High temperature structural foam", Proc. International SAMPE Symposium and Exhibition **1998**, *43*, 730-744.
24. Weiser, E. S.; Johnson, T. F.; St. Clair, T. L.; Echigo, Y.; Kaneshiro, H.; Grimsley, B. W. "High temperature polyimide foams for aerospace vehicles", International SAMPE Symposium and Exhibition **1999**, *44*, 511-525.
25. Weiser, E. S.; Johnson, T. F.; St. Clair, T. L.; Echigo, Y.; Kaneshiro, H.; Grimsley, B. W. "Polyimide foams for aerospace vehicles", High Performance Polymers **2000**, *12*, 1-12.
26. Williams, M. K.; Nelson, G. L.; Brenner, J. R.; Weiser, E. S.; St. Clair, T. L. "Cell surface area and foam flammability", Recent Advances in Flame Retardancy of Polymeric Materials **2001**, *12*, 137-150.
27. Williams, M. K.; Nelson, G. L.; Brenner, J. R.; Weiser, E. S.; St. Clair, T. L. In *Fire and Polymers*, 2001, p 49-62.
28. Weiser, E. S., Ph.D. Dissertation, 2004. College of William and Mary, Williamsburg, VA, 229 pp.
29. Williams, M. K.; Melendez, O.; Palou, J.; Holland, D.; Smith, T. M.; Weiser, E. S.; Nelson, G. L. "Characterization of polyimide foams after exposure to extreme weathering conditions", J. Adhesion Sci. Technol. **2004**, *18*, 561-573.
30. Williams, M. K.; Weiser, E. S.; Fesmire, J. E.; Grimsley, B. W.; Smith, T. M.; Brenner, J. R.; Nelson, G. L. "Effects of cell structure and density on the properties of high performance polyimide foams", Polym. Adv. Technol. **2005**, *16*, 167-174.
31. Williams, M. K.; Holland, D. B.; Melendez, O.; Weiser, E. S.; Brenner, J. R.; Nelson, G. L. "Aromatic polyimide foams: factors that lead to high fire performance", Poly. Degrad. Stab. **2005**, *88*, 20-27.

## Chapter 11

# Advanced Flame-Retardant Epoxy Resins for Composite Materials

**Michael Ciesielski, Jan Diederichs, Manfred Döring, and Alexander Schäfer**

**Institute for Technical Chemistry, Karlsruhe Research Center, Hermann-von-Helmholtz-Platz 1, D–76344, Eggenstein-Leopoldshafen, Germany**

The present article will highlight two different approaches to obtaining phosphorus-modified epoxy materials which are expected to be qualified for high performance printed wiring boards (PWB). The first method involves the incorporation of novel non-reactive derivatives of 9,10-dihydro-9-oxa-10-phosphaphenanthrene-10-oxide (DOPO) in an epoxy novolac (DEN 438) and subsequent curing with dicyandiamide (DICY) and fenuron. An alternative process for the manufacture of phosphorus-modified epoxy materials is the reactive introduction (preformulation) of aldehyde adducts of DOPO and 2,8-dimethyl-phenoxaphosphine-10-oxide (DPPO) into the backbone of the DEN 438 novolac, followed by curing with 4,4`-diaminodiphenylmethane (DDM). The epoxy materials obtained in this way reached the UL 94 V-0 rating at low phosphorus concentrations (0.6-1.4% P) and high $T_g$ values (180-190 °C). Comparison of the results obtained by both methods, however, revealed a slight superiority of the application of non-reactive additives in the epoxy system used. Whereas all formulations containing aldehyde adducts of DOPO and DPPO exhibited an insignificant drop of $T_g$, this parameter was maintained at the high level of the unmodified material when DOPO-based additives bearing bridging and nitrogen-containing substructures were applied. These novel

DOPO derivatives are the first non-reactive additives that do not influence the $T_g$ of an epoxy material. DEN 438 samples containing commercially available and various other already known organo-phosphorus compounds were manufactured in an analogous manner. All of them showed lower $T_g$ values and poorer flame retardant efficiencies than the novel phosphorus-modified materials. To explain the superior flame inhibition activity of phosphacyclic compounds, the decomposition behaviors of DOPO and DPPO, including their sulfur derivatives, were investigated by means of thermal desorption mass spectroscopy (TDMS) and high-resolution mass spectroscopy (HRMS). PO radicals as well as PS radicals were identified as gas-phase active species. However, such fragments could not be detected when investigating samples containing open-chained phosphorus compounds.

## Introduction

Electronic devices play an increasing role in our daily life and virtually all electronic items contain a printed wiring board (PWB). Most PWBs are manufactured from copper-clad epoxy-based laminates. However, the flammability of these materials is a crucial disadvantage in electronic and electrical applications.

Consequently, many efforts have been undertaken to impart fire retardance to epoxy resins. A widely used approach to rendering epoxies flame-retardant is the incorporation of bromine atoms in their polymeric backbone. Such flame-retardant epoxy resins are prepared by a fusion reaction (preformulation) of a brominated aromatic phenol – in most cases tetrabromobisphenol A – with an epoxy compound.

Due to ecological and health concerns, however, a general tendency towards banning halogenated fire-retardants is observed. This gives rise to extensive research activities that are aimed at replacing these substances. An alternative consists in the use of phosphorus-containing additives (*1,2*), preformulations (*3,4*) and curing agents (*5-11*) as flame retardants for PWB materials. Of these, 9,10-dihydro-9-oxa-10-phosphaphenanthrene-10-oxide (DOPO) and its derivatives proved to be very powerful flame retardants in epoxy thermosets (*1-3, 5-7,12*).

Normally, non-reactive additives seriously deteriorate the glass transition temperatures ($T_g$s) and other material properties of the epoxy thermosets and tend to leach out. The reactive approach solves the migration problem, but in many cases, the drop of $T_g$ can not be prevented and additional processing difficulties may occur.

This article will start with a survey of phosphorus-containing flame retardants suitable for common PWB-relevant epoxy resins and available today (*13*). The main part of this study will describe the flame-retardant properties of novel phosphacyclic additives (*14*) and preformulations (*15,16*) synthesized by our working group. These derivatives of DOPO and the structural analog compound 2,8-dimethyl-phenoxaphosphine-10-oxide (DPPO) do not or hardly suffer from the abovementioned disadvantages, such that they are expected to be qualified for epoxy resins commonly used in high-performance PWBs. The performances of these DOPO- and DPPO-based additives and preformulations will be compared to those of already reported DOPO-based compounds and commercially available phosphorus-containing flame retardants. The influence of structural parameters of the phosphorus-containing unit on the fire behavior, thermal stabilities, and $T_g$ values of epoxy materials will be discussed. Crucial importance of tailored chemical structures will be empasized.

Furthermore, the results of studies (*15,16*) of the flame retardant-action of DOPO- and DPPO-derivatives will be presented here. The detection of phosphorus-containing species like the PO radical by means of thermal desorption mass spectroscopy (TDMS) and high-resolution mass spectroscopy (HRMS) will be described and the importance of such radicals to the flame inhibition mechanism will be discussed.

This article will be confined to the flame-retardant properties of epoxy samples without glass fibers. The fire behavior of reinforced laminates containing the novel DOPO derivatives will be the subject of investigations in future.

## Halogen-free epoxy materials used in printed wiring boards

The National Electrical Manufacturers' Association (NEMA) has specified several classes of fire-retardant laminate materials used for PWBs. With approximately 80%, PWBs of the FR-4 classification, made of epoxy resins and reinforced by glass fibers, are most commonly used in the electronic industry. Such materials have to meet special requirements. In particular, they have to obtain the UL 94 V-0 classification that is the most widely used specification for the fire resistance of PWBs.

The increasing use of lead-free soldering has changed the PWB base material market. Due to the higher process temperatures of lead-free soldering, there is an increasing demand for PWB laminates of enhanced thermal stability. Often, epoxy-novolac resins are employed for such high-performance base materials of higher glass transition temperatures (Tg > 165 °C). These epoxy-novolac resins need less flame retardants compared to the normally used diglycidyl ether of bisphenol-A (DGEBA) resins, which makes it easier to meet the technical requirements with halogen-free flame retardants. Use of base materials of higher thermal resistance requires a reformulation of the base material recipes. Many manufacturers have taken this opportunity to investigate halogen-free FRs when developing new materials. Apart from inorganic hydroxides, several organo-phosphorus compounds have already been established as flame retardants for high-performance epoxy materials and represent a growing niche. The chemical structures of commercially available phosphorus-containing fire-retardants suitable for epoxy resins are shown in Figure 1.

Of these compounds, aluminium diethylphosphinate is a new non-halogenated flame retardant that is not hygroscopic, not toxic, has an extremely low solubility in water and common solvents and does not hydrolyze in the presence of water. The latter is most important, since the release of phosphoric acid cannot be tolerated in E&E applications.

9,10-dihydro-9-oxa-10-phosphaphenanthrene-10-oxide (usually denoted as DOPO) is a commercially available phosphacyclic flame retardant for epoxy resins. Today, DOPO may be regarded as the major building block used to

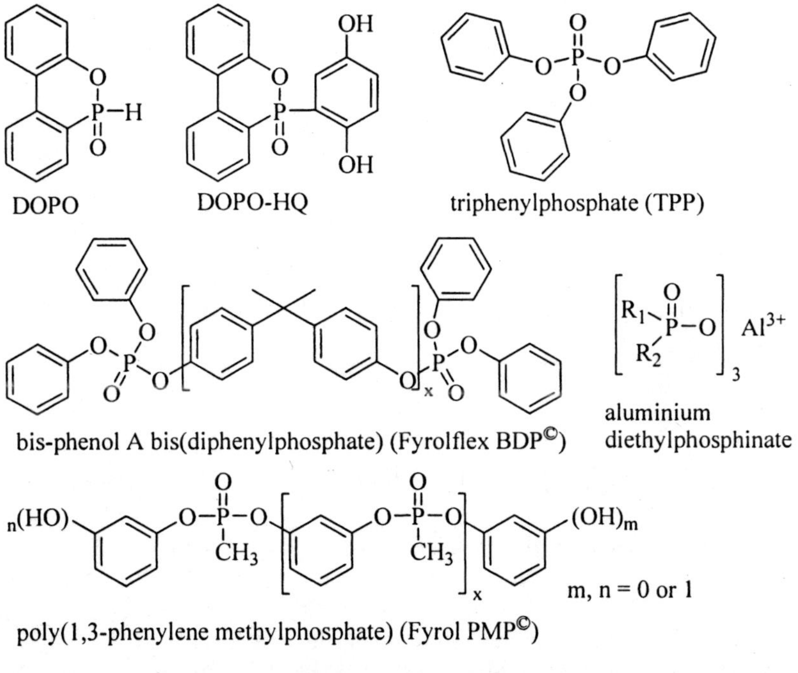

DOPO DOPO-HQ triphenylphosphate (TPP)

bis-phenol A bis(diphenylphosphate) (Fyrolflex BDP©)

aluminium diethylphosphinate

m, n = 0 or 1

poly(1,3-phenylene methylphosphate) (Fyrol PMP©)

*Figure 1. Selected commercially available, phosphorus-containing flame retardants for epoxies*

introduce phosphorus-containing units into epoxy resins ($T_g$ up to around 160 °C). DOPO is commercially available from different suppliers and global capacities increased over the last five years to meet the increasing market demand for PWBs. DOPO itself is mono-functional towards epoxy groups, but several modifications are commercially available, in particular its bifunctional benzoquinone adduct DOPO-HQ.

Poly(1,3-phenylene methylphosphonate) (Fyrol PMP®) also is a flame retardant especially developed for application in epoxy systems. Due to its hydroxyl groups, it can act as a curing agent for epoxies. This flame retardant is characterized by a high temperature stability. Table I lists the minimum amounts of commercially available flame retardants necessary to achieve the UL 94 V-0 classification for a typical PWB epoxy formulation, consisting of epoxy novolac resin DOW DEN 438, dicyandiamide (DICY) as hardener, and fenuron as accelerator. The experiments were performed with resins without glass fiber reinforcement. Therefore, these values only indicate the range in a final commercial FR 4 material.

**Table I. DOW DEN 438 resin hardened with DICY / fenuron (no glass fiber). Amount of flame retardant necessary for UL 94 V0 classification (*13*)**

| *Flame Retardant* | *Phosphorus-content (%)* | *FR-content (%)* | *UL 94 (4 mm)* | *Tg (DSC) (°C)* |
|---|---|---|---|---|
| – | – | – | V2 | 182 |
| aluminium diethylphosphinate | 3.2 | 16.7 | V0 | 167 |
| Fyrol PMP® | 3.2 | 23.5 | V0 | 165 |
| DOPO-HQ | 1.4 | 17.0 | V0 | 161 |
| DOPO | 0.93 | 6.5 | V0 | 158 |
| DOPO + 30 % boehmite | 0.42 | 2.9 | V0 | 168 |

(Reproduced from reference 13. Copyright 2007.)

## Novel Efficient DOPO-Based Flame Retardants

### Novel DOPO-Based Additives for High-$T_g$ Epoxy Resins

DOPO and DOPO-HQ induce excellent flame retardant properties, but they decrease the glass transition temperatures of the epoxy resins. DOPO derivatives which do not show this undesired effect were produced by a novel synthesis route (*14*) presented in Scheme 1.

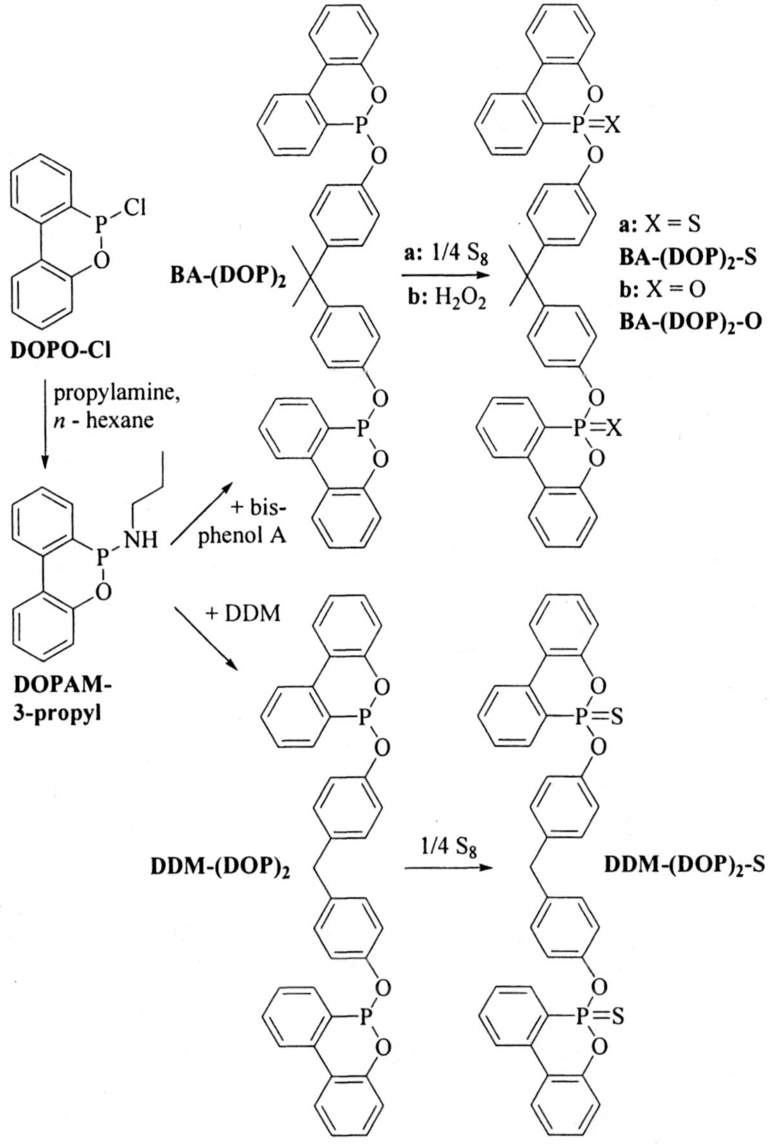

DOPO-Cl

propylamine, *n* - hexane

DOPAM-3-propyl

+ bis-phenol A

BA-(DOP)$_2$

a: 1/4 $S_8$

b: $H_2O_2$

a: X = S

BA-(DOP)$_2$-S

b: X = O

BA-(DOP)$_2$-O

+ DDM

DDM-(DOP)$_2$

1/4 $S_8$

DDM-(DOP)$_2$-S

*Scheme 1. Synthesis of the DOPO-based flame retardants. (Reproduced from reference 14. Copyright 2008 John Wiley & Sons, Ltd.)*

Using this pathway, the already described phosphorus-containing compounds BA-(DOP)$_2$ and BA-(DOP)$_2$-O (17) as well as the new DOPO derivatives BA-(DOP)$_2$-S, DDM-(DOP)$_2$, and DDM-(DOP)$_2$-S were synthesized.

These flame retardants were incorporated in a high-$T_g$ resin, consisting of an epoxy novolac (DEN 438), DICY as hardener, and fenuron. The fire-retardant properties of the DOPO-based additives in this PWB-relevant epoxy resin were evaluated using the UL 94 vertical burning test (*14*). The fire behavior and thermal properties of triphenylphosphate (TPP) and Fyrolflex BDP® were determined as well and compared to those of the DOPO derivatives.

The minimal phosphorus concentrations necessary to achieve V-0 are summarized in Table II together with the corresponding $T_g$ values determined by DSC measurements.

**Table II. Overview of the minimal phosphorus contents necessary to achieve Ul 94 V0 and the corresponding glass transition temperatures (*14*)**

| *Additive in DEN438/DICY*[a] | *Min. P content to achieve V0 (% P)* | $T_g$ *(°C)*[b] |
|---|---|---|
| TPP | 1.6 | 136 |
| BDP | 2.0 | 157 |
| BA-(DOP)$_2$ | 1.0 | 164 |
| BA-(DOP)$_2$-S | 1.4 | 167 |
| DDM-(DOP)$_2$ | 0.8 | 180 |
| DDM-(DOP)2-S | 1.2 | 184 |

[a] DOW DEN 438 resin hardened with DICY/fenuron (no glass fiber). The preparation and curing of the samples is described in (14).
[b] $T_g$ of the pure DEN 438 sample cured with DICY: 182°C.

The UL 94 tests revealed that some DOPO-based additives act as superior flame retardants in the DEN 438/DICY resin. The following order of flame-retardant efficiencies of the various additives was obtained:

DDM-(DOP)$_2$ > BA-(DOP)$_2$ > DDM-(DOP)$_2$-S > BA-(DOP)$_2$-S > TPP > BDP > BA-(DOP)$_2$-O

The flame-retardant properties were found to be strongly influenced by the structural features of the additives used. The results of the UL 94 tests showed that nitrogen-containing substituents at the phosphorus atom increase the fire-retarding efficiency. Furthermore it was found that DOPO derivatives with trivalent phosphorus are more effective than those with pentavalent phosphorus. Hence, the trivalent DOPO amide DDM-(DOP)$_2$ is the best fire retardant in the

DEN 438/DICY resin which required only 0.8% phosphorus to reach V-0. Replacing an oxygen atom by a sulfur atom also has a favorable effect on the flame-retardant properties. These results indicate that the flame-retarding effectiveness of DOPO-based compounds is enhanced with increasing electron density at the phosphorus atom.

The influence of the additive and its concentration on the $T_g$ values for DICY-cured DEN 438 resins is presented in Figure 2.

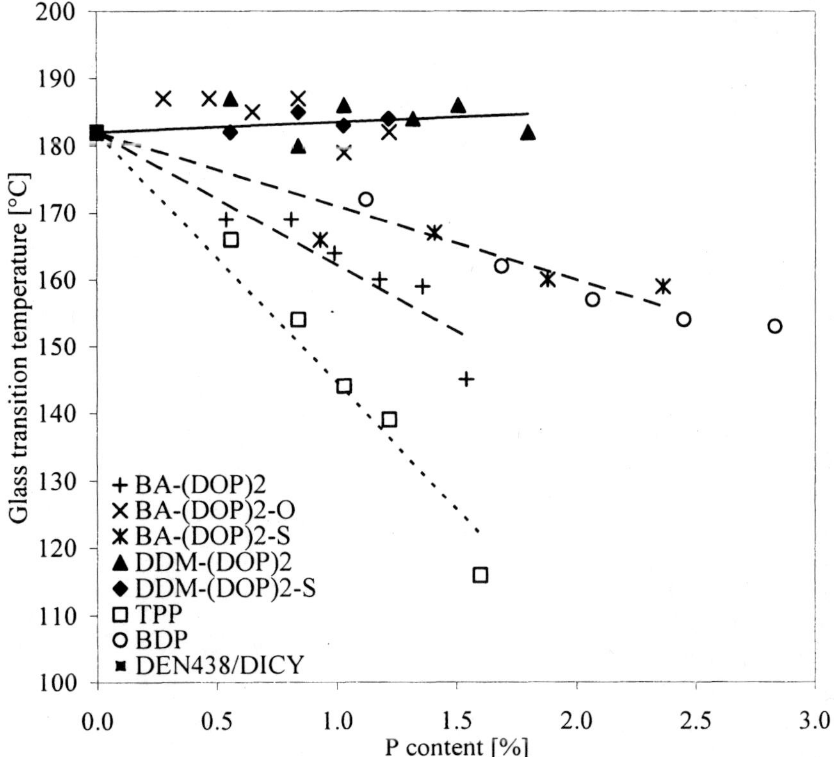


*Figure 2. Glass transition temperature versus phosphorus content of DEN 438 samples cured with DICY (Reproduced from reference 14. Copyright 2008 John Wiley & Sons, Ltd.)*

TPP strongly deterioates the $T_g$ in the DICY-cured epoxy resin. This result is not surprising, because TPP is a typical non-reactive additive having small molecules with a compact geometry. Fyrolflex BDP® and some DOPO-based additives reduce the $T_g$ in the epoxy system to a less significant extent only. The molecules of these compounds are elongated compared to those of TPP and have a rod-like geometry. The moderate plasticizing effect of these additives shows that the glass transition temperature is an important material property of epoxy materials that can be maintained at a considerably higher level, if non-reactive additives having oligomeric, bridging, and rather rigid molecular structures are used.

However, DDM-$(DOP)_2$, DDM-$(DOP)_2$-S and BA-$(DOP)_2$-O exhibited a quite unusual behavior that cannot be expected for non-reactive additives: they did not exhibit any decrease of the glass transition temperature even at high loadings.

Figure 3 allows for a direct assessment of the overall performances of the additives used in this study.

The minimal phosphorus concentrations necessary to achieve V-0 and the corresponding $T_g$ values were chosen to be crucial parameters in this

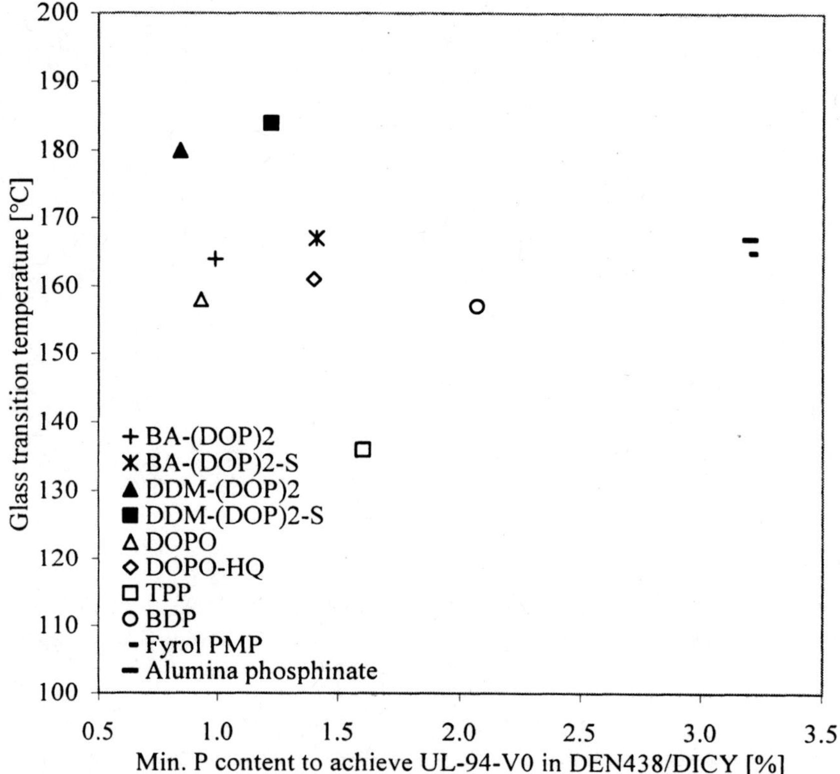


*Figure 3. Comparison of the overall performances of the DICY-cured samples flame-retardant with various compounds. (Reproduced from reference 14. Copyright 2008 John Wiley & Sons, Ltd.)*

comparison. For a better evaluation, the corresponding values of commercially available phosphorus-containing flame retardants are indicated as well. The DOPO-based additives DDM-$(DOP)_2$ and DDM-$(DOP)_2$-S show the highest performances in the epoxy systems used, because they are efficient at very low phosphorus concentrations and do not deteriorate the $T_g$ values. These novel flame retardants not only surpass the capabilities of all other additives used in this work, but also those of DOPO, DOPO-HQ, and all other commercially available products.

## DOPO-Based Preformulations for High-$T_g$ Epoxy Resins

Various flame-retardant epoxy thermosets with high $T_g$s were also obtained by the application of novel reactive DOPO- and DPPO-based flame retardants (*15,16*): aldehyde adducts of DOPO and DPPO were incorporated in the backbone of a novolac resin (DEN 438) by fusion reactions according to the synthetic paths shown in Schemes 2 and 3.

Preformulations with diethylphosphite$_2$-TDA and diphenylphosphite were synthesized in an analogous manner. The phosphorus-containing resins obtained

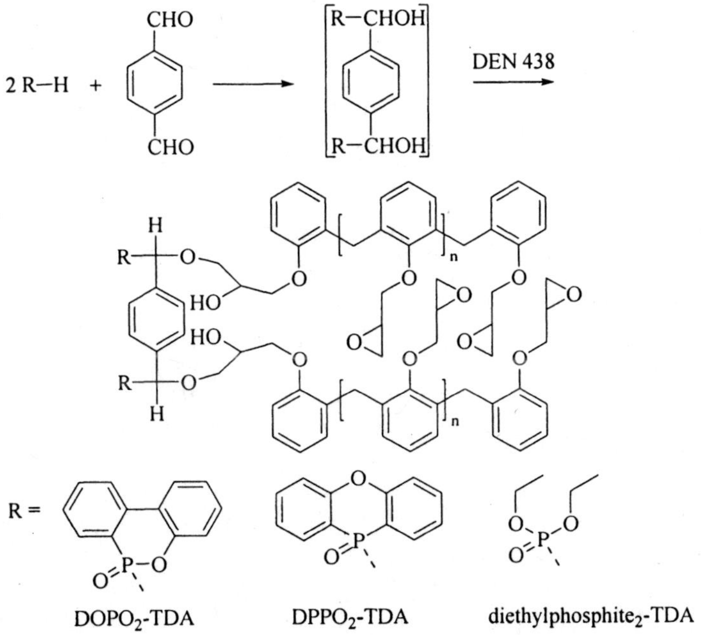

*Scheme 2. Synthesis of phosphorus-containing terephthalaldehyde adducts and subsequent fusion reactions with DEN 438 epoxy novolac.*

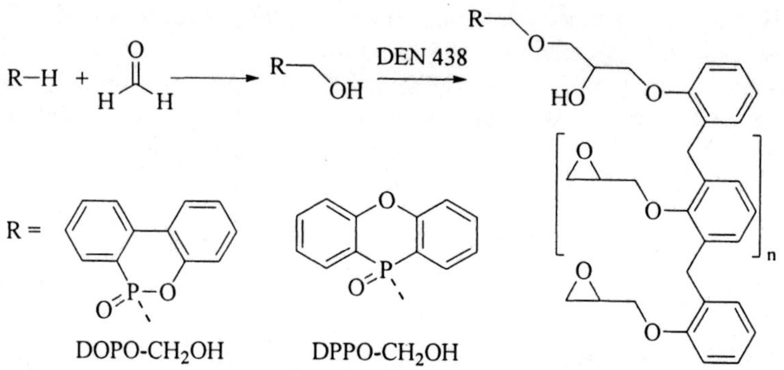

*Scheme 3. Phosphorus-containing formaldehyde adducts and subsequent fusion reactions with DEN 438 epoxy novolac.*

in this way were cured with 4,4'-diaminodiphenylmethane (DDM). Flammability and burning behaviors of the resultant epoxy thermosets were characterized by UL 94 and LOI and the $T_g$s were determined by DSC measurements. The minimal phosphorus concentrations necessary to pass the V-0 rating are summarized in Table III together with the corresponding glass transition temperatures and LOI values at a phosphorus content of 1.7%.

All DDM-cured DEN 438 resins with covalently integrated diethylphosphite$_2$-TDA and diphenylphosphite units were not classified by the UL 94 vertical burning test up to a phosphorus content of 2%. However, samples with DOPO or DPPO incorporated in the epoxy backbone exhibited superior fire-retardant efficiencies: To achieve the V-0 classification, only 1.4% of phosphorus was necessary in the DPPO$_2$-TDA based resin. The DOPO$_2$-TDA-containing resin required an even lower phosphorus content (1.0%) to reach the same UL 94 rating. The highest fire-retardant effects, however, were observed in epoxy thermoset samples based on preformulations containing the formaldehyde adducts DOPO-CH$_2$OH and DPPO-CH$_2$OH. These epoxy resins fulfilled the V-0 classification at phosphorus concentrations of 0.8 and 0.6%, respectively.

LOI results confirmed the outstanding fire-retarding effectiveness of covalently bound DOPO and DPPO derivatives in the epoxy system investigated. DEN 438 resins incorporating DOPO-based flame retardants showed high oxygen indices of up to 33% $O_2$ at a phosphorus content of 1.7%, while DPPO-based resins exhibited LOI values of up to 31% at comparable phosphorus contents.

All formultions containing aldehyde adducts of DOPO and DPPO exhibited an insignificantly drop of $T_g$. This parameter is maintained at a high level of 180-200 °C at the phosphorus contents necessary to achieve V-0.

**Table III. Overview of the fire behaviors and glass transition temperatures of phosphorus-containing DDM-cured DEN438 epoxy novolac resins**

| *Formulation*[a] | *Min. P content to achieve V0 (% P)* | *$T_g$ (°C) at the min P content to achieve V0* | *LOI (% $O_2$) at 1.7% P* |
|---|---|---|---|
| DEN438/DDM | not rated | – | 26.5 |
| $DOPO_2$-TDA DEN438/DDM | 1.0 | 189 | 33.1 |
| $DPPO_2$-TDA DEN438/DDM | 1.4 | 185 | 27.2 |
| diethylphosphite$_2$-TDA DEN438/DDM | not rated | – | 25.7 |
| DPPO-$CH_2OH$ DEN438/DDM | 0.6 | 199 | 31.1 |
| DOPO-$CH_2OH$ DEN438/DDM | 0.8 | 183 | 31.5 |
| diphenylphosphite DEN438/DDM | not rated | – | 27.3 |

[a] Neat resin (no glass fiber). The formulation and curing procedures are described in (15, 16).

## Mechanistic Studies to the Flame Retarding Process in Epoxy Resins

These experiments were aimed at identifying the active species resonsible for the flame-retarding process and explaining the differences in the fire-redardant efficiencies of heterocyclic and open-chained phosphorus compounds. Possible modes of action of flame retardants include charring, intumescence (both in the solid phase), and flame inhibition by radical species (in the gas phase). To detect gas-phase active decomposition products, the pyrolysis gases of cured resins with and without flame retardants were analyzed by means of TDMS (*16*). MS investigations may provide an idea of possible products formed during the thermal decomposition of a substance.

Powdered samples of the cured resins (10-20mg) were heated in a high-vacuum system ($10^{-8}$ hPa) from room temperature to 460 °C at a heating rate of 10K/min and the pyrolysis gases were analyzed by TDMS. The results of these investigations are presented in Figure 4.

HRMS measurements revealed that the decomposition of DOPO and DPPO starts with the cleavage of the P-H bond (see Scheme 4) (*15*). In the following

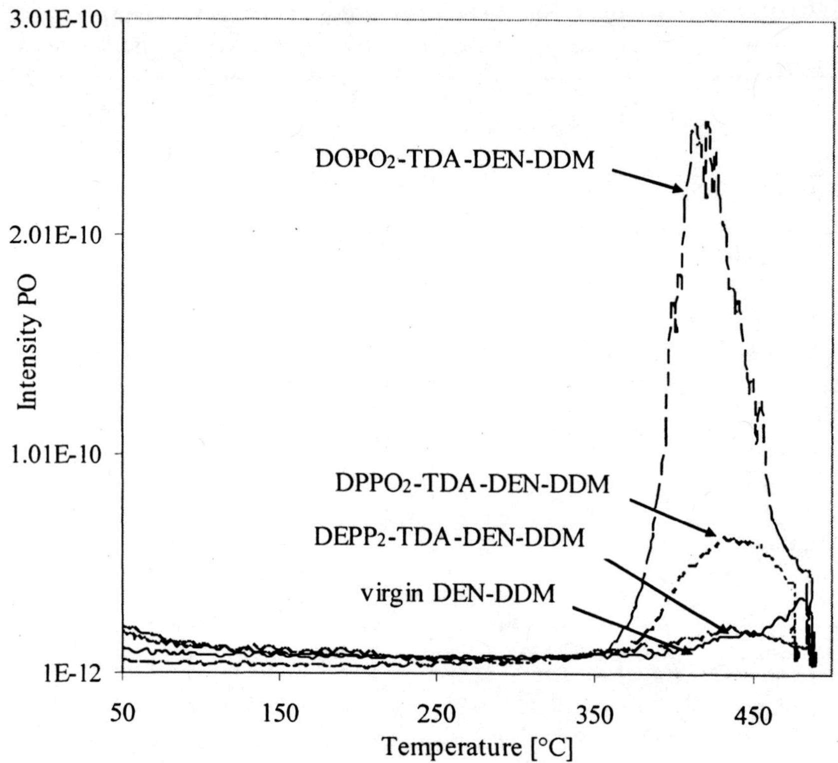


*Figure 4. Thermal desorption mass spectrometry of epoxy novolac samples preformulated with different P-containing terephthaldialdehyde derivatives, cured with DDM – PO detection. (Reproduced from reference 16. Copyright 2007 Wiley Periodicals, Inc.)*

fragmentation step, a dibenzofuran system is formed and the PO radical is released.

Results of density functional theory (DFT) calculations confirmed this decomposition route shown in Scheme 5 and excluded an alternative route which would involve the direct break-up of the phosphacycle into dibenzofuran and a HPO fragment.

The thermal decomposition behavior of epoxy samples containing sulfur-substituted DOPO derivatives were investigated in the same manner. In this case, PS radicals as the possible gas phase-active species were detected instead of PO-fragments.

Until now, PO and PX radicals have only been detected in the decomposition gases of DOPO- and DPPO-containing polymers. Further studies

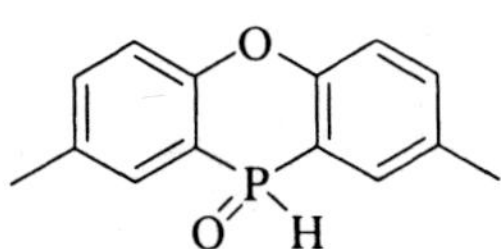

HRMS: found: 244.0596 u;
cal.: 244.0653 u.

H•

HRMS: found: 243.0540 u;
cal.: 243.0575 u.

PO•

HRMS: found: 196.0890 u;
cal.: 196.0888 u.

HRMS: found: 216.0348 u;
cal.: 216.0340 u.

H•

HRMS: found: 215.0306 u;
cal.: 215.0262 u.

PO•

HRMS: found: 168.0635 u;
cal.: 168.0575 u.

*Scheme 4. Possible decomposition behaviors of DOPO and DPPO*

| | |
|---|---|
| HPO ⟶ H· + PO· | 278.8 kJ/mol |
| DOPO-H· ⟶ PO· + DBF | 88.6 kJ/mol |
| DOPO ⟶ HPO + DBF | 163.9 kJ/mol |
| DOPO ⟶ H· + DOPO-H· | 354.2 kJ/mol |
| DOPO + H· ⟶ $H_2$ + DOPO-H· | -105.1 kJ/mol |
| DOPO + OH· ⟶ $H_2O$ + DOPO-H· | -148.8 kJ/mol |
| DOPO + H· ⟶ PO· + $H_2$ + DBF | -16.5 kJ/mol |
| DOPO + OH· ⟶ PO· + $H_2O$ + DBF | -60.3 kJ/mol |

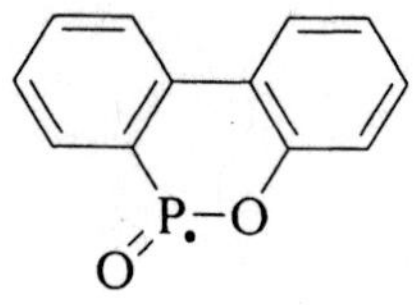

DOPO-H•

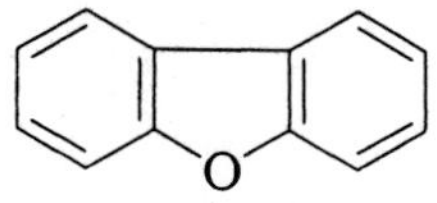

DBF

*Scheme 5. Calculated reaction energies (DFT)*

are now in progress to reveal if such or related fragments are also responsible for the flame inhibition activity of other phosphorus compounds.

## Conclusions

Two approaches to obtaining phosphorus-modified epoxy materials with glass transition temperatures around 180 °C and the UL 94 rating V-0 were presented.

The first involved the addition of non-reactive DOPO derivatives to an epoxy novolac (DEN 438) and the subsequent curing with DICY/Fenuron. The second approach was based on the reactive introduction (fusion reaction) of aldehyde adducts of DOPO and the structurally similar phosphacycle DPPO into the backbone of the epoxy novolac. DDM was used as curing agent in this case.

Surprisingly, the comparision of the results obtained revealed the superiority of the former approach in the epoxy system used: the application of non-reactive additives having tailored chemical structures. Especially DOPO amides with two DOPO units per molecule exhibited outstanding properties. These novel additives were found to be powerful fire retardants in the DEN 438/DICY resin. Only 1% phosphorus was necessary to obtain the UL 94 V-0 classification. The most important fact, however, is that these novel DOPO derivatives are the first non-reactive additives known which do not deteriorate the glass transition temperature. Hence, the $T_g$ as a crucial material parameter can be maintained at the high level of the pure resin, if non-reactive phosphorus compounds having bridging and rather rigid structures are used.

High-performance epoxy materials with the UL 94 V-0 rating at low phosphorus concentrations (0.6-1.4% P) were also obtained by the preformulation of aldehyde adducts of DOPO and DPPO with DEN 438 and curing with DDM. However, a slight drop of the $T_g$ could not be prevented by this approach.

Comparison of the results for our novel compounds with those of samples containing open-chained and commercially available fire retardants revealed that some of the new DOPO and DPPO derivatives have an outstanding performance. Furthermore, a general superiority of phosphacyclic flame retardants compared to open-chained ones was observed in the novolac-based epoxy matrix used.

Investigations by means of TDMS and HRMS provided an idea of the reason of this difference in flame-retardant efficiencies. Whereas the phosphacyclic compounds showed a significant release of small phosphorus-containing fragments (PO and PS radicals) which can inhibit the burning process, such species could not be detected if the open-chained compound and the virgin resin was investigated.

Due to their outstanding properties some of the novel phosphacyclic fire retardants described in this study are expected to be qualified for resins in advanced composites commonly used in high-performance PWB.

## References


1. Perez, R. M.; Sandler, J. K. W.; Altstädt, V.; Hoffmann, T.; Pospiech, D.; Ciesielski, M.; Döring, M. *J. Mater. Sci.* **2005**, *40*, 341-353.
2. Perez, R. M.; Sandler, J. K. W.; Altstädt, V.; Hoffmann, T.; Pospiech, D.; Ciesielski, M.; Döring, M.; Braun, U.; Knoll, U.; Schartel, B. *J. Mater. Sci.* **2006**, *41*, 4981-4984.
3. Wang, C.-S.; Lee, M.-C. *Polymer* **2000**, *41*, 3631-3638.
4. Jain, P.; Chaudary, V.; Varma, I. K. *J. Makromol. Sci.-Polym. Rev.* **2002**, *42*, 139-183.
5. Lu, S.-Y.; Hamerton, I. *Prog. Polym. Sci* . **2002**, *27*, 1661-1712.
6. Lin, C. H.; Cai, S. X. *J. Polym. Sci. Al.* **2005**, *43*, 5971-5986.
7. Artner, J.; Ciesielski, M.; Walter, O.; Doering, M.; Perez, R. M.; Sandler, J. K. W.; Altstaedt, V.; Schartel, B. *Macromol. Mat. and Eng.* **2008**, *293*, 503-514.
8. Liu, Y.-L.; Hsiue, G.-H.; Lee, R.-H.; Chiu, Y.-S. *J. Appl. Polym. Sci.* **1997**, *63*, 895-901.
9. Liu, Y.-L.; Hsiue, G.-H.; Chiu, Y.-S. *J. Polym. Chem.* **1997**, *35*, 565-571.
10. Shieh, J.-Y.; Wang, C.-S. *J. Appl. Polym. Sci.* **2000**, *78*, 1636-1644.
11. Hergenrother, P. M.; Thompson, C. M.; Smith, J. G.; Connell, J. W.; Hinkley, J. A.; Lyon, R. E.; Moulton, R. *Polymer,* **2005**, *46*, 5012-5024.
12. Levchic, S. V.; Weil, E. D. *J. Fire Sci.* **2006**, *24*, 345.
13. Döring, M.; Diederichs, J. *Halogen-free flame retardants in E&E applications,* Karlsruhe Research Center, Karlsruhe **2007**, www.halogenfree-flameretardants.com.

14. Ciesielski, M.; Schäfer, A.; Döring, M. *Polym. Adv. Technol.* **2008**, *19*, 507-515.
15. Schäfer, A.; Seibold, S.; Lohstroh, W.; Walter, O.; Döring, M. *J. Appl. Polym. Sci.* **2007**, *105*, 685-696.
16. Seibold, S.; Schäfer, A.; Lohstroh, W.; Walter, O.; Döring, M. *J. Appl. Polym. Sci.* **2008**, *108*, 264-271.
17. Su, W.-C.; Sheng, C.-S. U.S. Patent 0,101,793 A1, 2005.

# Conventional Fire Retardants

# Chapter 12

# Fire Retardancy of Polypropylene Composites Using Intumescent Coatings

**S. Duquesne[1], N. Renaut[1], P. Bardollet[2], C. Jama[1], M. Traisnel[1], and R. Delobel[1]**

**[1]Lab. Procédés d'Elaboration des Revêtements Fonctionnels (PERF), LSPES–UMR/CNRS 8008, ENSCL, Avenue Dimitri Mendeleïev - Bât. C7a, B.P. 90108, 59652 Villeneuve d'Ascq Cedex, France**
**[2]Schneider Electric, Industries SAS, Direction Scientifique et Technique, Direction de Recherche Matériaux 35800 Grenoble, France**

The aim of this study is to evaluate potential application of an intumescent coating in order to fire retard polypropylene composites. Using intumescent coatings and specific surface treatments such as cold plasma treatment prior to making the deposit, PP composites presenting a UL94 V-0 classification at low thickness have been obtained. Moreover, PP composites reinforced by $CaCO_3$ show a classification in the Glow Wire Index test at 960 °C. The surface modifications due to the cold plasma treatment are investigated using X-ray photoelectron spectroscopy, atomic force microscopy and contact angle measurement. It is observed that oxidation and etching occurs during the treatment resulting in an increased in the wetting behavior and thus to an improvement of the adhesion of the intumescent paints on the substrates. This good adhesion enables one to achieve good fire retardant properties.

## Introduction

Polyolefins, and in particular polypropylene, are increasingly used in building (flooring), transport and electrical applications (cables and wires) due to their low cost, properties and easy processing. Along with the good properties of polyolefins, such as chemical and electrical properties, their flammability is poor. They burn with a hot and clean flame, accompanied by melting and subsequent dripping or flowing of the molten polymer, limiting the widening of their application domain. To flame retard polymers in general and polyolefins in particular, fire retardant additives can be added to the matrix using an extrusion process. This methodology is widely used in the case of polyolefins because it is a low cost and well known technology *(1)*. However, in a number of applications, the required level of fire retardant properties is more and more difficult to reach. As an example, in electrical applications, whereas UL94 classification was desired by the companies for a thickness of sample of 3.2mm a few years ago, a UL94 classification for 1.6mm and even 0.8mm thick samples is now required *(2)*. At such thickness, when the FR additives are added to the matrix at a level of 20-30% (typical loading for phosphorus-based compounds), the quantity of available additives in case of fire is very low and the FR properties of materials are not high enough *(3)*. When the FR content is increased, the mechanical performance of the materials rapidly degrade. As a consequence, two approaches can be followed to achieve acceptable performance: (i) the development of more effective FR systems and/or the development of synergistic systems and (ii) the use of fire retardant coatings. The second approach has been followed in this study.

In the field of fire retardant coatings, intumescent coatings are well known. Their efficiency in the field of fire protection, in particular for the protection of steel structures is well established *(4-5)*. Intumescent coatings are composed of a resin (polyacrylic, polyvinyl acetate, epoxy...) in which intumescent additives (typically ammonium polyphosphate, pentaerythritol and melamine) are embedded. When exposed to a heat source, they develop a foamed char structure limiting the heat transfer between the heat source and the substrate *(6)* resulting in a higher resistance to fire of the substrate.

Dealing with surface coatings on polymer substrates in general, and in polyolefins in particular, leads to the problem linked with their poor surface properties and, in particular, to their low wetting behavior. Cold plasma surface treatment is known as a powerful process used to modify surface characteristics of polymeric materials *(7)*. Therefore, it represents an efficient, clean and economic alternative to activate polymeric surfaces. A plasma is an ionized gas containing both charged and neutral particles, such as electrons, ions, atoms, molecules and radicals. There are two kinds of plasma: cold and thermal. The difference is the temperature of the heavy species: in a thermal plasma, the temperature is similar to that of electrons and of the other species in a thermal plasma ($10^4$–$10^5$ K) *i.e.*, this medium is in an equilibrium state, while it is lower than 773 K in a cold plasma (or non-equilibrium plasma).

The aim of this study is to evaluate potential application of intumescent coatings in order to fire retard polypropylene composites. In thefirst part, the fire retardant properties of the coated materials with and without surface treatment are compared. Then, in the second part, the surface of the treated and untreated samples will be analyzed. Changes in surface wetting behavior of the treated surfaces will be investigated by water contact angle measurements. XPS will then be used to investigate the chemical compositional changes of the surface (region that is responsible for the improvement in wettability after exposure to plasma), the level of oxygen incorporation providing a means to monitor the extent of surface functionalization. The morphology of the modified surfaces will be finally analyzed by Atomic Force Microscopy.

## Experimental

### Materials

Raw materials are isotactic polypropylene (PPH 7060, MFI=12g/min) supplied by Total Petrochemicals and a maleic anhydride modified polypropylene (Orevac CA100, MFI=5-15g/min) supplied by Arkema; talc (A3) supplied by Rio Tinto Minerals and calcium carbonate (Socal 312) supplied by Solvay. Table I reports the composition of the PP composites. The intumescent coating used in this study was supplied by Nullifire (commercial name S607) and was applied by brush on the substrate to obtain a coating thickness varying from 100 to 150 μm.

**Table I. Composition of the Materials**

| Sample | PP (wt.-%) | PP-g-MA (wt.-%) | Talc (wt.-%) | $CaCO_3$ (wt.-%) |
|---|---|---|---|---|
| PPc | 67 | 3 | _ | 30 |
| PPt | 83,5 | 1,5 | 15 | _ |

### Plasma treatment

A capacitive cold radio frequency plasma (13.56 MHz) was generated by an EUROPLASMA set-up. The treatments were performed at a ratio of $O_2$/Ar equal to 2/1 at constant total flow of 1200 sccm ($O_2$/Ar =800/400sccm) and at a pressure of 550mTorr. The treatments of the samples were realized at constant electric power of 100W for 10min. After plasma treatment, samples were exposed to ambient atmosphere, prior to analyses.

### Surface characterization

The adherence of the film on the substrate was evaluated according to ASTM D 3359-02 "Standard Test Methods for Measuring Adhesion by Tape Test". X-ray photoelectron spectroscopy (XPS) measurements were performed with a VG Escalab 220 XL system, using non-monochromatic Mg Kα-radiation (hν = 1253.6 eV) operated at 15 kV and 20 mA. The peaks were resolved using peak analysis software (Peakfit of Jandel Scientific) assuming a Lorentzian/Gaussian line shape. Contact angle measurements were carried out with a GBX Digidrop. Contact angles were measured from the profile of liquid drops of distilled water placed on the polymer plaques at room temperature. AFM (Nanoscope III system, Digital Instruments) was used to study the surface morphology. Scanning Electron microscopy was carried out on the Hitachi S4700 SEM.

### Fire Testing

UL-94 tests were carried out on (125x12.5x1.6) $mm^3$ sheets according to the UL-94 testing *(8)*. The sample rod is placed in a holder in a vertical position and the lower end of the rod is contacted by a flame for 10 s thus initiating burning. A second ignition is made after self-extinguishing of the flame at the sample. The burning process is characterized by the times t1 and t2 pertaining to the two burning steps. The parameters t1 and t2 denote the time between removing the methane flame and self-extinguishing of the sample. Moreover, it is always noted whether drips from the sample are released or drips make absorbent cotton flame during the burning times t1 and t2.

The glow wire tests were carried out on (60x60x3) $mm^3$ sheets according to the international standard IEC 60695-2 *(9)*. The glow wire is heated via electrical resistance to a specified elevated temperature. A test specimen is held for 30 seconds against the tip of the glow wire with a force of 1 N. After the glow wire is removed, the time for the flames to extinguish is noted along with details of any burning drops. Cotton is placed beneath the specimen during the test to determine the effects of burning drops. The Glow Wire Flammability Index (GWI) is the highest temperature which satisfies one of the following conditions in three successive tests: there is no flame and no glowing (no ignition) or burning or the glowing time is less than 30 seconds after removal of the glow wire and the cotton does not ignite. These sizes correspond to the size of the uncoated samples.

## Results and Discussion

### Fire retardant properties

Table II reports the fire retardant properties of the untreated and plasma treated samples coated with the intumescent paint. It is observed that both the untreated PPt and PPc materials coated with the intumescent paint fail in both UL94 and GWI tests. To the contrary, both PPt and PPc achieve a V-0 classification in UL94 tests when they have been treated by plasma before the deposition of the intumescent coating. Moreover, the plasma treated and coated PPc succeed in the GWI tests at 960°C. The difference between the behavior of PPc and PPt regarding this test could be attributed to differences in term of mechanical properties of the material. Indeed, during GWI test, a force is applied on the material via the wire and the deformation of the material will affect the results of the test. Moreover, $CaCO_3$ can decompose leading to the formation of $CO_2$ that will have a positive effect on the fire performance.

**Table II. Fire retardant properties of the intumescent coated materials**

| Sample | Untreated samples | | Plasma treated samples | |
|---|---|---|---|---|
| | **UL-94 rating (1.6mm)** | **GWI test (960°C)** | **UL-94 rating (1.6mm)** | **GWI (960°C)** |
| PPt | NC | Failed | V-0 | Failed |
| PPc | NC | Failed | V-0 | Pass |

Observation of the materials after the test (Figure 1) reveals that the intumescent coating develops a protective layer that interrupt the combustion cycle leading to flame extinguishing.

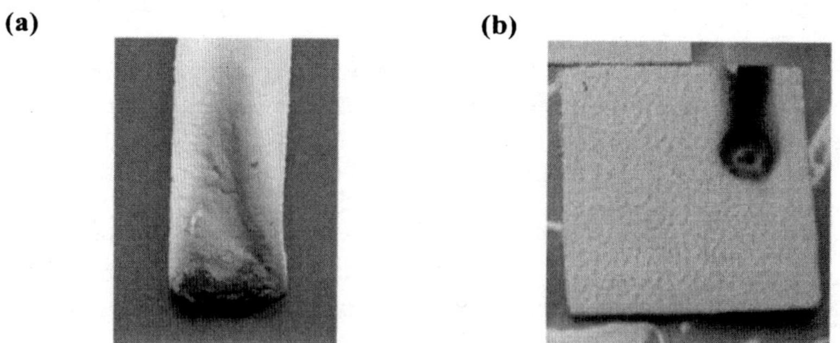


*Figure 1. Plasma treated coated PPc after UL94 test (a) and GWI test (b).*

When the materials are not cold plasma treated prior to the deposition of the intumescent coatings, low adhesion of the coating is observed (Figure 2). Intumescent coating acts in developing a multicellular and expanded layer when exposed to heat. This protective barrier limits at the same time the heat and mass (oxygen/fuel) transfer between the material and the heat source. If there is not a good adhesion between the coating and the substrate, the coating may fall down due to gravity forces creating cracks into which mass and heat transfer may occurs.

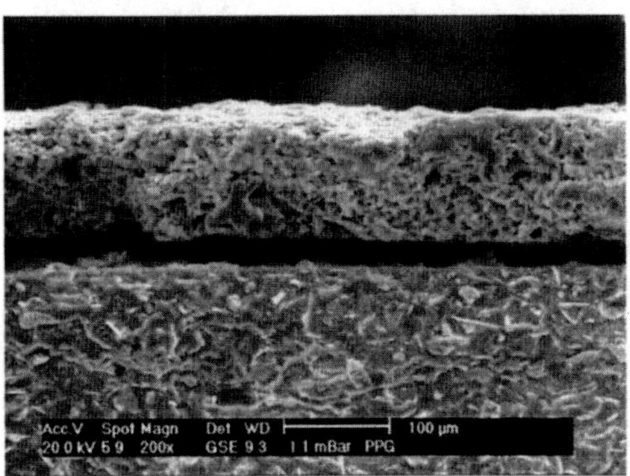


*Figure 2. Scanning electron microscopic picture of the interface between the intumescent coating and the substrate for untreated PPc.*

**Surface properties of untreated materials**

The surface properties of the untreated materials (virgin PP and PP composites) have been investigated in order to better understand the phenomenon linked with the improvement of the adhesion of the paint on the substrate.

Table III to Table V V report the binding energy values of the different elements detected on the surface of the sample for the virgin PP and for the two PP composites as well as the full width at maximum half-height (FWMH) and the composition of the surface (at.-%). Whatever the materials, the C1s spectra

**Table III. Binding energy values and elemental composition of virgin PP**

| Assignement | El(eV) | FWMH (eV) | at.-% |
|---|---|---|---|
| C1s | 285.0 | 1.4 | 90.6 |
| O1s | 532.4 | 1.8 | 4.7 |
| Si2p | 102.2 | 1.8 | 4.4 |

presents a sharp peak centred at 285eV that can be assigned to the aliphatic carbon composing the polymer chain in PP *(10)*.

On the other hand, contamination of the surface is observed since oxygen as well as silicon are detected at the surface of all materials. Their atomic concentration varies between 4.7 and 6.9at.-% and between 1.5 and 4.4at.-%, respectively, for oxygen and silicon. As a consequence, they can not be considered as negligeable.

**Table IV. Binding energy values and elemental composition of PPt composites**

| Assignement | El(eV) | FWMH (eV) | at.-% |
|---|---|---|---|
| C1s | 285.0 | 1.4 | 88.8 |
| O1s | 532.5 | 2.0 | 6.9 |
| Si2p | 102.4 | 2.4 | 3.8 |

**Table V. Binding energy values and elemental composition of PPc composites**

| Assignement | El(eV) | FWMH (eV) | at.-% |
|---|---|---|---|
| C1s | 285.0 | 1.4 | 93.1 |
| O1s | 532.4 | 2.4 | 5.2 |
| Si2p | 102.5 | 2.8 | 1.5 |

The Si2p spectra can be decomposed into two components centered at 104.1 and 102.3 eV (Figure 3). The peak at around 104.1 eV may be assigned to inorganic silicon species, typically silica (Q4 configuration) whereas the peak around 102.3 eV is attributed to organic silicone (as example in D2 configuration) (*10*). It has to be noticed that PPt contains talc, a hydrated magnesium sheet silicate with the chemical formula $Mg_3Si_4O_{10}(OH)_2$. Its elementary sheet is composed of a layer of magnesium-oxygen/hydroxyl octahedra, sandwiched between two layers of silicon-oxygen tetrahedra. As a consequence, the presence of silicon in PPt is not surprising. However, this is not the case for virgin PP or for PPc in which no silicon was found. Since the silicon is detected in the virgin PP, also in the granule coming from the supplier (results not shown), the contamination can not be attributed to pollution occurring during the extrusion process or during thermoforming of the material but is rather attributed to the presence of an additive in the virgin polymer.

Silicone based additives (such as for example polydimethylsiloxane) are widely used as mold-releasing agent and as a consequence it is reasonable to assume that the silicon comes from this type of additives. However, the chemical composition of this agent is not clearly identified because 1) this is not the aim of this study and 2) since the XPS by itself can not clearly identify the chemical composition.

The surface properties of untreated materials in term of contact angle measurements, surface roughness and adhesion are presented in Table VI. The high values of the contact angle measured for all the materials demonstrate that their surface are hydrophobic. The higher value obtained for virgin PP can be correlated with the higher contribution of the Si2p peak centred around 102.3eV ("organic" part of the silicone based species). The higher concentration of the mold-releasing agent at the surface of the virgin PP compare to PP composites leads to higher hydrophobicity of its surface. It can be proposed that the presence of filler limits the diffusion of the mold-releasing agent into the polymer matrix.

**Table VI. Surface properties of untreated materials**

| Material | Contact angle (°) | Surface Roughness (nm) | Adhesion (ASTM D 3359-02) |
|---|---|---|---|
| PP | 106.6 | 3.8 | 0B |
| PPt | 97.6 | 20.8 | 0B |
| PPc | 95.8 | 14.0 | 0B |

The surface roughness of the three materials are very different. Since all the materials have been processed in the same way, it is possible to assume that the presence of filler in the PP matrix modifies the surface properties of the matrix (for example its crystallinity), resulting in an increase in the surface roughness of the thermoformed plaques. All materials present poor adhesion with the intumescent paints which is not surprising according to their surface properties.

## Surface properties of plasma treated materials

The C1s spectrum of the treated PPt is presented in Figure 4. Whereas the C1s spectrum of untreated PPt is composed of one main peak at 285.0 eV, C1s spectra for the plasma treated material presents three additional bands at 286.8eV, 287.9eV and 289.2eV respectively assigned to C-O, C=O and O-C=O (or $CO_3^{2-}$ in the case of PPc for which similar results were obtained).

These results show the formation of functional groups on the surface of PP composites after the plasma treatment. It demonstrates that oxidation occurs when PP composites are exposed to plasma treatment. This functionalization occurs by grafting of oxygen containing species in the free radicals generated during the treatment and also after plasma treatment, when plasma-treated samples are exposed to ambient atmosphere leading to an intensification of surface functionalization phenomenon *(11)*. Most of the papers related to

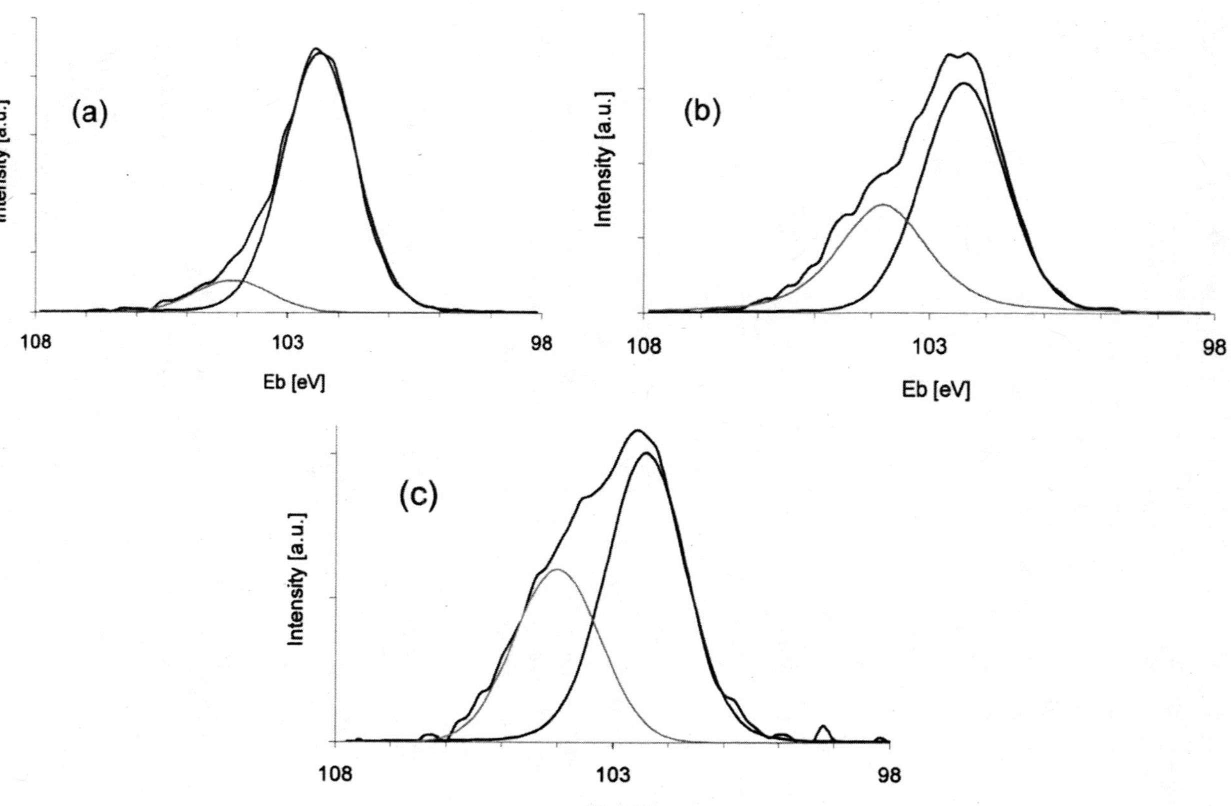


*Figure 3. XPS Si2p spectra of untreated materials (a) virgin PP, (b) PPt and (c) PPc*

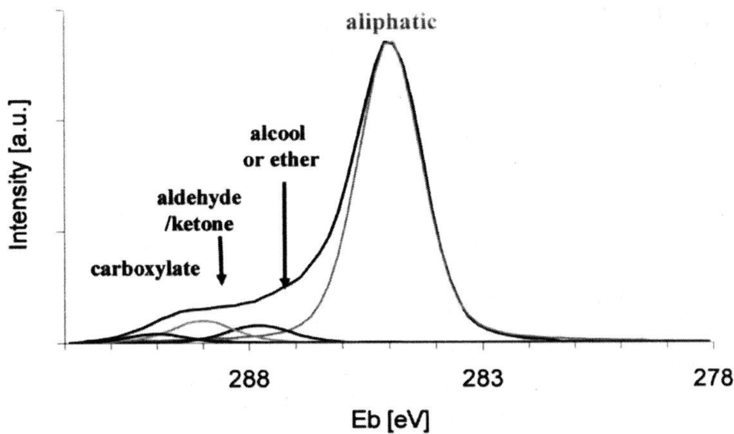


*Figure 4. C1s spectrum of plasma treated PPt*

adhesion improvement by plasma treatment are unanimous in attributing an important role to the chemical functionalization of the polymer surface by polar chemical groups *(12)*.

Si2p spectra of plasma treated composites are presented in Figure 5. It is obvious that the nature of the silicon based components observed at the upper surface of the composite after plasma treatment differ from those observed for untreated materials (comparison of Figure 3 and Figure 5). The intensity of the peak attributed to silica (around 104.1eV) increases when compared with that attributed to silicone (around 102.3eV). As a consequence, degradation of the mold removing agent into silica is suspected.

The AFM images of untreated and plasma treated PPc (Figure 6) reveal the surface structures of the material, showing the change in surface morphology of the composites before and after plasma treatment. The surface of the untreated PPc material presents a smooth structure whereas it is obviously roughened after the plasma treatment since aggregate structures with various sizes can be seen on the surface. Those results are confirmed by the surface roughness measured from the AFM software. The roughness sharply increases from 17.0 nm before plasma treatment to 37.8 nm after the treatment. This effect is even more pronounced in the case of PPt (Tables VI and VII).

Depending on the gas composition and treatment conditions, plasma treatments involve several phenomenon such as etching, activation and/or cross-linking (14). The increase in surface roughness demonstrate that not only functionnalization but also etching is observed in our experimental conditions.

Finally, the contact angle of the PP composites is found to change from 97.6° and 95.8° respectively for the untreated PPt and PPc (Table VI) to 41.0° and 37.4° after treatment (Table VII). Contact angle measurement characterizes the surface free energy of solids that determines most of the surface properties

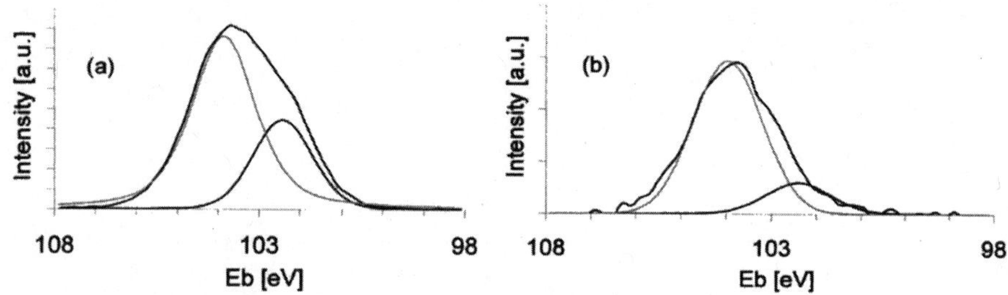


*Figure 5. Si2p spectra of plasma treated composites ((a) PPt and (b) PPc)*

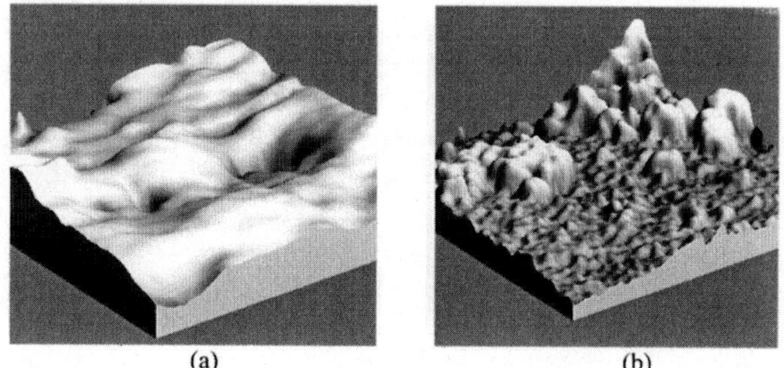


*Figure 6. AFM images of untreated (a) and plasma treated (b) PPc*

**Table VII. Surface properties of treated materials**

| Material | Contact angle (°) | Surface Roughness (nm) | Adhesion (ASTM D 3359-02) |
|---|---|---|---|
| PPt | 41.0 | 76.6 | 5B |
| PPc | 37.4 | 37.8 | 5B |

such as adsorption, wetting or adhesion. It is obvious that the plasma treatment leads to a significant decrease in contact angle and thus to an improvement of the wettability of the PP composites.

## Conclusion

This study demonstrates that intumescent coatings can be used advantageously to flame retard PP composites. However, good adhesion between the coating and the substrate is required. It also demonstrates that the plasma treatment significantly alters the surface wetting behavior of the PP composites. It significantly improves the wetting behavior of the materials, introducing polar groups on surfaces, reducing the contact angles and increasing the surface roughness. The chemical functionalization significantly improves adhesion between the treated polymer and the other component by interactions ranging from the rather weak Van der Waals or acid/base interactions to strong covalent bonding. As a consequence, it is reasonable to assume that this effect is observed in our study. This results in an improvement of the adhesion of the intumescent paint on the composites andthis good adhesion enables one to achieve a V-0 rating and a classification at the GWI test at 960 °C for PPc material.

## References

1. Zhang, S.; Horrocks, A.R. *Progress in Polymer Science* **2003**, *28(11)*, 1517-1538.
2. Babrauskas, V.; Simonson, M. *Fire and Materials*, **2006**, *31 (1)*, 83-96.
3. Duquesne, S.; Samyn, F.; Bourbigot, S.; Amigouet, P.; Jouffret, F.; Shen, K. *Polymers for Advanced Technologies*, **2008**, *19(6)*, 620-627.
4. Duquesne, S.; Magnet, S.; Jama, C.; Delobel, R. *Surface and Coatings Technology*, **2004**, *180-181*, 302-307.
5. Jimenez, M.; Duquesne, S.; Bourbigot, S. *Surface and Coatings Technology*, **2006**, *201 (3-4)*, 979-987.
6. Bourbigot, S.; Le Bras, M.; Duquesne, S.; Rochery, M. *Macromolecular Materials and Engineering*, **2004**, *289 (6)*, 499-511.
7. Mühlan, C.; Nowack, H. *Surface and Coatings Technology*, **1998**, *98(1-3)*, 1107-1111.
8. Tests for flammability of plastics materials for part devices and appliances, Underwriters Laboratories. Northbrook, IL: ANSI//ASTM D-635/77, 1977.
9. International Standard IEC 60695-2-12 - Fire hazard testing –Part 2-12: Glowing/hot-wire based test methods – Glow-wire flammability test method for materials.
10. Benoit, R. XPS database, http://www.lasurface.com/accueil/index.php

11. France, R.M.; Short, R.D. *Langmuir*, **1998**, *14*, 4827-4835.
12. Schultz, J.; Nardin, M. In *Adhesion Promotion Techniques*; Mittal K.L.; Pizzi, A. Eds.; Series: Materials Engineering Volume: 14; CRC Press: Washington DC, 1999; pp. 1-26.
13. Liston, E.M. Martinu, L. Wertheimer, M.R. *Journal of Adhesion Science and Technology*, **1993**, *7 (10)*, 1091-1127.

## Chapter 13

# Combustion and Thermal Properties of Polylactide with an Effective Phosphate-Containing Flame-Retardant Oligomer

**Jing Zhan[1,2], Lei Song[1], and Yuan Hu[1,*]**

**[1]State Key Laboratory of Fire Science, University of Science and Technology of China, 96 Jinzhai Road, Hefei, Anhui 230026, People's Republic of China**
**[2]Department of Polymer Science and Engineering, University of Science and Technology of China, 96 Jinzhai Road, Hefei, Anhui 230026, People's Republic of China**

A phosphate-polyether copolymer poly (polyethylene glycol spirocyclic pentaerythritol bisphosphate) (PPGPB) has been synthesized and its structure was characterized by Fourier transformed infrared spectrometry (FTIR) and $^1$H nuclear magnetic resonance ($^1$HNMR). A series of polylactide (PLA)-based flame retardant composites containing PPGPB were prepared by melt blending. The combustion properties of PLA/PPGPB composites were evaluated through UL-94 tests and microscale combustion calorimetry (MCC) experiments. It is shown that PPGPB significantly improved the flame retardancy of PLA and only 5wt% additive allowed the composites to achieve UL-94 V-0. The possible flame retardance mechanism of PPGPB in PLA was substantiated by scanning electron microscopy (SEM) and in situ FTIR. In addition, thermogravimeric analysis (TGA) showed the weight loss rate was decreased by introduction of PPGPB, and it is found from the differential scanning calorimetry (DSC) results that PPGPB can facilitate the crystallization of PLA.

# Introduction

With the quickly decrease of petroleum energy sources and environmental pollution brought by the petroleum-derived plastics, renewable polymers have aroused much interest recently in preparing substitutes (*1*). Polylactide (PLA) is a biodegradable polymer derived from agricultural resources through bioconversion and polymerization of lactide with good biocompatibility, mechanical properties, thermal plasticity, and gas permeability (*2*). Even when burned, it produces no nitrogen oxide gases and only one third of the combustible heat generated by polyolefin (*3*). Therefore it is a promising polymer for various end-use applications such as medicine, household, engineering, packaging fields and so on (*4*). However, just like other plastics, PLA is easily to burn which limits its application and development to some extent. As far as we are aware, there are few reports about the flame retardancy of PLA up to now.

Although traditional halogen flame retardants show good flame retardancy in polymer, the release of some toxic gases during their decomposition results in some ecological and physical problems, thus leading to globally increasing attention to halogen-free flame retardant. Adding organophosphorus flame-retardant to polyester is an efficient way to obtain an environmentally friendly flame-retardant system (*5-7*), among which intumescent flame retardants (IFR) containing phosphorus shows its potential in flame retardance with the advantages of outstanding effectivity, low smoke and toxicity, long life, and halogen-free. The IFR system is usually composed of three components: acid source, carbonization agent, and blowing agent, and on heating, such a system can experience an intense expansion and form protective foamed cellular charred layers on their surface, thus effectively protecting the underlying material from the action of the heat flux or flame during combustion (*8-10*). Réti and Casetta (*11*) studied the efficiency of different intumescent formulations to flame retardant PLA, and evaluated the flammability properties when pentaerythritol was substituted by bioresources such as lignin and starch. The quantity of the additives has been optimized thanks to a mixture design and so has to decrease the quantity of ammonium polyphosphate in the formulation.

However, the poor miscibility between IFR and PLA matrix will deteriorate the mechanical properties of PLA. As is well known, the lower molecular weight poly(ethylene glycol) (PEG) which is often used as a polymeric plasticizer has good miscibility with PLA (*12-14*). Therefore PEG was introduced into the flame retardant, and a halogen-free flame retardant poly(polyethylene glycol spirocyclic pentaerythritol bisphosphate) (PPGPB) was synthesized through polycondensation of spirocyclic pentaerythritol bisphosphorate disphosphoryl chloride (SPDPC) and PEG. SPDPC contains an acid source and a carbon source in one molecule, is a very effective flame retardant for polyolefin and polyesters and a series of phosphates, phosphate amides, polyphosphates, and polyphosphate amides have been synthesized from it (*15-17*). In addition, the advantages of the introduction of PEG can be summarized as follows: PEG can eliminate the chlorine atom of the SPDPC realizing halogen free flame retardance; transforms the flame retardant into polymer in order to prevent the migration of the small molecule in the composites; in a way, serves as a gas source to form a protective foamed cellular; slightly improves the toughness of the PLA as reported by many researchers (*18, 19*).

In this article, the main aim is to study the flame retardancy of PLA containing synthetic PPGPB. The combustion properties of PLA composites were investigated by UL-94 tests and microscale combustion calorimetry experiments. It is found that PPGPB can significantly improve the flame retardancy of PLA and low filler concentration enabled the composites to achieve UL-94 V-0. The SEM and in situ FTIR were used to evaluate the possible flame retardant mechanism and the results show that the introduction of PPGPB has entirely changed the degradation process of PLA. In addition, thermal stability and crystallization behavior of PLA composites containing PPGPB were studied.

## Experimental

### Materials

PLA was supplied by Cargill Dow. N,N-Dimethylformamide(DMF), pyridine, PEG400 and pentaerythritol were all provided by Shanghai Chemicals Co.. Phosphorus oxychloride was purchased from Tianjin Guangfu fine chemical research institute. SPDPC was synthesized by the reaction of phosphorus oxychloride with pentaerythritol (Scheme 1) as previously reported (*17*). PEG and all the solvents were purified before used.

### Synthesis of PPGPB

A 500-ml four-necked round bottom flask was equipped with a mechanical stirrer, reflux condenser connected with a calcium chloride tube, and a thermometer. The flask was charged with 300 ml DMF, 20 ml pyridine, 8.91 g (0.03 mol) SPDPC, 12 g (0.03 mol) PEG400, and then the mixture was strongly stirred and gradually heated until reflux and keep under reflux. The reaction was

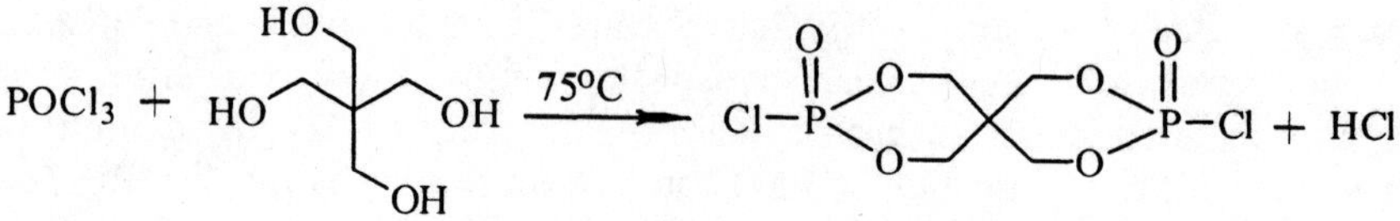

Scheme 1. Synthesis of SPDPC.

maintained for about 24 h, and then filtered after the solution cooled. The solution was vacuum distilled to remove part of the solvent and the product was precipitated in ether, finally it was dried at 80 °C under vacuum to a constant weight. The synthesis route is illustrated in Scheme 2.

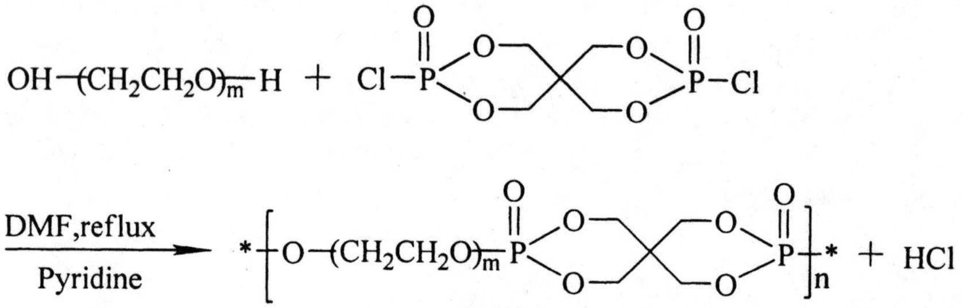

Scheme 2. Synthesis of PPGPB.

### Preparation of PLA/PPGPB composites

PLA pellets were dried under vacuum at 80 °C overnight before use. All the samples were prepared on a two-roll mixing mill (XK-160, Jiangsu, China ) at the temperature of 165 °C and the roll speed was maintained at 35 rpm. PLA was first added to the mill at the beginning of the blending procedure. After PLA was molten, PPGPB was then added to the matrix and processed for about 10 min until a visually good dispersion was achieved. The resulting samples were compressed and molded into sheets (3 mm thickness). Table I shows the composition of the samples.

### Characterization

The $^1$H NMR spectrum was recorded with an AVANCE 300 Bruker spectrometer using tetramethylsilane as an internal reference. The solvent for SPDPC and PPGPB is DMSO-*d6*, and PEG is dissolved in $D_2O$.

**Table I. The composition of the samples.**

| *Samples* | *PLA (wt%)* | *PPGPB (wt%)* |
|---|---|---|
| PLA | 100 | -- |
| PLA/1PPGPB | 99 | 1 |
| PLA/5PPGPB | 95 | 5 |
| PLA/10PPGPB | 90 | 10 |

The Fourier transform infrared (FTIR) spectra were recorded with a MAGNA-IR 750 spectrometer (Nicolet Instrument Co., USA).The in situ FTIR spectra were recorded using the apparatus equipped with a ventilated oven having a heating device. Tablet samples were placed into the oven and then the temperature of the oven was raised to 450 °C at a heating rate of 10 °C /min.

UL-94 vertical burning tests were performed with samples of dimensions $130\times13\times3$ $mm^3$, suspended vertically above a cotton patch. The classifications are defined according to the American National Standard UL-94.

A Govmark MCC-2 microscale combustion calorimetry (MCC) was used to determine the flammability characteristics of PLA composites according to ASTM D 7309-07. About 5 milligram specimens were thermally decomposed in an oxidizing (aerobic) environment at a constant heating rate of 1 K/s.

TGA experiments of PLA composites were performed using a TA Instruments Q5000 IR thermoanalyzer instrument under an air flow of 25 $ml\cdot min^{-1}$. The specimens (about 10 mg) were heated from room temperature to 600 °C at a linear heating rate of 10 °C/min.

Differential scanning calorimetry (DSC) measurements were performed using a Perkin-Elmer DSC-7 calorimeter under nitrogen flow. The samples were first heated at 50 °C/min to 180 °C and held for 1.0 min to eliminate the different thermal history, and then rapidly cooled to the isothermal crystallization temperature (120 °C) where they remained for 60min. In order to record the events of interest and compare crystallization properties, finally, thermal scanning at 10 °C/min up to 180 °C were performed after fast cooled to room temperature at 50 °C /min.

The morphologies of the char residue of PLA composites after MCC experiments were observed using the scanning electron microscope (PHILIPS XL30ESEM) after samples were coated with a thin layer of gold.

# Results and Discussion

## Characterization of PPGPB

Figure 1 is the $^1H$ NMR spectrum, in which it can be found that the $^1H$ of -$OCH_2$- from SPDPC appeared at δ=4.2 and 4.1 ppm with equal intensity. The peak corresponding to-$CH_2$- from PEG is observed at 3.6 ppm and the broad peak at 6.1 ppm can be assigned to the –OH from the end of PPGPB. From the FTIR spectrum (Figure 2), the strong absorption peaks are observed at 2871 and 1465 $cm^{-1}$ ascribed to the stretching and bending vibration of -$CH_2$- mostly from PEG. Meanwhile, the absorption peak of P=O is observed at 1290 $cm^{-1}$, with in-depth analysis there exists a shoulder at 1250 $cm^{-1}$ which can be assigned to its rotational isomer. The absorption band between 980 and 1200 $cm^{-1}$ associated with P–O–C becomes stronger and broader owing to the introduction of C-O-C from PEG. In addition, the absorption peaks at 3456 $cm^{-1}$ can be assigned to O–H stretching vibration, and no absorption of P-Cl at 576 $cm^{-1}$ means that chlorine atom was totally replaced.

## Flammability of PLA/PPGPB composites

The UL-94 vertical burning test is a common testing method for flame retardancy that determines the upward burning properties of a polymer. Table II gives UL-94 vertical burning test results of PLA composites containing different amount of PPGPB. It is clear that the burning time of PLA decreases as the PPGPB content increases, and only 5wt% can endow PLA with a V-0 classification in UL-94 protocol, which shows PPGPB can significantly improve the flame retardancy of PLA. Certainly, PLA still drips and this has not been improved, but it could not ignite the cotton.

Microscale combustion calorimetry (MCC) is a new test method to evaluate the efficiency of flame-retardant additives in polymers developed by Lyon (*20*). The test method is a thermal analysis method that improves upon previous methods by directly measuring the heat of combustion of the gases evolved during controlled heating of milligram-sized samples. It can quickly and easily measure the key fire parameters of plastics, wood textiles and composites. From just a few milligrams of specimen a wealth of information on material combustibility and fire hazard is obtained in minutes. The measured thermal combustion properties include the heat of complete combustion of the pyrolysis gases per unit mass of original solid *h*c (J/g), heat release rate (HRR), the temperature at maximum pyrolysis rate $T_{max}$ (°C) and heat release capacity

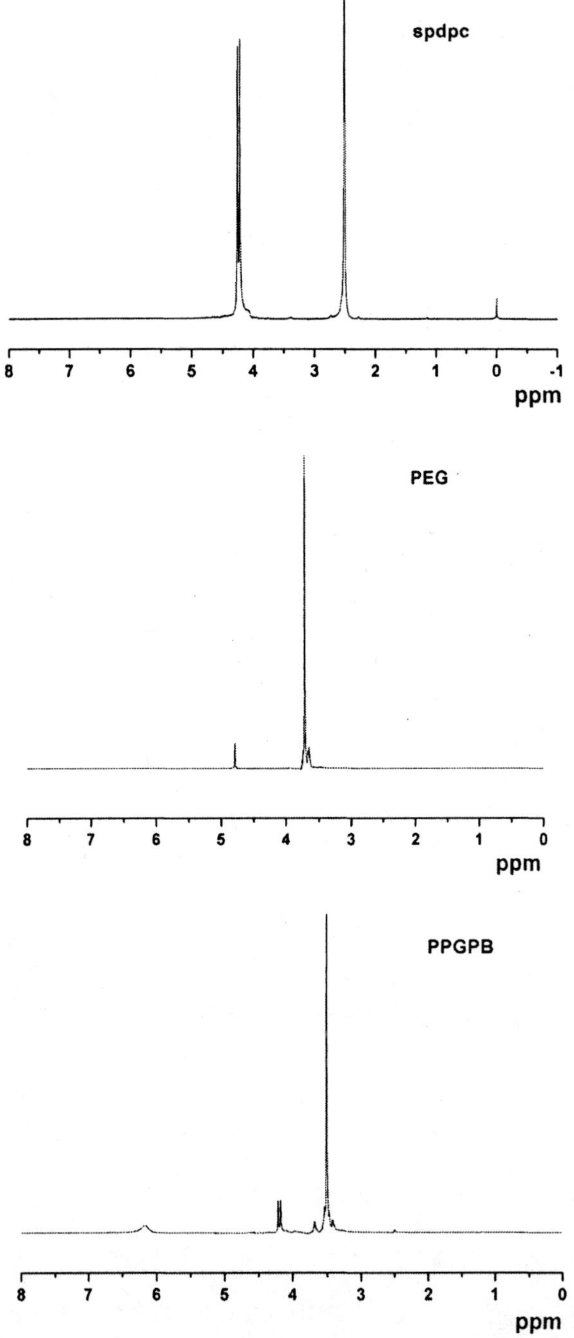


*Figure 1. [1]H NMR spectrum of SPDPC, PEG and PPGPB.*

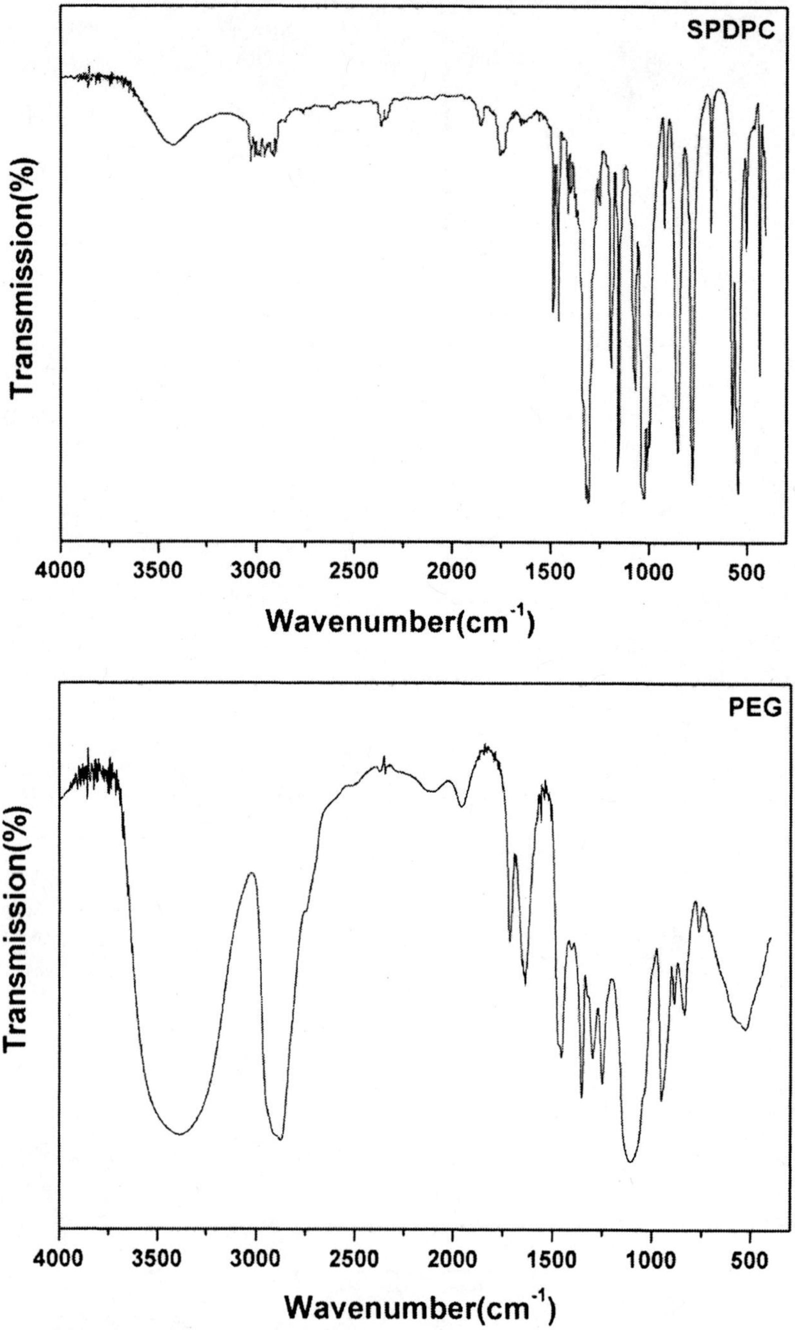

SPDPC

Transmission(%)

4000 3500 3000 2500 2000 1500 1000 500

Wavenumber($cm^{-1}$)

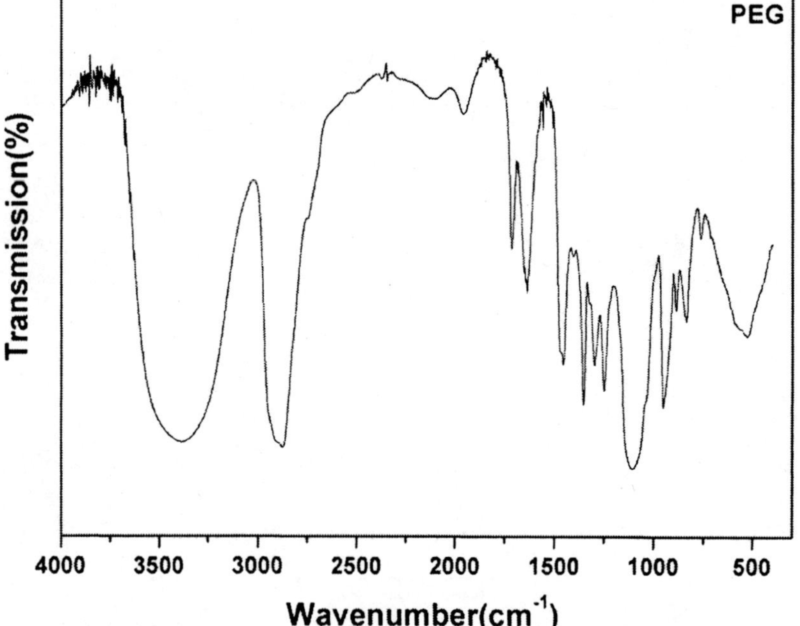


*Figure 2. FTIR spectrum of SPDPC, PEG and PPGPB.*

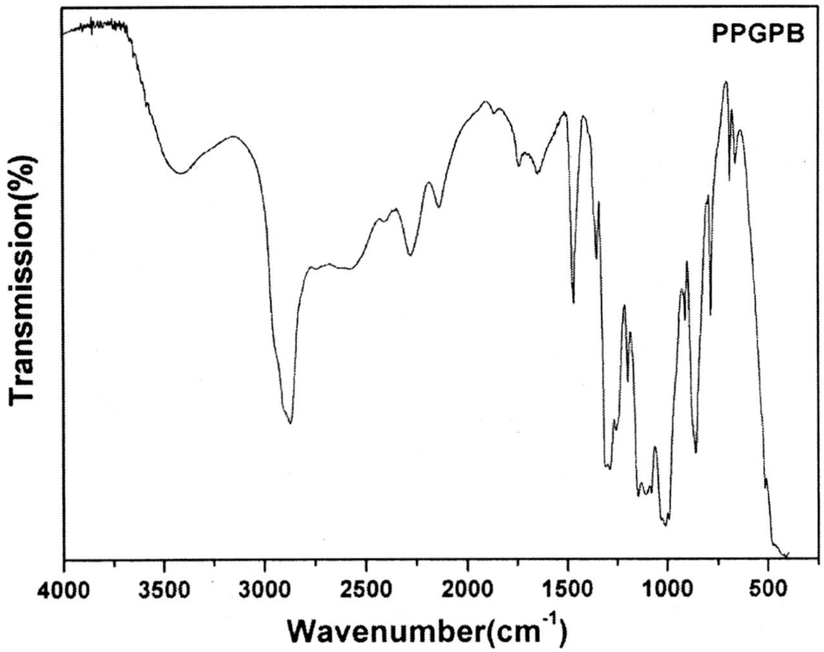


*Figure 2. Continued.*

**Table II. UL-94 test results and part data recorded in MCC experiments**

| *Composition (wt%)* | *burning time (s)* | *UL-94 rating* | *ηc (J/g-K)* | *pHRR (W/g)* | *hc (kJ/g)* | $T_{max}$ *(°C)* |
|---|---|---|---|---|---|---|
| PLA | N.R.[a] | N.R. | 484 | 491.1 | 16.2 | 386.5 |
| PLA/1PPGPB | 10.8 | V-2 | 463 | 464.8 | 15.8 | 388.7 |
| PLA/5PPGPB | 4.2 | V-0 | 372 | 374.0 | 15.1 | 391.8 |
| PLA/10PPGPB | 2.3 | V-0 | 259 | 260.1 | 14.2 | 383.9 |

N.R[a]. denotes no record due to the complete combustion of the material.

(the maximum specific heat release rate during a controlled thermal decomposition divided by the heating rate in the test) $\eta c$ (J/g-K).

Figure 3 shows the HRR curves of PLA composites with different PPGPB content and the corresponding combustion data are presented in Table II. From Figure 3, it is found that the peak HRR of PLA composites reduce with increasing PPGPB content. Compared to pure PLA, the peak HRR of the sample containing 10wt% PPGPB decreases from 491 to 231 W/g, and the reduction is 5.4% and 23.8% for the composites containing 1wt% and 5wt% PPGPB. The small peak between 250 °C and 300 °C corresponding to PLA/10PPGPB can be assigned to the degradation of PEG chain from PPGPB. Meanwhile, the total heat released hc is slightly decreased as the flame retardant additives increases. The heat release capacity $\eta c$ is a relatively good predictor of the heat release rate in flaming combustion and the propensity for ignition (*21, 22*). Low values of $\eta c$ are indicative of low flammability in the PLA test and low full scale fire hazard. It can be seen from Table III that the heat release capacity $\eta c$ decreases with flame retardant additives content increasing, and the $\eta c$ of the PLA/5PPGPB sample is 372 J/g-K (112 J/g-K lower than pure PLA). These results show the PPGPB is very efficient on a weight basis at improving the flame retardancy of PLA.

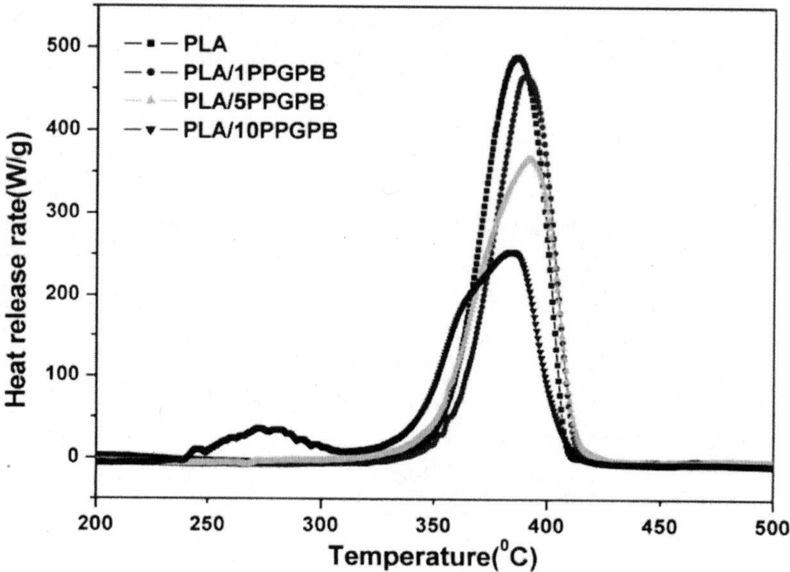


*Figure 3. The HRR curves of PLA and its composites at 1 K/s heating rate.*

## In situ FTIR

The thermal degradation process of PPGPB was characterized by in situ FTIR which provided information about the mass transfer and the decomposed sample. As shown in the FTIR spectra in Figure 4, in the temperature range 200-310 °C, the peaks at about 3417 $cm^{-1}$ assigned to –OH quickly decrease and disappear at 270 °C. The absorbance peaks of C-H at 2871 $cm^{-1}$ and 1465 $cm^{-1}$ weaken substantially. Furthermore the broad absorbance band between 980-1200 $cm^{-1}$ gets narrower due to the disappearance of some ether bond. All the above changes show the degradation of the chain from PEG. During this process, a new peak at 1730 $cm^{-1}$ associated with C=O appears and attains its highest absorbance at 250 °C, and then with the temperature increasing, it weakens at 310 °C and almost disappears by about 350 °C. The change of this peak is likely ascribed to the formation and loss of oxidation products of the –$CH_2$-$CH_2$-O- chain.

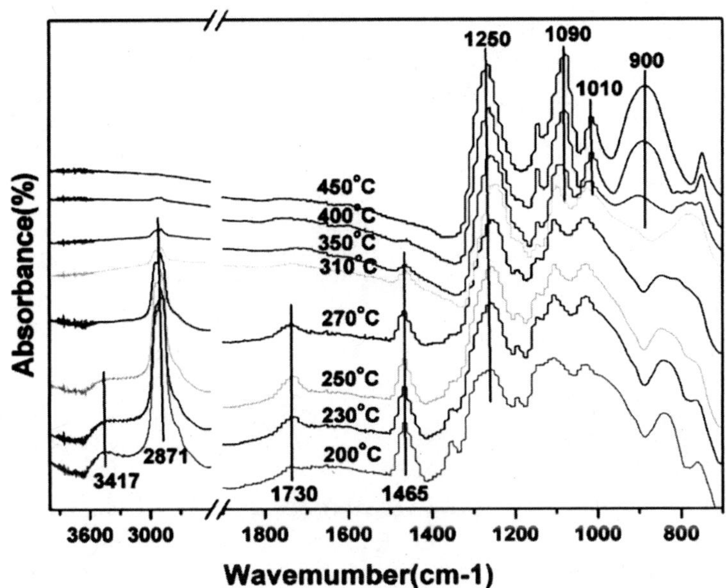


*Figure 4. In situ FTIR spectra for the degradation products of PPGPB at different temperatures.*

As the temperature increases, the two peaks due to the rotational isomers of P=O merge at 1250 $cm^{-1}$ and the peak gets stronger and stronger during the heating process. As the chain from PEG decomposes, some phosphate groups link and form P-O-P bonds which appear at 1010 $cm^{-1}$. Two peaks at 1090 $cm^{-1}$

and 900 $cm^{-1}$ assigned to the stretching vibration of $PO_2/PO_3$ in phosphate carbon complexes appear and further intensify at higher temperature (*23,24*).The new peaks are induced by the pyrolysis of the spirocyclic structure which will further release phosphoric acid as reported before (*10*).

In order to give a more complete picture of the flame retardance mechanism of PPGPB acting on PLA, the in situ FTIR spectra monitoring the degradation process of both PLA and its composite with 5wt% PPGPB at different temperatures were undertaken and this is shown in Figure 5. It can be found that the addition of PPGPB has entirely changed the degradation process of PLA. At the temperature 330 °C, almost all of the ester bonds of the pure PLA have been pyrolyzed along with the degradation of the C-H, based on the decrease of the peaks at 1750 $cm^{-1}$corresponding to C=O, and to the peaks at 1380 $cm^{-1}$,1460

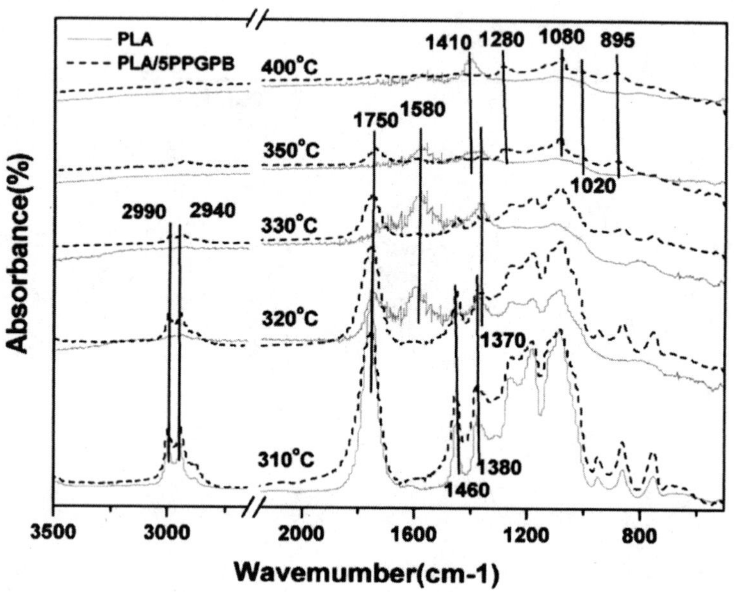


*Figure 5. In situ FTIR spectra for the degradation of the pure PLA and its composites with 5wt% PPGPB at different temperatures*

$cm^{-1}$, 2940 $cm^{-1}$and 2990 $cm^{-1}$ ascribed to the stretching and bending vibration of C-H. The new peaks, appearing at 1370 $cm^{-1}$, 1410 $cm^{-1}$ and 1580 $cm^{-1}$, can be assigned to the olefinic ketone, aromatic ring or some complicated carboxylate. There is no new peak appearing before 350 °C, therefore the pyrolysis process of the composites with the flame retardant is mainly the ester exchange reaction due to the catalysis of phosphoric acid derived from the degradation of PPGPB. If the molecular weight of an oligomer is sufficiently low to evaporate, it is removed from the mixture of esters in the molten polymer and this leads to a weakening of the peaks associated with the polyester. In the spectra at 400 °C, the peak at 1280 $cm^{-1}$ is the stretching vibration of P=O. Peaks at 1080 $cm^{-1}$and 895 $cm^{-1}$can be assigned to the stretching vibration of $PO_2/PO_3$ in phosphate carbon complexes and the peak at 1020 $cm^{-1}$ is assigned to the stretching vibrations of P-O-P bond. The possible degradation process of the flame retardancy PLA composites with PPGPB is shown in scheme 3.

## Morphology of char

SEM was used to observe and study the morphological structure of the char formed after MCC experiments. From Figure 6, after PPGPB was introduced, the surface of the char forms many cellular bubbles, and as the content of the flame retardant increases, the bubbles become coarser, tighter and stronger. As is shown in Figure 6(e), no bubble exists when the PLA matrix is modified with 3wt%SPDPC, and this composite can not reach UL-94 V-0 rating with a little more phosphorus content than PLA/5PPGPB sample. This phenomenon implies that there exists some synergism between PEG chain and phosphorus-containing segment in PPGPB. It can be explained that the pyrolysis of PEG from PPGPB happens first during the heating process from in situ FTIR results. The burning of the gaseous pyrolytic product has low pHRR as MCC shows and forms $CO_2$, $H_2O$ etc, and then these less combustible gases dilute the flammable pyrolysates from PLA. On the other hand, the formation of $H_2O$ induces the formation of phosphoric acid which will change the degradation process of PLA, catalyze carbonization and act as a surface coating (*25*).

Therefore PPGPB is analagous with intumescent flame retardants. The phosphorus-containing segment serves as an acid source to catalyze carbonization and change the degradation process of PLA, and it is also a char-forming agent due to its pentaerythritol component. The effect of the PEG chain is similar to a foaming agent that produces less combustible gases to dilute the flammable pyrolysates and allows an insulating cellular carbonaceous layer to form. All the above endow the PLA/5PPGPB sample with a UL-94 V-0 rating, and a great decrease of $\eta c$ and pHRR in MCC tests.

## Thermogravimetric analysis

Figure 7 shows the TGA and the derivative thermograms (DTG) curves of PLA with different amount of PPGPB. It is found from Figure 7 (a) that the onset degradation temperature ($T_{-5\%}$) of PLA/PPGPB decreased with the increasing

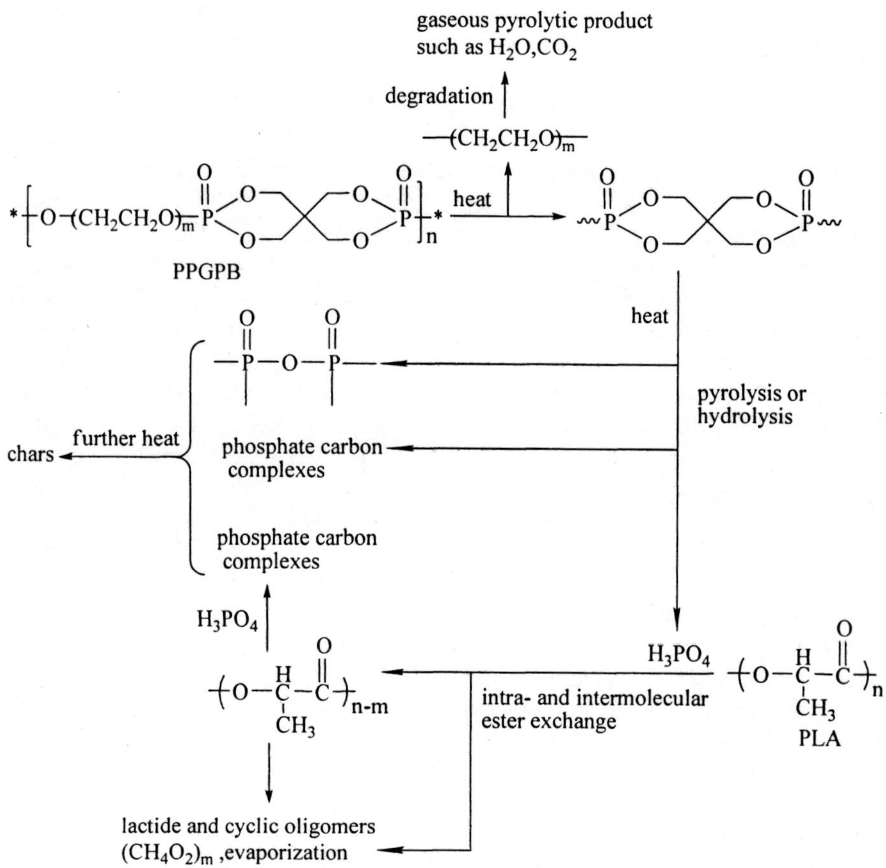


*Scheme 3. The possible degradation process of the flame retardant PLA composites with PPGPB.*

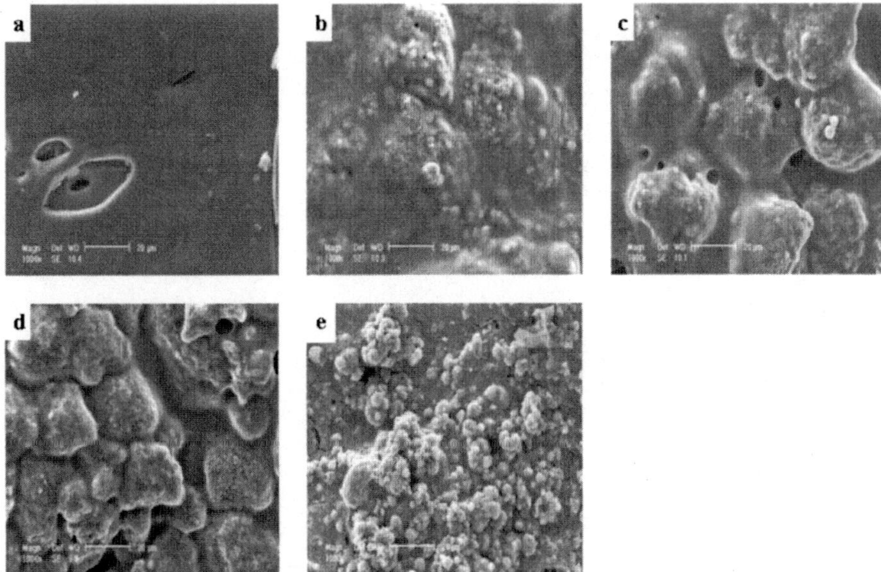

*Figure 6. SEM micrographs of char residue from PLA and its composites: (a) PLA, (b) PLA/1PPGPB, (c) PLA/5 PPGPB, (d) PLA/10 PPGPB, (e) PLA/3 SPDPC.*

additive content, which is mainly due to the lower thermal stability of PPGPB. $T_{-5\%}$ of PPGPB starts at 195 °C which is mainly ascribed to by the degradation of the PEG chain in PPGPB in accordance with the analysis of the in situ FTIR spectra. The increase of the char residue is realized with adequate flame retardant at 10wt%PPGPB. Fig.7 (b) shows the derivative thermograms (DTG), which gives the weight loss rates in the heating process. It can be seen that the weight loss rate obtained from the samples containing PPGPB is smaller than pure PLA, and it decreases as PPGPB content increases. This is mainly due to the change of the pyrolysis process of PLA according to the in situ FTIR.

### Crystallization behavior

PLA is a partially crystalline polymer with low glass transition temperature ($T_g$). The thermal and crystalline properties of the PLA composites determined from the second heating DSC scans are presented in Figure 8, and the DSC data are given in Table III. The presence of PEG in PPGPB, which can be used as plasticizer, tends to decrease the $T_g$ (*26*), cold crystallization temperature ($T_c$) and melting temperature ($T_m$) of PLA. It enhances segmental mobility of the

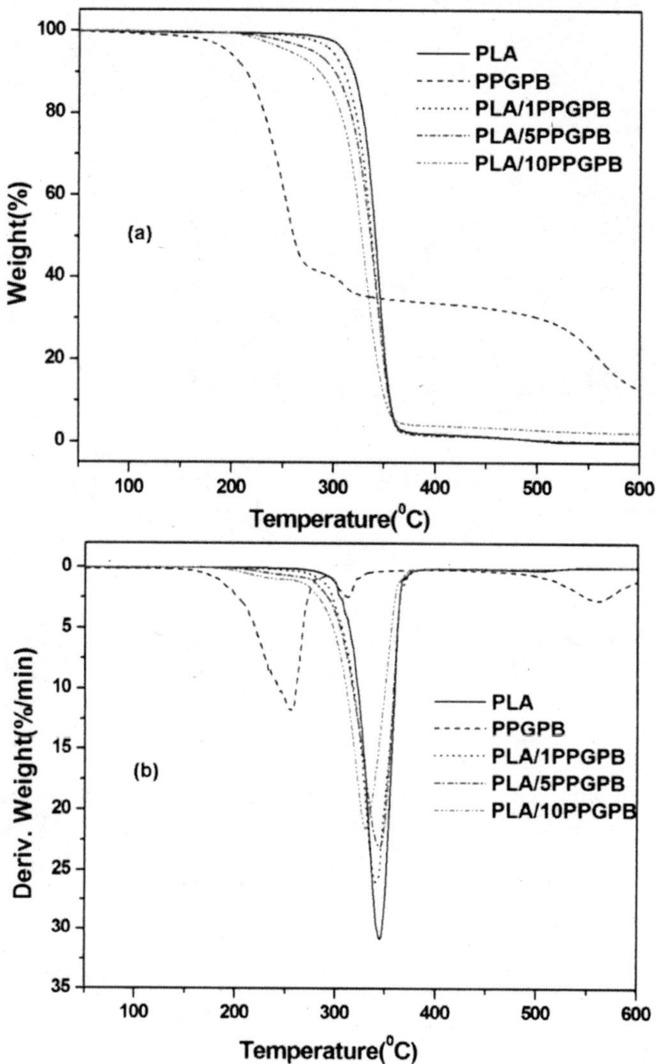


*Figure 7. The TG curves of (a) and DTG (b) of PLA/PPGPB composites.*

PLA chains as reported, so as Table III shows, Tg for PLA/10PPGPB decreases from 58.9 °C of the pure PLA to 51.2 °C. In-depth analysis of these curves reveals an endothermic peak associated with the glass transition, typically attributed to stress relaxation on heating (*27*). The plasticizer also facilitates the process of crystallization. As is known, the degree of crystallinity (θ) after the DSC heating scan can be calculated as follows:

$$\theta = \Delta H_m / \Delta H_\infty$$

Where $\Delta H_m$ and $\Delta H_\infty$ represent the relative melting enthalpy of the sample and of the 100% crystalline PLA. Data in Table III show that the bigger value of $\Delta H_m$ is obtained with an increasing amount of PPGPB, and for the sample containing 10wt% PPGPB, the $\Delta H_m$ is 15.85 J/g, approximately 15 J/g higher than pure PLA, indicating the achievement of high crystallinity. As shown in Figure 8, the cold crystallization exotherm appears when 5wt% or more plasticizer is added, which can be regarded as evidence that illustrates the increase of the molecular mobility and the crystallization ability of PLA.

**Table III. DSC data of PLA and its composites with PPGPB derived from the second heating scan.**

| *Samples* | $T_g$ (°C) | $T_c$ (°C) | $T_m$ (°C) | $\Delta H_m$ (J/g) |
|---|---|---|---|---|
| PLA | 58.9 | / | 149.7 | 0.89 |
| PLA/1PPGPB | 53.9 | / | 148.5 | 2.98 |
| PLA/5PPGPB | 53.0 | 127.2 | 148.1 | 9.76 |
| PLA/10PPGPB | 51.2 | 123.7 | 147.5 | 15.85 |

## Conclusions

A novel phosphorus-containing flame retardant PPGPB was successfully synthesized, and it is found that PPGPB could effectively improve the flame retardancy of PLA. Microscale combustion calorimetry experiments showed a great decrease of $\eta c$ and pHRR of PLA composites with increasing PPGPB content, and UL-94 V-0 was achieved when only 5wt% PPGPB was added. The flame retardant mechanism of PPGPB in PLA was mainly due to the change of the degradation process of PLA according to the analysis of the in situ FTIR, and some synergism occurs between PEG chain and phosphorus-containing part in PPGPB. From the thermal decomposition behavior, it was found that the onset decomposition temperature of PLA was reduced due to the degradation of the

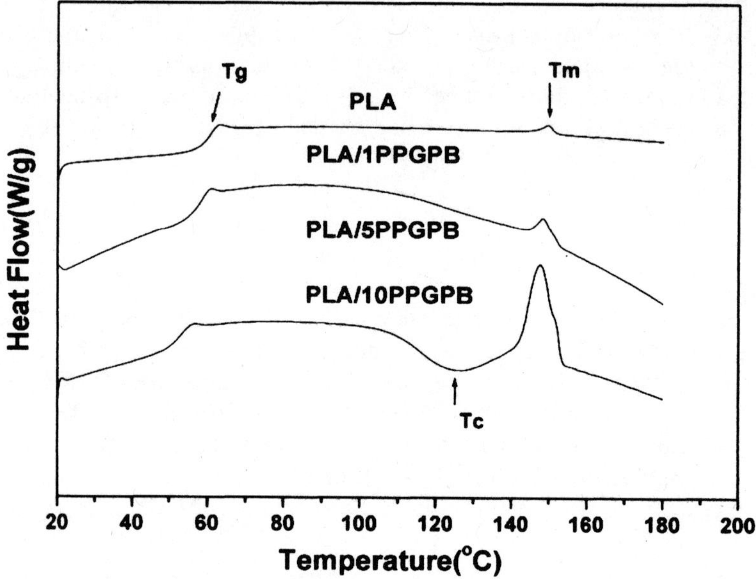


*Figure 8. DSC heating thermograms recorded at 10 °C/min for PLA and its composites with PPGPB.*

PEG chain, and the weight loss rate decreases as PPGPB content increases. The DSC investigation showed that the introduction of PPGPB decreases the glass transition temperature and facilitates the process of crystallization.

## Acknowledgements

The work was financially supported by the Program for New Century Excellent Talents in University, and National 11th 5-year Program (No. 2006BAK01B03, 2006BAK06B06, and 2006BAK06B07).

## References

1. Yu L.; Dean K.; Li L. *Prog. Polym. Sci.* **2006**, *31*, 576-602.
2. Fang Q.; Hanna MA. *Ind .Crop. Prod.* **1999**, *10*, 47-53.
3. Ray S.S.; Okamoto M. Macromol. *Mater. Eng.* **2003**, *288*, 936-944.
4. Meinander K.; Niemi M.; Hakola J.S.; Selin J.F. *Macromolecular Symposia.* **1997**, *123*, 147-153.

5. Wang C.S.; Shieh J.Y.; Sun Y.M. *Eur. Polym. J.* **1999**, *35*, 1465-1472.
6. Wang C.S.; Lin C.H. *Polymer.* **1999**, *40*, 5665-5673.
7. Wu B.; Wang Y.Z.; Wang X.L.; Yang K.K.; Jin Y.D.; Zhao H. *Polym. Degrad. Stab.* **2002**, *76*, 401-409.
8. Duquesne S.; Magnet S.; Jama C.; Delobel R. *Surf. Coat. Tech.* **2004**, *180*, 302-307.
9. Le Bras M.; Camino G.; Bourbigot S.; Delobel R.; Lewin M. *Fire Retardancy of Polymer*; The Royal Society of Chemistry: Cambridge, UK, 1997.
10. Camino G.; Martinasso G.; Costa L.; Gobetto R. *Polym. Degrad. Stab.* **1990**, *28*, 17-38.
11. C. Reti; M. Casetta; S. Duquesne; S. Bourbigot; R. Delobel. *Polym. Adv. Technol.* **2008**, *19*, 628-635.
12. Baiardo M.; Frisoni G.; Scandola M.; Rimelen M.; Lips D.; Ruffieux K.; Wintermantel E. J. *Appl. Polym. Sci.* **2003**, *90*, 1731-1738.
13. Jacobsen S.; Fritz H. G. *Polym. Eng. Sci.* **1999**, *39*, 1303-1310.
14. Martin O.; Averous L. *Polymer.* **2001**, *42*, 6209-6219.
15. Horrocks A.R.; Zhang S. *Polymer.* **2001**, *42*, 8025-8033.
16. Ma Z.L.; Zhao W.G.; Liu Y.F.; Shi J.R. *J. Appl. Polym. Sci.* **1997**, *66*, 471-475.
17. Chen D.Q.; Wang Y.Z.; Hu X.P.; Wang D.Y.; Qu M.H.; Yang B. *Polym. Degrad. Stab.* **2005**, *88*, 349-356.
18. Jacobsen S.; Fritz H.G. *Polym. Eng. Sci.* **1999**, *39*, 1303-1310.
19. Kulinski Z.; Piorkowska E. *Polymer.* **2005**, *46*, 10290-10300.
20. Lyon R.E.; Walters R.N.; Stoliarov S.I. *Eng. Sci.* **2007**, *47*, 1501-1510.
21. Lyon R.E. *Seventh International Symposium on Fire Safety Science*; Worcester Polytechnic Institute, Worcester, MA, 2002.
22. Walters R.N.; Lyon R.E. *Final report DOT/FAA/AR-01/31*; U.S. Department of Transportation, Washington, DC, 2001.
23. Liang H.B.; Shi W.F. *Polym. Degrad. Stab.* **2003**, *84*, 525-532.
24. Liu W.; Chen D.Q.; Wang Y.Z.; Wang D.Y.; Qu M.H. *Polym. Degrad. Stab.* **2007**, *92*, 1046-1052.
25. Zhang S.; Horrocks A.R. *Prog. Polym. Sci.* **2003**, *28*, 1517–1538.
26. Pillin I.; Montrelay N.; Grohens Y. *Polymer.* **2006**, *47*, 4676-4682.
27. Pluta M.; Galeski A.; Alexandre M.; Paul M.A.; Dubois P. *J. Appl. Polym. Sci.* **2002**, *86*, 1497-1506.

## Chapter 14

# Effects of Boric Acid on Flame Retardancy of Intumescent Flame-Retardant Polypropylene Systems Containing a Caged Bicyclic Phosphate

**Hua-Qiao Peng and Yu-Zhong Wang***

**Center for Degradable and Flame-Retardant Polymeric Materials, College of Chemistry, Sichuan University, Chengdu 610064, China**

The synergistic effects of boric acid (BA) as a synergist in polypropylene (PP) flame-retarded by a novel intumescent flame-retardant system based upon a caged bicyclic phophate, bis (2, 6, 7- trioxa- 1- phosphabicyclo [2, 2, 2] octane- 1- oxo- 4- hydroxymethyl) phenylphosphonate, were investigated using the limiting oxygen index (LOI), the UL-94 test, thermogravimetric analysis (TGA), Fourier transform infrared spectroscopy, X-ray photoelectron spectroscopy (XPS) and scanning electronic microscopy (SEM). BA added to intumescent flame-retardant PP (IFR-PP) systems has a synergistic flame retardancy with the intumescent flame retardant, which gives an LOI of 34.2 and UL-94 V-0 rating when only 1 wt. % BA is used. The IFR-PP/BA has a high yield of residual char at high temperature, and the FTIR analysis of the residues has proven that the matrix of IFR-PP/BA system is protected well compared to that of IFR-PP system. The degraded products were investigated by XPS analysis. The morphology observed by SEM have demonstrated that BA can improve the quality of charred layers.

## Introduction

Polypropylene (PP) as a synthetic material could be used in a lot of fields, such as housing, wire and cables, cars and architectural materials because it is easily processed into products with different shapes. However, the flammability of PP restricts its applications. This shortage could be solved effectively by incorporating halogen-containing compounds with antimony trioxide into PP, but they are limited in use because of the evolution of toxic gases and corrosive smoke during combustion (*1*). Over the past few years, more and more studies have been carried out in intumescent flame retardants (IFR) (*2–9*), which work mainly by a condensed-phase mechanism. IFRs are regarded as promising flame retardants because of their environment-friendly properties. When being ignited or heated, the intumescent systems start to swell, and a foamed charred layer on the surface is formed which protects the underlying matrix. The greatest benefit to be obtained in this way is a dramatic decrease in the heat produced due to the exothermic combustion of polymers (*7*).

Recently, the synergistic effect has been widely described and discussed on the flame-retardant properties of polymers. It has been found that the presence of some metal-containing (*7, 10–12*) or silicon-containing (*13*) compounds has an excellent synergistic effect on the flame retardancy of polymers when used together with IFR. It is well-known that boron-containing compounds could be used as flame-retardant additives (*14–19*), which act in the condensed phase by redirecting the decomposition process in favor of carbon formation rather than CO or $CO_2$ (*20*), but they have not been investigated as extensively as phosphorous, nitrogen and halogen-containing flame retardants. Therefore, there is little literature about boric acid (BA) used in intumescent flame-retardant PP systems (*21*).

In this article, a novel charring agent, bis (2, 6, 7- trioxa- 1- phosphabicyclo [2, 2, 2] octane- 1- oxo- 4- hydroxymethyl) phenylphosphonate (BCPPO) (see Scheme 1) was used to prepare IFR together with ammonium polyphosphate and melamine. BA was used as a synergistic agent in the IFR-PP system. Flame retardancy and thermal properties were investigated using LOI, UL-94 test, and TGA. The synergism between IFR and BA in PP was studied, and the residue was investigated by FTIR, XPS and SEM.

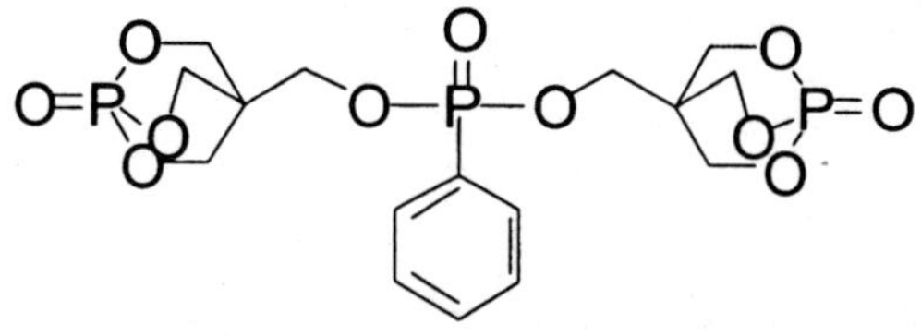

*Scheme 1. The structure of BCPPO*

# Experimental

## Materials

The caged bicyclic phosphate, bis(2, 6, 7- trioxa- 1- phosphabicyclo[2, 2, 2]octane- 1- oxo- 4- methoxyl) phenylphosphine oxide (BCPPO) was synthesized according to our previous report (*22*). Its chemical structure is shown in Scheme 1. Boric acid (BA) was supplied by Changzheng Chemical Reagent Corp. (Chengdu, China). Ammonium polyphosphate (APP, Type I ($(NH_4PO_3)_n$, n>50)) was supplied by Changfeng Chemical Corp. (Shifang, China). Melamine (MA) was provided by Sichuan Chemical Group Co. Ltd. (Chengdu, China). Polypropylene (1300) was provided by Yanshan Petroleum Chemical Company (Beijing, China).

## Preparation of Flame-retardant PP with Different Formations

The weight proportions of APP, MA and BCPPO in IFR were kept at 3:1:1 according to our previous study on PP. BA, IFR and PP were mixed according to the formulations shown in Table I. All samples were prepared on a two-roll mill in the temperature range of 180–185 °C for 15 min. After mixing, the samples were hot-pressed under 10 MPa for 5 min at 190 °C into sheets of suitable thickness and size.

## Measurements

### *LOI and UL-94*

The limiting oxygen index (LOI) values were measured on a HC-2C oxygen index meter (Jiangning Analysis Instrument Company, China) with sheet dimensions of 130×6.5×3 mm according to ASTM D2863-97.

The vertical burning tests were carried out on a CFZ-2-type instrument (Jiangning Analysis Instrument Company, China) with sheet dimensions of 130×13×3 mm according to ASTM D3801.

### *Thermogravimetric Analysis (TGA)*

Thermogravimetric analysis (TGA) was conducted on a TA Q500 thermogavimetric analyzer (TA, USA). 10 mg samples (platinum pan) were heated from room temperature to 600 °C at a heating rate of 10 °C / min under nitrogen and air, respectively, at a flow rate of 60 mL / min.

*Fourier transform infrared (FTIR) spectroscopy*

Residue samples of intumescent flame-retardant polypropylene (IFR-PP) and intumescent flame-retardant polypropylene containing boric acid (IFR-PP/BA) for the measurement of FTIR spectrum were obtained after the samples are heated in air at different temperatures for 20 minutes. The FTIR spectroscopy was recorded using a Nicolet FTIR 170SX infrared spectrophotometer.

*X-ray photoelectron spectroscopy (XPS)*

X-ray photoelectron spectroscopy (XPS) was recorded on a XSAM80, using Mg $K_\alpha$ excitation radiation, operated at 210 W (14 kV and 15 mA) under a high vacuum between $10^{-5}$ and $10^{-6}$ Pa. High resolution spectra of carbon 1s (C1s), nitrogen 1s (N1s), oxygen 1s (O1s) and phosphorus 2p (P2p) core level for samples were obtained. Binding energies were referenced to the carbonaceous carbon at 285.0 eV.

*Scanning electronic microscopy (SEM)*

Scanning electronic microscopy (SEM) observed on a JEOL JSM-5900LV was used to investigate the outer and inner surface of residucs of the IFR-PP and IFR-PP/BA systems. The residue samples for SEM were obtained after LOI test. SEM graphs of the residual char samples were recorded after gold coating surface treatment.

## Results and Discussion

### Effect of Boric Acid (BA) Concentrations on LOI and UL-94 Value of IFR-PP Systems

It is known that the structure and formation of char layers have an important effect on the flame retardancy of IFR. Table I presents the LOI values and UL-94 ratings for IFR-PP system with or without BA as a synergistic agent when the total amount of additives is kept to 30 wt. %. It can be seen that PP is an easily flammable polymeric material, accompanied by melt dripping, and its LOI value is only 18.3 and it is not classified by the UL-94 test. However, when the IFR containing APP, MA and BCPPO was incorporated into PP, a satisfactory result was obtained: the LOI value was 30.3 and UL-94 test reached a V-0 rating. Moreover, when only 1 wt. % BA was added, the LOI value increased from 30.3 to 34.2. It is proposed that due to the dehydration of BA, $B_2O_3$ forms, and leads to the formation of a "glass-like" material which increases the viscosity of the melt and prevents the gaseous decomposition products from escaping to feed the flame. Furthermore, the water of hydration is lost by endothermic decomposition and, therefore, both dilutes and cools, by absorbing the thermal energy from the flame (*23, 24*). Unfortunately, after that, the LOI values decrease quickly with increasing the content of BA. The LOI value reached a minimum of 29.9, and UL-94 rating was V-1 rating when 3 wt. % BA was adopted in formulation 5. Furthermore, when the loading of IFR-PP/BA is 25 wt. %, according to the proportion of formulation 3, the LOI value is 30.4 and UL-94 test reaches a V-0 rating.

**Table I. Effect of IFR Composition on Flame Retardance of PP with a 30wt.-% Total Loading Level of Additives**

| *No.* | *Composition of IFR-PP* | | | *LOI* | *UL-94* | |
|---|---|---|---|---|---|---|
| | *PP* | *IFR* | *BA* | | *Rating* | *Dripping* |
| 1 | 100 | 0 | 0 | 18.3 | Fail | Yes |
| 2 | 70 | 30 | 0 | 30.3 | V-0 | No |
| 3 | 70 | 29 | 1 | 34.2 | V-0 | No |
| 4 | 70 | 28 | 2 | 32.6 | V-0 | No |
| 5 | 70 | 27 | 3 | 29.9 | V-1 | No |

### Thermal Behavior of PP, IFR-PP and IFR-PP/BA System

To evaluate the thermal properties of polymers, TGA curves under nitrogen (a) and air (b) atmosphere were obtained and analyzed. From Figure 1 (a), it can be seen that both formulation 2 and formulation 3 show an earlier decomposition behavior than PP. When the temperature is higher than 426 °C, larger amounts of residue are generated for formulation 3. However, the corresponding temperature of formulation 2 is 454 °C, which confirms the higher thermal stability of formulation 3.

Figure 1 (b) presents the TGA curves of PP, IFR-PP (formulation 2) and IFR-PP/BA (formulation 3) under air flow, respectively. It is shown that pure PP begins to decompose at 278 °C, and there is no char residue over 390 °C, which may be responsible for the poor flame retardancy of PP due to insulating

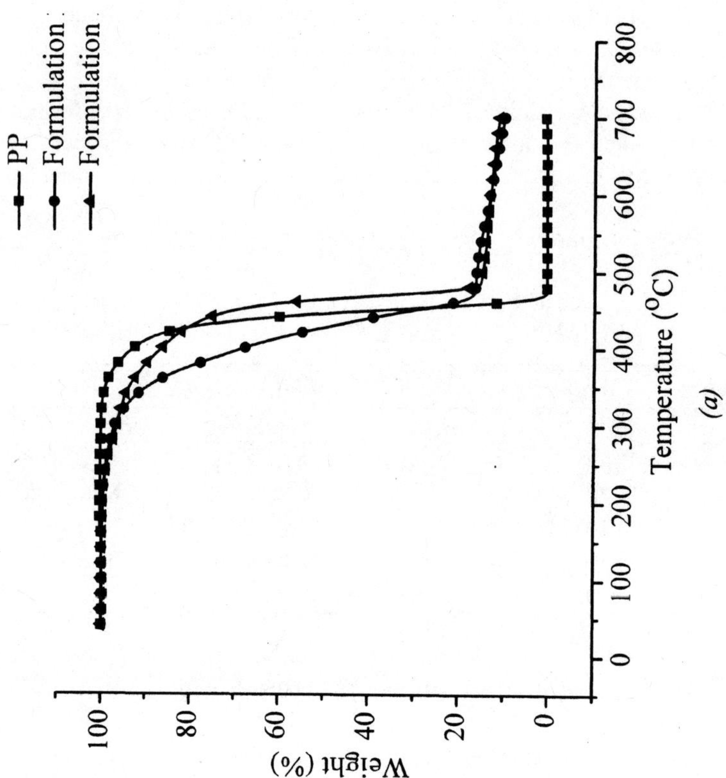

PP
Formulation
Formulation
Weight (%)
100
80
60
40
20
0
0
100
200
300
400
500
600
700
800
Temperature (°C)


*(a)*

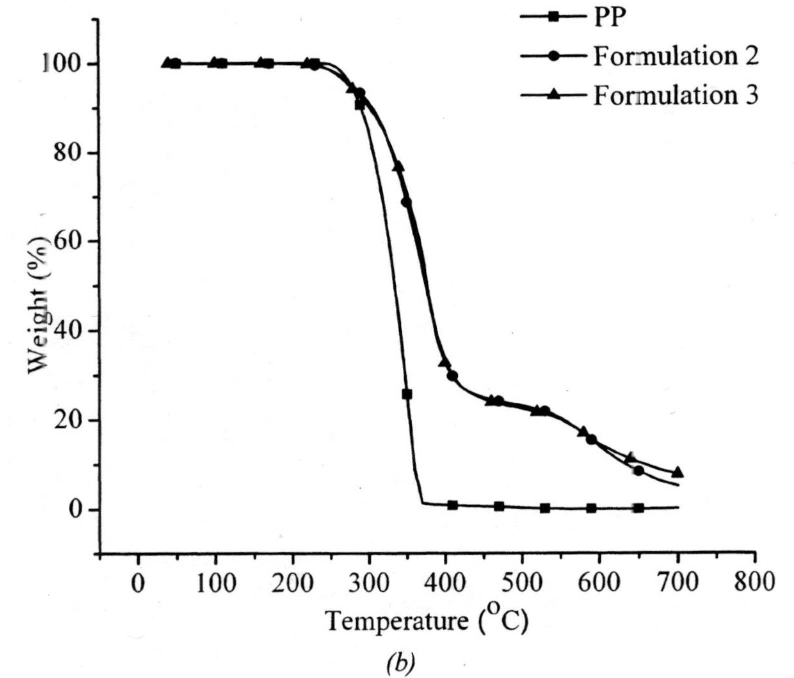


*(b)*

*Figure 1. TG curves of pure PP, IFR-PP and IFR-PP/BA samples under a flow of nitrogen (a) and air (b), respectively.*

char or residue formed. However, both the IFR-PP and IFR-PP/BA exhibit an enhanced thermal property in comparison with the virgin PP, due to high amount of residues generated from 282 °C to 700 °C. The two samples show similar TGA behavior under the same experimental conditions, and the main difference between them is that the charred layers from the IFR-PP/BA sample have more char residue than those from IFR-PP in the range of 450-500 °C under air. This means that BA can promote the formation of carbonaceous materials at high temperature because BA can be decomposed into metaboric acid ($HBO_2$), and the dehydration of $HBO_2$ occurs with further degradation into boron oxide ($B_2O_3$). As noted before, the glass-like material, $B_2O_3$, is very hard and highly thermally stable residue, which provides a "glue" to hold the combustion char together and provide structural integrity to the char (*25*). So IFR-PP/BA outperformed IFR-PP for flame retardance of PP, and the higher stability and char yield at higher temperature may account for this.

## Chemical Structural Analysis of Degradation Residue of IFR-PP and IFR-PP/BA by FTIR Spectrum at Different Temperatures

Figures 2 and 3 show the FTIR spectra of residues of IFR-PP and IFR-PP/BA systems, respectively. From Figure 2 we can see that the intensities of the peaks at 2829-2955 $cm^{-1}$ are assigned to the $CH_2$ or $CH_3$ asymmetric and symmetric vibrations (*26*), which disappear after heating for 20 min at 450 °C. When the temperature is further increased, a peak at 1455 $cm^{-1}$ from the asymmetric deformation vibration of $CH_2$ or $CH_3$ groups disappears at 520 °C. These results indicate that the C-H and C-C main chains of PP are being broken (*27*). The absorption of C=C is observed at 1631 $cm^{-1}$ resulting from the cleavage of the main chains after thermal oxidation (*28*). The absorption of P=O is observed at 1249 $cm^{-1}$ (*5*) in the phosphate complexes (*29*). The peak at 1062 $cm^{-1}$ is attributed to the stretching vibration of P-O-C (*5*) which is in the phosphocarbonaceous complexes (*29*). The peak at 997 $cm^{-1}$ (P-O-P) (*30*) becomes gradually evident as temperature increases, indicating that the char formation occurs during the decomposition of IFR-PP.

The FTIR of the residue from IFR-PP/BA is similar to that of IFR-PP under the same experimental conditions. However, the biggest difference is that the obvious absorption of $CH_2$ or $CH_3$ ranged from 2838 to 2955 $cm^{-1}$ can be still observed when heated for 20 min at 450 °C, proving that the matrix of IFR-PP/BA system is protected much better than that of IFR-PP system. However, With a further increase in the degradation temperature, the peaks disappear due to cleavage of the main chain.

All the above assignments confirm that the carbon layers are formed during heating, which serve to protect the underlying matrix so that the flame retardancy of PP is improved.

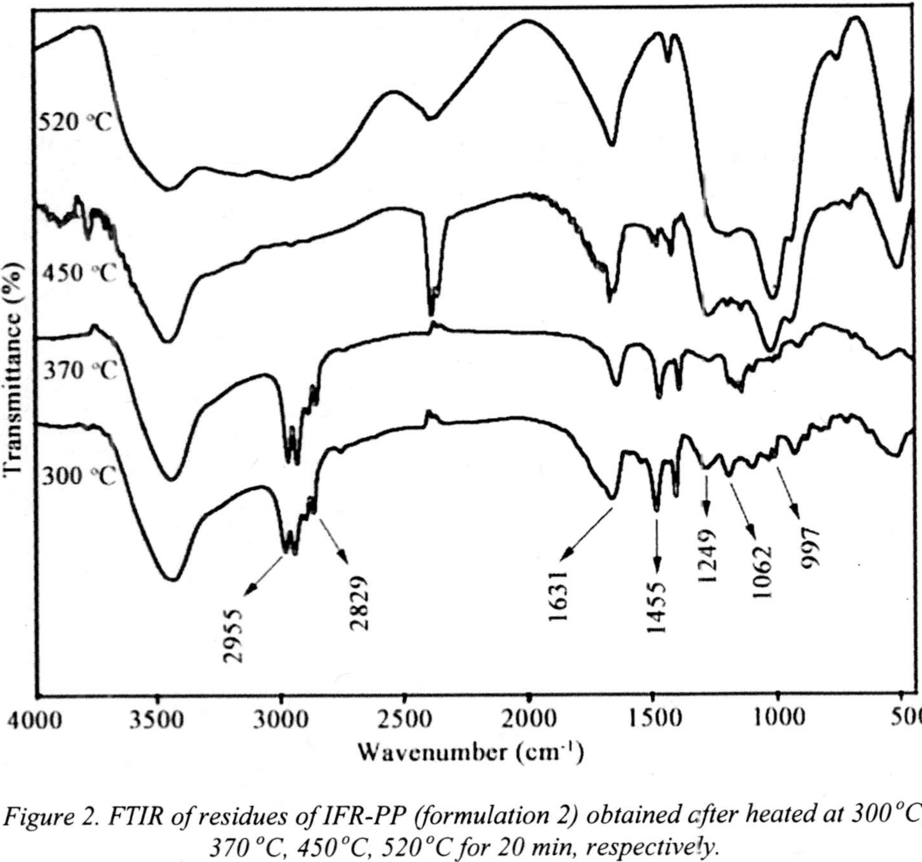


*Figure 2. FTIR of residues of IFR-PP (formulation 2) obtained after heated at 300°C, 370°C, 450°C, 520°C for 20 min, respectively.*

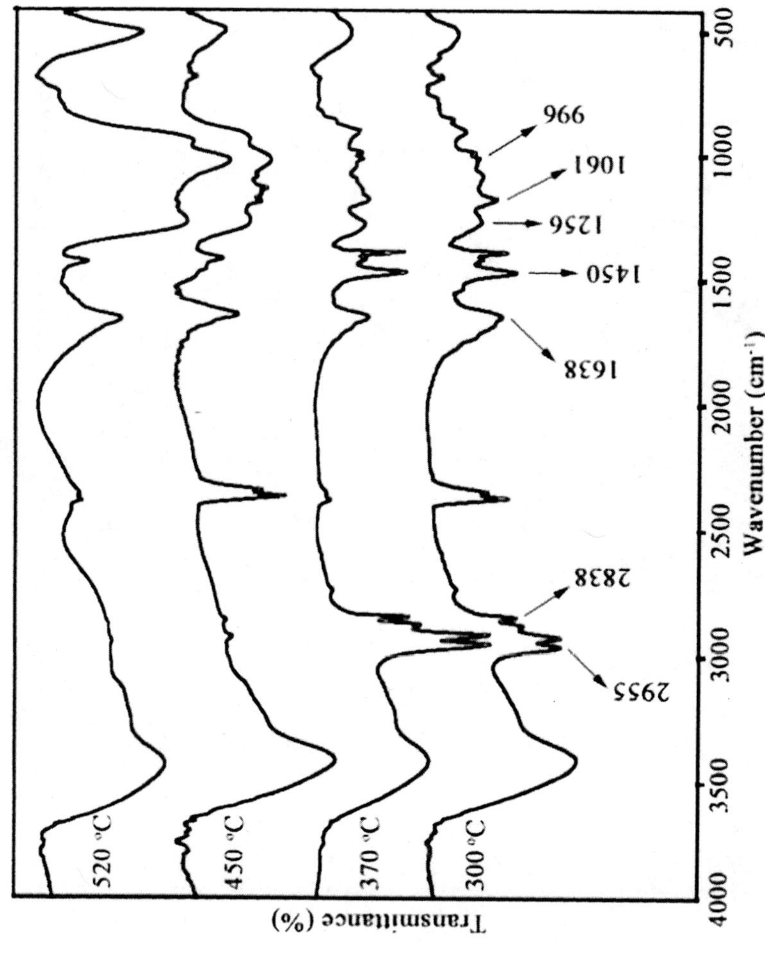


*Figure 3. FTIR of residues of IFR-PP/BA (formulation 3) obtained after heated at 300°C, 370 °C, 450°C, 520°C for 20 min, respectively.*

## X-rayPhotoelectron Spectroscopy (XPS) Analysis

XPS provides detailed information about elemental composition and content of the blend and charred layers, which complements the results from the FTIR spectra. Table II and Figure 4 show the atomic concentrations of C1s, N1s, O1s, P2p and P/C (a), O/C (b), and N/C (c) ratios of formulation 2 and formulation 3 after treatment at 300 °C, 370 °C, 450 °C and 520 °C for 20 min, respectively,as measured by XPS.

The atomic concentration of C1s is decreased with increasing temperature, due to the escape of carbon-containing gas, however, the atomic concentration of O1s and P2p are increased as a result of producing more phosphorus-containing materials (Table II). It should be pointed out that the atomic concentrations of C1s both in formulation 2 and formulation 3 decrease rapidly when the temperature is between 370 °C and 450 °C, which may indicate that esterification, dehydration and carbonization are occurring. During this stage, much char is crosslinked and more char layers are produced. From graphs (a) and (b) in Figure 4, we can see that the P/C and O/C curves have a similar behavior. Consequently, O is mainly linked to P in the material (*31*). From 300 °C to 370 °C, both the ratios of P/C and O/C increase slowly.

However, when the temperature is between 370 °C and 520 °C, a significant enhancement is observed: the P/C ratio of formulation 2 increases from 0.07 to 0.30, and the P/C ratio of formulation 3 increases from 0.06 to 0.34, respectively. The curves of N/C ratio are quite different from those of P/C and O/C. From Figure 4 (c), we can see that the N/C ratio of formulation 2 increased from 300 °C to 370 °C, and reaches a maximum of 0.05, while the maximum value of formulation 3 is 0.03. Above this temperature, the N/C ratios of both formulations 2 and 3 decrease rapidly, but the decrease of formulation 2 is greater than that of formulation 3. At 520 °C, the N/C ratios of formulation 2 and

**Table II. Atomic Concentrations of Formulation 2 and Formulation 3 at Different Heat Treated Temperatures**

| *No.* | *Temperature °C* | *C1s %* | *N1s %* | *O1s %* | *P2p %* | *P/C* | *O/C* | *N/C* |
|---|---|---|---|---|---|---|---|---|
| 2 | 300 | 90.10 | 2.05 | 5.14 | 2.70 | 0.030 | 0.057 | 0.023 |
| | 370 | 82.47 | 3.75 | 8.21 | 5.57 | 0.068 | 0.100 | 0.045 |
| | 450 | 59.72 | 2.19 | 23.85 | 14.24 | 0.238 | 0.400 | 0.037 |
| | 520 | 54.34 | 0.91 | 28.46 | 16.29 | 0.300 | 0.524 | 0.017 |
| 3 | 300 | 88.57 | 2.76 | 5.43 | 3.24 | 0.037 | 0.061 | 0.031 |
| | 370 | 84.44 | 2.92 | 7.55 | 5.10 | 0.060 | 0.089 | 0.035 |
| | 450 | 64.47 | 1.77 | 22.07 | 11.69 | 0.181 | 0.342 | 0.027 |
| | 520 | 53.65 | 1.17 | 26.95 | 18.23 | 0.340 | 0.502 | 0.022 |

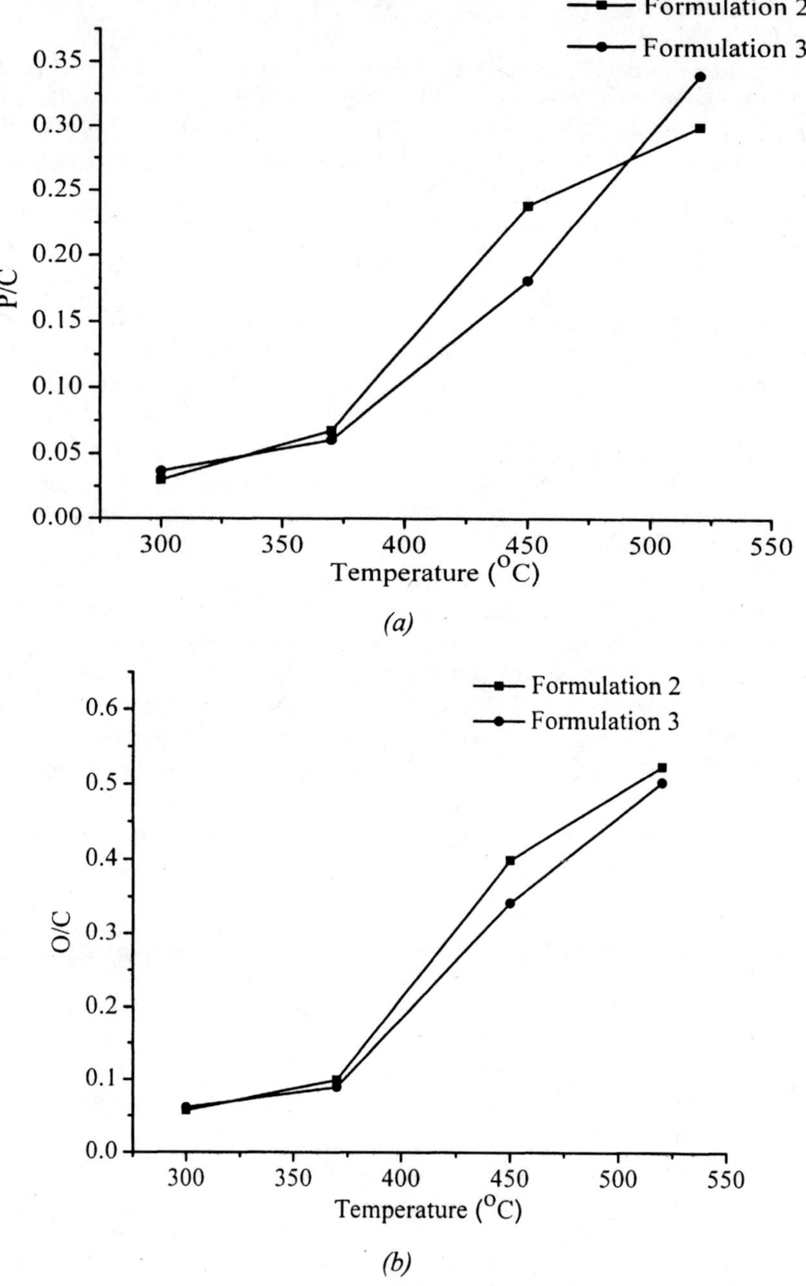


*Figure 4. Comparison of P/C (a), O/C (b), and N/C (c) ratios measured by XPS analysis.*

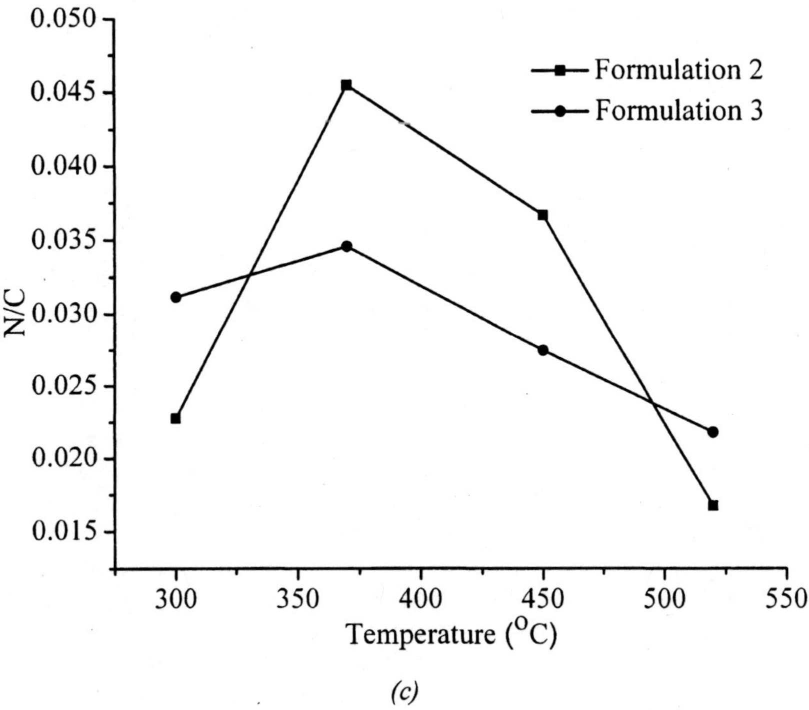


*(c)*

*Figure 4. Continued.*

formulation 3 are 0.017 and 0.022, respectively. Obviously, more nitrogen-containing compounds are preserved in the residue in formulation 3 due to its better protection of "char quantity" and "char quality".

The C1s spectra of formulation 2 (a) and formulation 3 (b) are shown in Figure 5. The peak of C1s around 284.8 eV is the contribution of C-C in aliphatic and aromatic species in the materials due to the dehydration during the heating (*32*).

The N1s XPS spectra of formulation 2 (a) and formulation 3 (b) are shown in Figure 6. From the Figure, we can see that these curves have similar behavior. In Figure 6 (a), the main peak shifts upward from 400.1 eV at 300 °C to 401.6 ev at 450 °C, while an additional weak peak at 399.0 eV appears. Then, the peak shifts downward to 400.4 eV at 520 °C. The band at 401.6 eV could be a result of the formation of quaternary nitrogen and some oxidized nitrogen compounds, as reported in the literature (*33*). In Figure 6 (b), the main peak shifts upward from 399.8 eV at 300 °C to 401.5eV at 450 °C.

The O1s XPS spectra of formulation 2 (a) and formulation 3 (b) are shown in Figure 7. Two bands around 531.2 and 532.5 eV are observed from the O1s spectra, as shown in Figure 7 (a) and (b). It is reported that the band at 531.2 eV can be assigned to double bonding in phosphate (P=O) or carbonyl (C=O) groups (*34*), while the band centered around 532.5 eV can be assigned to -O- in C-O-C, C-O-P and P-O-P groups (*35, 36*), which is in agreement with the FTIR analysis.

The P2p XPS spectra of formulation 2 (a) and formulation 3 (b) are shown in Figure 8. Usually the binding energy of P2p of inorganic phosphorus is 130 eV, whereas the P2p peak of sample appears around 134.0 eV, which can be assigned to C-O-P or $PO_3^-$ groups (*34*). The increase of binding energy demonstrates the formation of oxidation products of the phosphate species, (*29*) which confirms that the phosphorus in the blends has been converted into phosphocarbonaceous complexes in the char after burning, in good agreement with the results of the FTIR analysis (*33*).

## SEM Observation

The SEM micrographs of char from the outer surfaces of formulation 2 and formulation 3 are presented in Figure 9 (a) and (b), respectively. From the two micrographs, it can be seen that the outer surfaces of the two samples are compact. However, the inner surface char of formulation 2 and formulation 3 show different structure, presented in Figure 10 (a) and (b), respectively. Some closed cells are observed in graph (b), which can improve the thermal protection due to the charred layer (*13*), however, the cells in graph (a) are broken. These results show that formulation 3 has a better flame retardancy in PP than formulation 2.

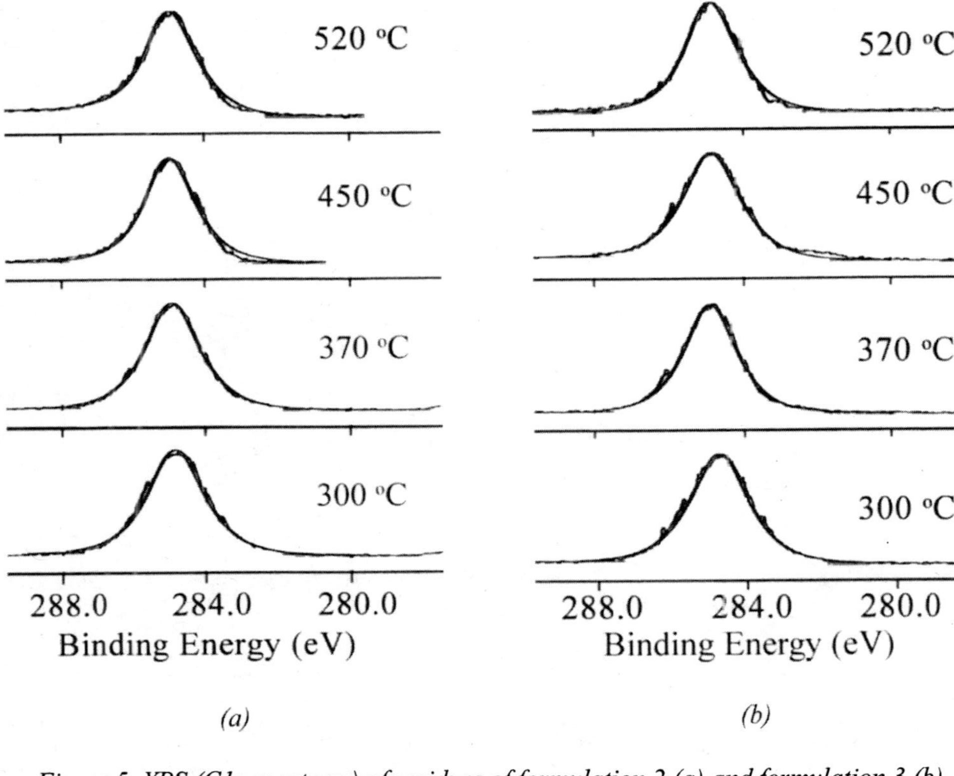


*Figure 5. XPS (C1s spectrum) of residues of formulation 2 (a) and formulation 3 (b) obtained after heated at 300°C, 370°C, 450°C, 520°C for 20 min, respectively.*

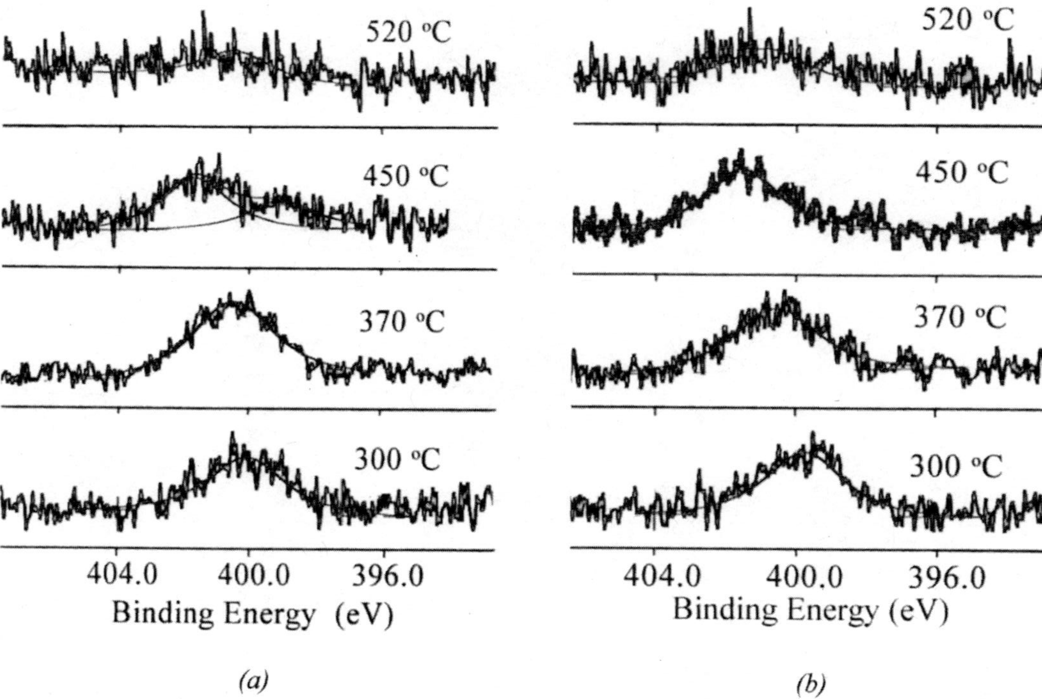


*Figure 6. XPS (N1s spectrum) of residues of formulation 2 (a) and formulation 3 (b) obtained after heated at 300°C, 370 °C, 450°C, 520°C for 20 min, respectively.*

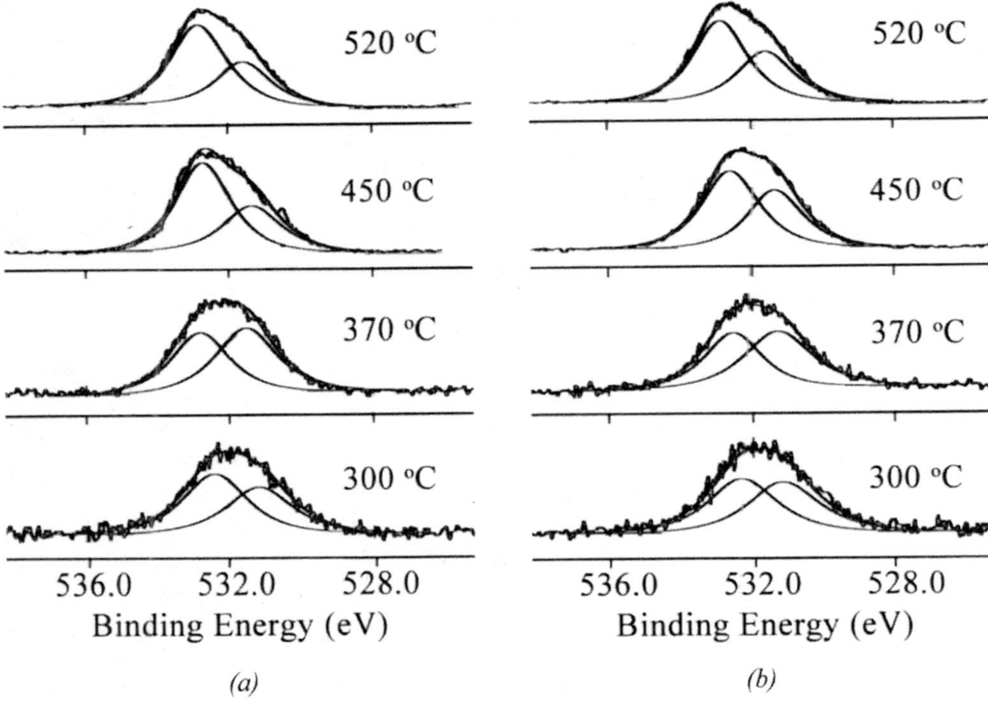


*Figure 7. XPS (O1s spectrum) of residues of formulation 2 (a) and formulation 3 (b) obtained after heated at 300°C, 370 °C, 450°C, 520°C for 20 min, respectively.*

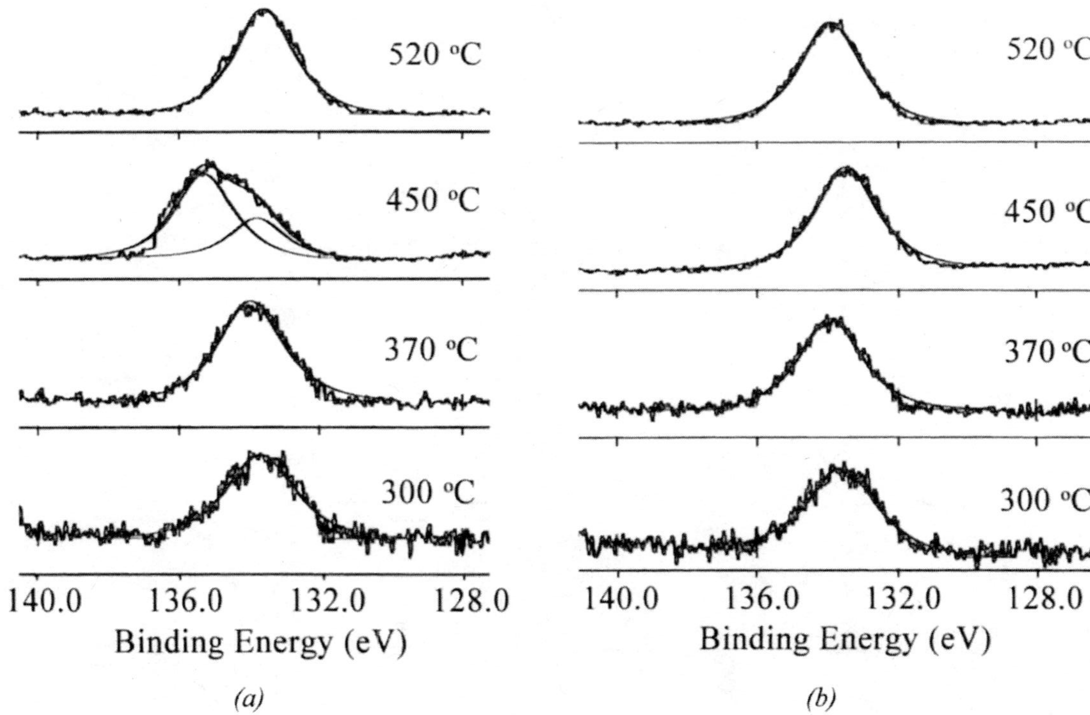


*Figure 8. XPS (P2p spectrum) of residues of formulation 2 (a) and formulation 3 (b) obtained after heated at 300°C, 370°C, 450°C, 520°C for 20 min, respectively.*

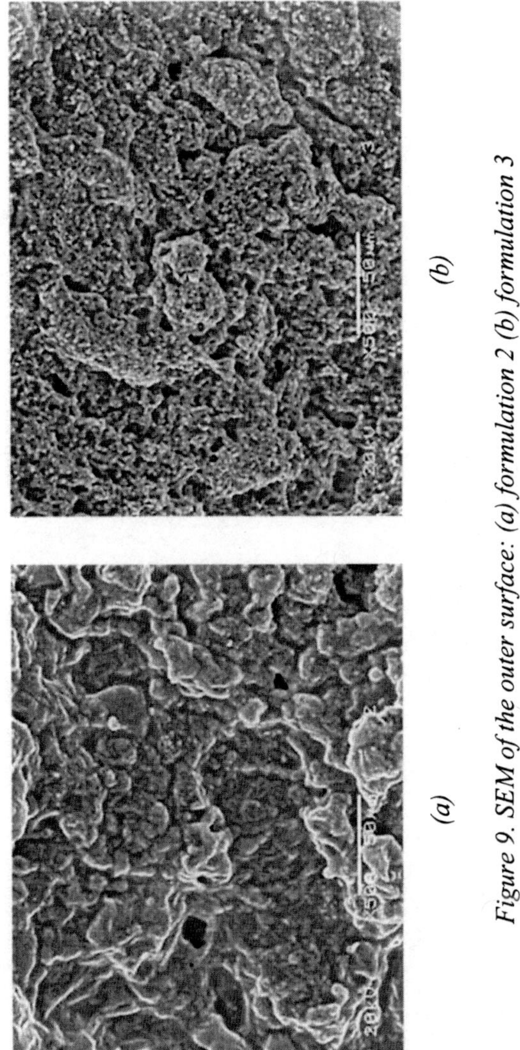

(a) (b)

Figure 9. SEM of the outer surface: (a) formulation 2 (b) formulation 3

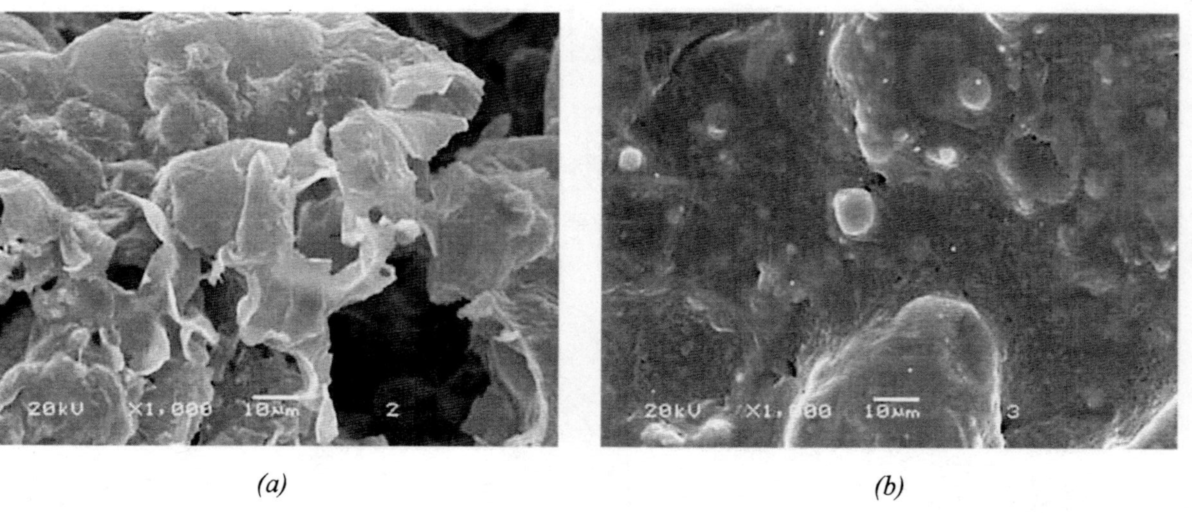


(a) (b)

*Figure 10. SEM of the inner surface: (a) formulation 2 (b) formulation 3*

## Conclusions

In summary, a small amount of boric acid (BA) incorporated into an intumescent flame retardant polypropylene (PP) can play a critical role in the enhancement of LOI value. The optimum flame retardant formulation is 29 wt. % of IFR and 1 wt. % of BA, which achieves an LOI value of 34.2 and aUL-94 V-0 rating. The addition of BA to IFR-PP composites can improve the thermal stability, especially at the high temperature, and apparently promote the formation of carbonaceous char layers. Both FTIR and XPS confirm that a phosphorus-containing carbon layer is formed during heating, which serves to protect the underlying matrix so that the flame retardancy of PP is improved. SEM results provide positive evidence that IFR-PP containing BA can promote the formation of compact char layer.

## Acknowledgments

This work was supported by the National Science Fund for Distinguished Young Scholars (50525309).

## References

1. Sen, A.K.; Mukherjee, B.; Bhattacharya, A. S.; Sanghi, L. K.; De, P. P.; Bhowmick, A. K. Preparation and characterization of low-halogen and nonhalogen fire-resistant low-smoke (FRLS) cable sheathing compound from blends of functionalized polyolefins and PVC. *J. Appl. Polym. Sci.* **1991**, *43*, 1673-1684.
2. Fontaine, G.; Bourbigot, S.; Duquesne, S. Neutralized flame retardant phosphorus agent: Facile synthesis, reaction to fire in PP and synergy with zinc borate. *Polym. Degrad. Stab.* **2008**, *93*, 68-76.
3. Xie, R. C.; Qu, B. J. Expandable graphite systems for halogen-free flame-retarding of polyolefins. I. Flammability characterization and synergistic effect. *J. Appl. Polym. Sci.* **2001**, *80*, 1181-1189.
4. Li, B.; Xu, M.J. Effect of a novel charring–foaming agent on flame retardancy and thermal degradation of intumescent flame retardant polypropylene. *Polym. Degrad. Stab.* **2006**, *91*, 1380-1386.
5. Lv, P.; Wang, Z. Z.; Hu, K. L.; Fan, W. C. Flammability and thermal degradation of flame retarded polypropylene composites containing melamine phosphate and pentaerythritol derivatives. *Polym. Degrad. Stab.* **2005**, *90*, 523-534.
6. Almeras, X.; Dabrowski, F.; Le Bras, M; Poutch, F.; Bourbigot, S.; Marosi, G.; Anna, P. Using polyamide-6 as charring agent in intumescent

polypropylene formulationsI. Effect of the compatibilising agent on the fire retardancy performance. *Polym. Degrad. Stab.* **2002,** *77*, 305-313.
7. Xie, F.; Wang, Y. Z.; Yang, B.; Liu, Y. A novel intumescent flame-retardant polyethylene system. *Macromol. Mater. Eng.* **2006**, *291*, 247-253.
8. Bourbigot, S.; Le Bras, M.; Duquesne, S.; Rochery, M. Recent advances for intumescent polymers. *Macromol. Mater. Eng.* **2004**, *289*, 499-511.
9. Kandola, B. K. ; Horrocks, A. R. ; Horrocks. S. Char formation in flame-retarded fibre-intumescent combinations. Part V. Exploring different fibre/intumescent combinations. *Fire Mater.* **2001**, *25*, 153-160.
10. Lewin, M.; Endo, M. Catalysis of intumescent flame retardancy of polypropylene by metallic compounds. Polym. Adv. Technol. **2003,** *14*, 3-11.
11. Chen, X. C.; Ding, Y. P.; Tang, T. Synergistic effect of nickel formate on the thermal and flame-retardant properties of polypropylene. *Polym. Int.* **2005**, *54*, 904-908.
12. Tang, Y.; Hu, Y.; Wang, S. F.; Gui, Z.; Chen, Z. Y.; Fan, W. C. Intumescent flame retardant-montmorillonite synergism in polypropylene-layered silicate nanocomposites. *Polym. Int.* **2003**, *52*, 1396-1400.
13. Wu, Q. ; Qu, B. Synergistic effects of silicotungistic acid on intumescent flame-retardant polypropylene. *Polym.Degrad. Stab.* **2001**, *74*, 255-261.
14. Morgan, A. B. ; Jurs, J. L. ; Tour, J. M. Synthesis, flame-retardancy testing, and preliminary mechanism studies of nonhalogenated aromatic boronic acids: A new class of condensed-phase polymer flame-retardant additives for acrylonitrile-butadiene-styrene and polycarbonate. *J. Appl. Polym. Sci.* **2000**, *76*, 1257-1268.
15. Myers, R. E.; Dickens, E. D. Licursi, E.; Evans, R. E. Ammonium pentaborate—an intumescent flame-retardant for thermalplastics polyurethanes. *J. Fire. Sci.* **1985**, *3*, 432-449.
16. Armitage, P.; Ebon, J. R.; Hunt, B. J.; Jones, M. S.; Thorpe, F. G. Chemical modification of polymers to improve flame retardance—I. The influence of boron-containing groups. *Polym. Degrad. Stab.* **1996**, *54*, 387-393.
17. Gao, M.; Yang, S. S.; Yang, R. J. Flame retardant synergism of GUP and boric acid by cone calorimetry. *J. Appl. Polym. Sci.* **2006**, *102*, 5522-5527.
18. Martín, C.; Ronda, J. C.; Cádiz, V. Novel flame-retardant thermosets: Diglycidyl ether of bisphenol A as a curing agent of boron-containing phenolic resins. *J. Polym. Sci. Part A: Polym. Chem.* **2006**, *44*, 1701-1710.
19. Martín, C.; Ronda, J. C.; Cádiz, V. Development of novel flame-retardant thermosets based on boron-modified phenol-formaldehyde resins. *J. Polym. Sci. Part. A: Polym. Chem.* **2006**, *44*, 3503-3512.
20. Lu, S. Y.; Hamerton, I. Recent developments in the chemistry of halogen[hyphen]free flame retardant polymers. *Prog. Polym. Sci.* **2002**, *27*, 1661-1712.

21. Sain, M.; Park, S. H.; Suhara, F.; Law, S. Flame retardant and mechanical properties of natural fibre–PP composites containing magnesium hydroxide. *Polym. Degrad. Stab.* **2004**, *83*, 363-367.
22. Peng, H. Q.; Wang, D. Y.; Zhou, Q.; Wang, Y. Z. A novel charring agent containing caged bicyclic phosphate and its application in intumescent flame retardant polypropylene systems. *J. Ind. Eng. Chem.* **2008**, *(in press)*
23. Jimenez, M.; Duquesne, S.; Bourbigot, S. Characterization of the performance of an intumescent fire protective coating. *Surf. Coat. Technol.* **2006**, *201*, 979-987.
24. Martín, C.; Hunt, B. J.; Ebdon, J. R.; Ronda, J. C.; Cádiz, V. Synthesis, crosslinking and fame retardance of polymers of boron-containing difunctional styrenic monomers. *React. Funct. Polym.* **2006**, *66*, 1047-1054.
25. Jimenez, M.; Duquesne, S.; Bourbigot, S. Intumescent fire protective coating: Toward a better understanding of their mechanism of action. *Thermochimica acta* **2006**, *449*, 16-26.
26. Li, Q.; Zhong, H. F.; Wei, P.; Jiang, P. K. Thermal degradation behaviors of polypropylene with novel silicon-containing intumescent flame retardant. *J. Appl. Polym. Sci.* **2005**, *98*, 2487-2492.
27. Xie, R. C.; Qu, B. J. Thermo-oxidative degradation behaviors of expandable graphite-based intumescent halogen-free flame retardant LLDPE blends. *Polym. Degrad. Stab.* **2001**, *71*, 395-402.
28. Pages, P.; Carrasco, F.; Saurina, J.; Colom, X. FTIR and DSC study of HDPE structural changes and mechanical properties variation when exposed to weathering aging during canadian winter. *J. Appl. Polym. Sci.* **1996**, *60*, 153-159.
29. Xie, R. C.; Qu, B. J. Expandable graphite systems for halogen-free flame-retarding of polyolefins. II. Structures of intumescent char and flame-retardant mechanism. *J. Appl. Polym. Sci.* **2001**, *80*, 1190-1197.
30. Bugajny, M.; Borbigot, S. The origin and nature of flame retardance in ethylene-vinyl acetate copolymers containing hostaflam AP 750. *Polym. Int.* **1999**, *48*, 264-270.
31. Duquesne, S.; Le Bras, M.; Jama, C.; Weil, E. D.; Gengembre, L. X-ray photoelectron spectroscopy investigation of fire retarded polymeric materials: application to the study of an intumescent system. *Polym. Degrad. Stab.* **2002**, *77*, 203-211.
32. Bourbigot, S.; Le Bras, M.; Delobel, R.; Gengembre, L. XPS study of an intumescent coating: II. Application to the ammonium polyphosphate/pentaerythritol/ethylenic terpolymer fire retardant system with and without synergistic agent. *Appl. Surface. Sci.* **1997**, *120*, 15-29.
33. Qu, B. J.; Xie, R. C. Intumescent char structures and flame-retardant mechanism of expandable graphite-based halogen-free flame-retardant linear low density polyethylene blends. *Polym. Int.* **2003**, *52*, 1415-1422.
34. Zhu, S. W.; Shi, W. F. Thermal degradation of a new flame retardant phosphate methacrylate polymer. *Polym. Degrad. Stab.* **2003**, *80*, 217-222.

35. Brown, N. M. D.; Hewitt, J. A.; Meenan, B. J. X-ray-induced beam damage observed during x-ray photoelectron spectroscopy (XPS) studies of palladium electrode ink materials. *Surf. Interf. Anal.* **1992**, *18*, 187-198.
36. Shih, P. Y.; Yung, S. W.; Chin, T. S. Thermal and corrosion behavior of $P_2O_5$-$Na_2O$-CuO glasses. *J. Non-Cryst. Solids.* **1998**, *224*, 143-152.

# Chapter 15

# Brominated Aryl Phosphorlane Flame Retardants

**Bob A. Howell* and Y.-J. Cho**

**Center for Applications in Polymer Science, Central Michigan University, Mt. Pleasant, MI 48859 0001**
***Corresponding author: bob.a.howell@cmich.edu**

For most applications polymeric materials must contain flame retardant additives. Traditionally, organohalogen compounds, particularly brominated aromatics, have been utilized in this application. They are generally readily available in relatively pure form at acceptable cost and are effective gas phase flame retardants. However, increasing concern about the accumulation of organohalogen compounds in the environment make the development of flame retardants that are effective at very low loading particularly attractive. Compounds that display two modes of action might impart sufficient flame retardancy at much lower levels than traditional organohalogen flame retardants. Brominated aryl phospholanes contain both bromine (for gas phase activity) and phosphorus (for solid phase activity). The carbon-carbon bond of these five-membered heterocycles undergoes thermolysis at modest temperatures to generate a diradical capable of initiating polymerization of vinyl monomers. The resulting polymer contains flame retarding units in the mainchain.

Decabromodiphenyl ether remains the largest selling organic flame retardant additive for polymeric materials (*1*). However, organohalogen flame retardants are coming under increasing pressure from environmentalists and regulatory agencies (*2*). There is a growing concern about the accumulation of halogen from all sources in the environment. Decabromodiphenyl ether is readily available, reasonably inexpensive, and highly effective as a flame retardant (*3,4*). It will likely retain a major position in U.S. markets for the foreseeable future. However, even here it is being displaced by organophosphorus in some applications; in ABS for computer housings, for example. In contrast to organohalogen flame retardants which are gas-phase active, organophosphorus compounds are solid-phase active, i.e., they promote char formation at the surface of the burning polymer. The formation of the char layer insulates the polymer from heat feedback from the flame so that polymer pyrolysis to generate fuel fragments is diminished. Compounds containing both phosphorus and bromine might behave as dual-functional flame retardants, i.e., they might be expected to display both types of flame retardant activity. In this case, a smaller amount of total halogen would be required to achieve a suitable level of flame retardancy. If a synergy of action could be achieved, a much lower level of additive would be required (*6-8*).

## Experimental

### Methods and Instrumentation

All new materials were fully characterized by spectroscopic, thermal and chromatographic methods. Nuclear magnetic resonance spectra ($^{1}H$ and $^{13}C$) were obtained using solutions in deuterochloroform, dimethyl sulfoxide-$d_6$ or toluene-$d_8$ containing tetramethylsilane (TMS) as internal reference and a Varian Mercury 300 spectrometer. Infrared spectra were obtained using solid solutions (1 %) in anhydrous potassium bromide (as pellets) or thin films between sodium chloride discs using a model 560 Nicolet MAGN-IR spectrophotometer. Melting points were determined by differential scanning calorimetry (DSC) using a TA Instruments Thermal Analyst model 2100 system equipped with a model 2910 MDSC cell. The sample compartment was subject to a constant purge of dry nitrogen at 50 mL/min. Thermal degradation characteristics were examined by thermogravimetry using a TA Instruments model 2950 TGA unit interfaced with the Thermal Analyst Control module. The TGA cell was swept with nitrogen at 50 mL/min during degradation runs. The sample size was 5-10 mg in a platinum sample pan. The temperature was ramped at a rate of 5 °C/min.

## Synthesis

In general, syntheses were carried out in oven-dried glassware which had been allowed to cool under a stream of dry nitrogen. Reaction mixtures were blanketed as necessary with a static atmosphere of dry nitrogen.

**Di(3,5-dibromophenyl)methanol.** A 2.5 M solution of butyllithium in hexane (36.0 mL, 90 mmol) was added dropwise to a stirred solution of 28.0 g (90 mmol) of 1,3,5-tribromobenzene in 700 mL of anhydrous diethyl ether maintained at -78 °C. The mixture was stirred for two hours and 2.70 ml (44 mmol) of methyl formate was added dropwise. The resulting mixture was allowed to slowly warm to room temperature and stirred overnight. Water (100 ml) was slowly added. The layers were separated and the aqueous layer was extracted with two 40-mL portions of diethyl ether. The ether layers were combined, washed with saturated aqueous sodium chloride solution, and dried over anhydrous sodium sulfate. The solvent was removed from the dry solution by rotary evaporation at reduced pressure to afford 18.7 g of brown residual solid. This material was crystallized from hexane to provide 13.6 g (60.5% yield) of the desired alcohol as white needles, mp 180 °C (DSC). [lit. (*9*) 169.5-170.5 °C]: $^1H$ NMR (δ, $CDCl_3$) 2.39 (hydroxyl proton), 5.65 (benzylic proton), 7.43, 7.62 (aromatic protons); $^{13}C$ NMR (δ, $CDCl_3$) 73.84 (benzylic carbon atom), 123.35, 128.24, 133.80, 146.11 (aromatic carbon atoms).

**Di(3,5-dibromophenyl)methanone.** A solution of 26.2 g (0.05 mol) of di(3,5-dibromophenyl)methanol in 100 mL of dichloromethane was added in a single portion to a suspension of 30.3 g (0.14 mol) of pyridinium chlorochromate (PCC) in 400 mL of dichloromethane. The resulting mixture was stirred at room temperature for 1.5 hr. Diethyl ether, 400 mL, was added and the mixture was filtered through a short alumina column. The solvents were removed by rotary evaporation at reduced pressure to afford a pale yellow crystalline solid. Crystallization from hexane afforded 25.6 g (98.3% yield) of di(3,5-dibromophenyl)methanone as white needles, mp 210 °C (DSC). [lit. (*10*) 208.5-209 °C]: $^1H$ NMR (δ, $CDCl_3$) 7.83, 7.80 (aromatic protons); $^{13}C$ NMR (δ, $CDCl_3$) 123.45, 131.28, 138.40, 139.31 (aromatic carbon atoms), 190.87 (carbonyl carbon atom).

**1,1,2,2-Tetra(3,5-dibromophenyl)-1,2-ethanediol.** A solution of 1.0 g (2.0 mmol) of di(3,5-dibromophenyl)methanone in 25 mL of tetrahydrofuran (THF) was added, dropwise, to a cold (0 °C) suspension of 5.0 g (76 mmol) of zinc powder and 1.0 g (7.0 mmol) of zinc chloride in 50 mL of 50% aqueous THF. The mixture was stirred for two hours at 0 °C and 10 mL of 3N aqueous hydrochloric acid solution was added dropwise. The resulting mixture was stirred 0.5 hr. The solids present were removed by filtration and the filtrate was extracted with three 30-mL portions of diethyl ether. The ether layers were combined, washed with saturated aqueous sodium chloride solution, and dried over anhydrous sodium sulfate. The solvent was removed by rotary evaporation at reduced pressure to obtain a pale yellow solid. This material could be purified by crystallization from hexane or silica chromatography (ethyl acetate: hexane, 20:80) to provide the expected diol as white needles, mp 167 °C (DSC). Proton and carbon-13 NMR spectra of this compound are displayed below in Figures 1 and 2. The carbon-13 spectrum, in particular, is diagnostic for the diol (absorption for the benzylic carbon atom near δ 80) and derivatives.

## Results and Discussion

A way of achieving enhanced flame retardant activity is to construct compounds which display more than a single mode of action or that are capable of a synergy of flame suppressant properties. Compounds containing both a high level of halogen, in particular bromine, and another element, which may be converted to a cross-linking, char-promoting agent during the combustion process, might display these characteristics. Organic compounds, which promote char formation, are those containing phosphorus, nitrogen, sulfur, phenoxy oxygen, and a few other heteroatoms. Many organophosphorus compounds are very effective solid phase flame retardants particularly when used with oxygen-containing polymers (*5*). Nitrogen compounds such as melamine, melamine phosphate, melamine polyphosphate, and ethylenediamine phosphate are frequently used as synergists to reinforce the effect of phosphorus compounds (*11-14*). Several approaches to the preparation of useful flame retardants containing both bromine and phosphorus have been explored.

A series of bromoaryl phosphates which could simultaneously promote char formation at the surface of a burning polymer and at the same time display effective gas-phase disruption of flame propagating reactions has been prepared from bromophenols and diethylphosphite (*15,16*). The thermal degradation characteristics of these compounds are shown in Table I. It might be noted that all undergo decomposition at approximately 250 °C and that two are crystalline solids.

The effectiveness of these compounds as flame retardants for poly(styrene) was evaluated using the UL-94 vertical burn test (ASTM D3801). Plaques for testing were prepared by dissolving the flame retardant in styrene monomer at the appropriate concentration, carrying out partial polymerization (AIBN-initiated) to generate a syrup, and curing the syrup in a mold of dimensions 2x13x127 mm. The results of the flame test are collected in Table II. As can be seen, all the additives are quite effective as flame retardants for poly(styrene) at the 5% level. Even at a low loading of 3% an appreciable flame retarding influence is observed. An enhancement of flame retarding effectiveness has been noted for compounds containing both bromine and phosphorus (*17*).

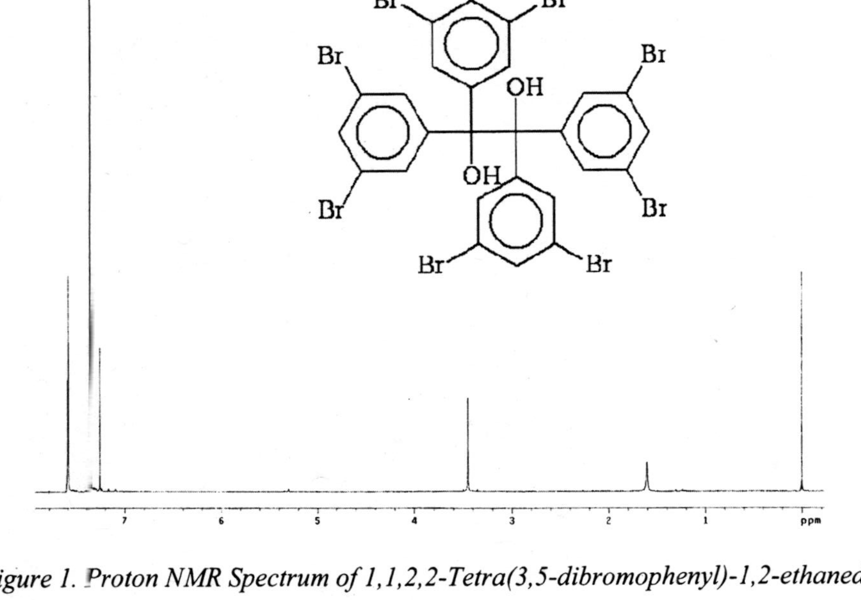


*Figure 1. Proton NMR Spectrum of 1,1,2,2-Tetra(3,5-dibromophenyl)-1,2-ethanediol.*

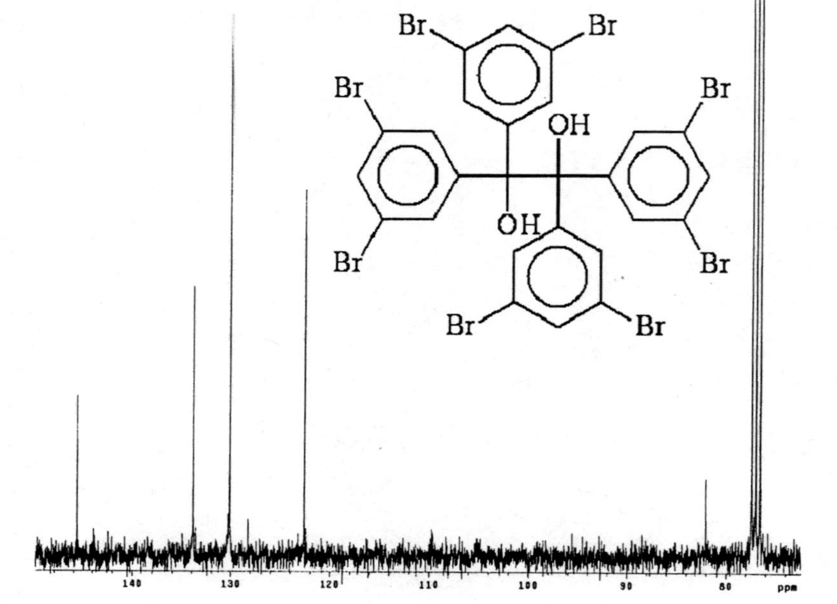


*Figure 2. Carbon-13 NMR Spectrum of 1,1,2,2-Tetra(3,5-dibromophenyl)-1,2-ethanediol.*

**Table I. Thermal Characteristics of [(Bromo)$_x$phenyl]diethylphosphates.**

| *Compound (Bromophenyl Substiuent)* | *Melting Point (°C), DSC)* | *Decomposition Temperature (°C, TGA)* |
|---|---|---|
| 4-Bromophenyl | | 231 |
| 2,4-Dibromophenyl | | 265 |
| 2,4,6-Tribromophenyl | 73 | 259 |
| Pentabromophenyl | 137 | 242 |

**Table II. Impact of the Presence of Brominated Aryl Phosphates on the Flammability of Poly(styrene).**

| *Flame Retardant* | *Loading (wt. %)* | *UL-94 Rating[a]* |
|---|---|---|
| (4-Bromophenyl)diethylphosphate | 3 | V-1 |
| (2,4-Dibromophenyl)diethylphosphate | | V-1 |
| (2,4,6-Tribromophenyl)diethylphosphate | | V-1 |
| (Pentabromophenyl)diethylphosphate | | V-0 |
| (4-Bromophenyl)diethylphosphate | 5 | V-0 |
| (2,4-Dibromophenyl)diethylphosphate | | V-0 |
| (2,4,6-Tribromophenyl)diethylphosphate | | V-0 |
| (Pentabromophenyl)diethylphosphate | | V-0 |

[a] Ul-94 rating of V-1 indicates that burning stops (without flaming drips) within 60 seconds after two applications of ten seconds each of a flame to a test bar; V-0 indicates that burning stops (without flaming drips) within 10 seconds after two applications of ten seconds each of a flame to a test bar and is the most desirable performance rating.

Poly(styrene)s containing phosphonate moieties in the mainchain have been prepared using dioxaphospholanes as polymerization initiators. Several suitable reactive dioxaphospholanes may be prepared from benzpinacol (*18*). The synthesis of one of these 2,4,4,5,5-pentaphenyl-1,3,2-dioxaphospholane, is outlined in Scheme 2.

The cyclic phospholane can be obtained in good yield as a white crystalline solid with a melting point of 72 °C (DSC) (19,20). The proton NMR spectrum contains absorption for aromatic protons and the phosphorus spectrum consists of a singlet at δ 22.1 relative to the absorption for 85% aqueous phosphoric acid as external reference. The infrared spectrum of this compound contains an aromatic absorption band at 1601 $cm^{-1}$ as well as strong bands for O-P-O absorption at 1059 and 759 $cm^{-1}$. This five-membered ring structure contains a sterically strained carbon-carbon bond that might be expected to undergo thermally induced homolysis at modest temperatures. This may be readily demonstrated using thermogravimetry. The compound readily undergoes thermal decomposition with a maximum rate of degradation at 120 °C and a degradation onset temperature of 70 °C. Because of the ease with which the central carbon-carbon bond of this compound undergoes homolysis, it should function as a polymerization initiator. This is illustrated in Scheme 3 for the polymerization of styrene.

If the compound functions as initiator, each polymer chain generated should contain one phosphorus moiety. Alternatively, it is possible that the phosphorus compound could also function as a comonomer in the polymerization, i.e., the propagating poly(styryl) radical could add to the strained carbon-carbon bond.

A series of mixtures was prepared in standard polymerization tubes. The first tube contained pure styrene monomer and was used as a control. The remaining tubes contained solutions of 1, 5 and 10% by mass of 2,4,4,5,5-pentaphenyl-1,3,2-dioxaphospholane in styrene monomer, respectively. Tubes were placed in an oil bath maintained at 70 °C. Progress of polymerization was followed by removing aliquots of the mixture as a function of time for viscosity measurement. The aliquots were diluted with benzene (1g of mixture/100 mL of benzene). The relative viscosity increased as a function of time reflecting the extent of polymerization. Size exclusion chromatographic data for the polymers formed are contained in Table III.

From the chromatographic data, it is apparent that thermal decomposition of 2,4,4,5,5-pentaphenyl-1,3,2-dioxaphospholane is effective in initiating styrene polymerization. In fact, when no phospholane was present in the monomer, no polymer was produced at 70 °C. In the presence of phospholane, polymer is readily produced under the same conditions. Further, the molecular mass of the polymer formed decreases regularly, as expected, as the concentration of initiator (phospholane) is increased. Analysis of the polymers by both infrared and phosphorus-31 NMR spectroscopy indicated that phosphorus units were incorporated into the polymer mainchain at a much higher level than can be accounted for by initiation, i.e., the phosphorus moiety probably functions as both initiator and comonomer.

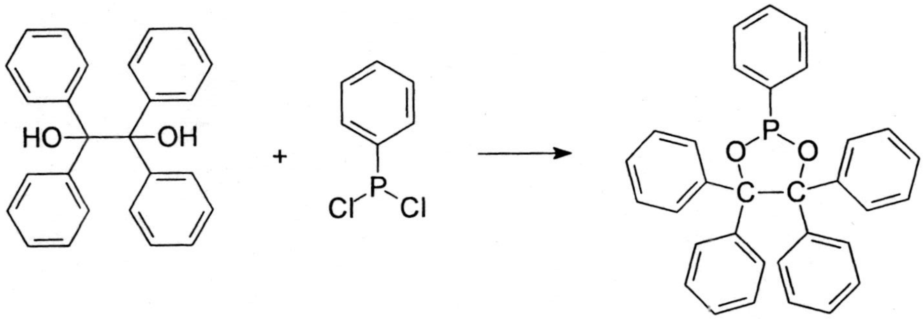

*Scheme 2. Synthesis of 2,4,4,5,5-pentaphenyl-1,3,2-dioxaphospholane.*

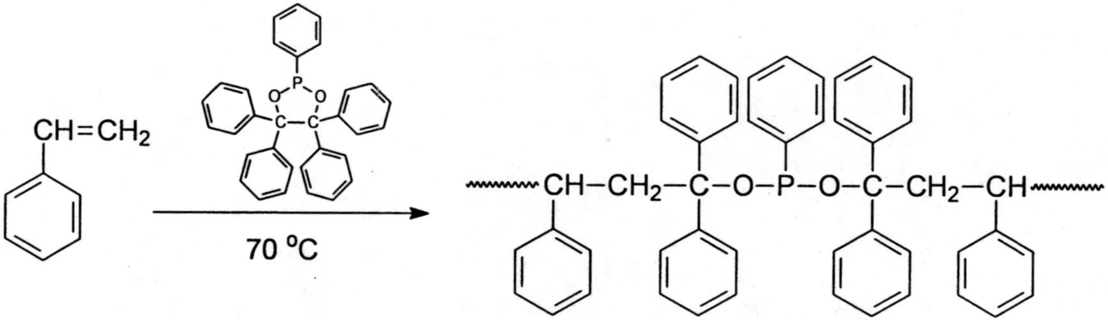

CH=CH$_2$

70 °C

CH–CH$_2$–C–O–P–O–C–CH$_2$–CH

*Scheme 3. Styrene Polymerization Initiated with 2,4,4,5,5-Pentaphenyl-1,3,2-dioxaphospholane.*

**Table III. Size Exclusion Chromatographic Characterization of Poly(styrene) Produced Using 2,4,4,5,5-Pentaphenyl-1.3.2-dioxaphospholane as Initiator at 70 °C.**

| *Initiator Present (Wt. %)* | $M_n$ | $M_w$ | *Polydispersity* |
|---|---|---|---|
| 0 | N[1] | N[1] | N[1] |
| 1 | 180000 | 365400 | 2.03 |
| 5 | 120000 | 302400 | 2.52 |
| 10 | 90000 | 273600 | 3.04 |

N[1]: No polymer formed.

**Table IV. Comparison of the Extrapolated Onset and Maximum Decomposition Temperature of Styrene Polymers Produced in the Presence of One, Five and Ten Percent 2,4,4,5,5-Pentaphenyl-1,3,2-dioxaphospholane.**

| *Phospholane in Polymerization Mixture (Wt. %)* | *Extrapolated Onset Temperature for Decomposition ( °C)* | *Maximum Decomposition Rate Temperature ( °C)* |
|---|---|---|
| 0 | 418.1 | 438.9 |
| 1 | 424.2 | 443 |
| 5 | 429.3 | 448.3 |
| 10 | 433.2 | 454.1 |

The thermal stability of the styrene polymers generated using one, five and ten mass percent 2,4,4,5,5-pentaphenyl-1,3,2-dioxaphospholane as initiator was examined using thermogravimetry. The relevant decomposition data are displayed in Table IV.

From these results, it would appear that the thermal stability of the polymers containing a phospholane unit is similar to that of ply(styrene), i.e., incorporation of the phospholane into the polymer mainchain does not diminish the thermal stability of poly(styrene).

The flammability of these polymers was evaluated using the UL 94 vertical burn test. Results are presented in Table V.

**Table V. Flammability Behavior of Styrene Polymers Generated by Initiation with 2,4,4,5,5-Pentaphenyl-1,3,2-dioxaphospholane at 70° C.**

| *Observation* | *Level of Phospholane in Polymerization Mixture (Wt. %)* | | | |
|---|---|---|---|---|
| | 0% | 1% | 5% | 10% |
| Total flaming combustion for each specimen | 30s | 30s | 30s | 30s |
| Total flaming combustion for all 5 specimens of any set | 250s | 250s | 250s | 80s |
| Flaming and glowing combustion for each specimen after second burner | 60s | 60s | 60s | 60s |
| Cotton ignited by flaming drips from any specimen | YES | YES | YES | NO |
| Glowing or flaming combustion of any specimen to holding clamp | YES | YES | NO | NO |
| **Classification** | | | **V-1** | **<V-1** |

The results presented in Table V suggest that, at the level of incorporation, the phospholane imparts a modest flame retardancy to poly(styrene).

This approach may be exploited to introduce the flame retarding influence of bromine as well as phosphorus. Bromine atoms may be introduced into the phenyl groups of the phospholane. Using the phospholane for initiation of vinyl polymerization would lead to the formation of a polymer containing both bromoaromatic and phosphorus functionality. This is illustrated in Scheme 4. An approach to the synthesis of the phospholane may be based on commercially available 1,3,5-tribromobenzene as starting material. Lithium-halogen exchange generates 3,5-bromophenyllithium which may be treated with methyl formate to form di(3,5-dibromophenyl)methanol (21). Mild oxidation with pyridinium chlorochromate generates the corresponding ketone. Metal-promoted reductive

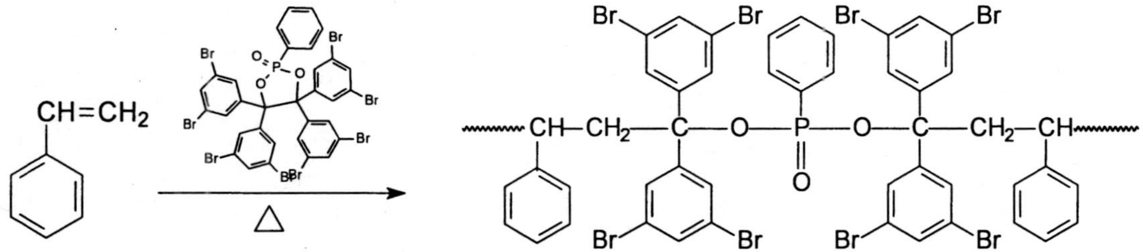

*Scheme 4. Styrene Polymerization Initiated with 2-Oxo-2-phenyl-4,4,5,5-tetra(3,5-dibromophenyl)-1,3,2-dioxaphospholane.*

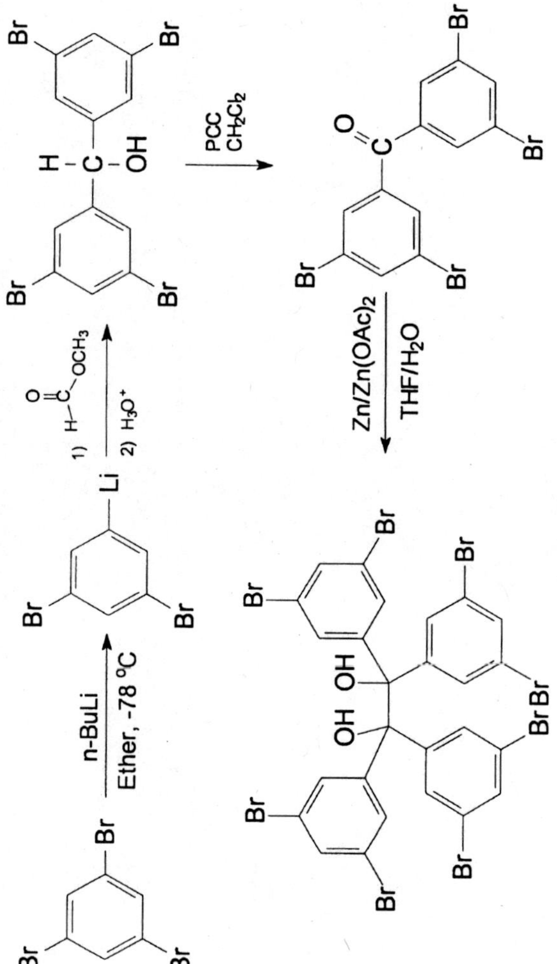
n-BuLi
Ether, -78 °C
1) H–C(=O)–OCH3
2) H3O+
PCC
CH2Cl2
Zn/Zn(OAc)2
THF/H2O

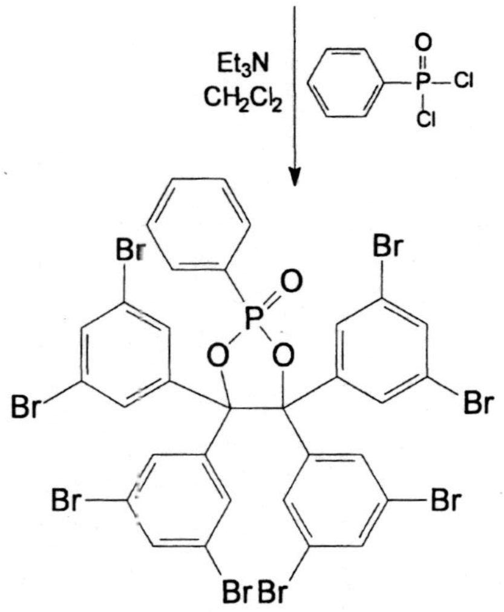

*Scheme 5. A Route to 2-Oxo-2-phenyl-4,4,5,5-tetra(3,5-dibromophenyl)-1,3,2-dioxophospholane.*

coupling of the ketone generates the diol precursor to the phospholate (22,23). The proposed synthesis is illustrated in Scheme 5. A variety of conditions for the conversion of the diol to the phospholane are being explored. The five-membered ring containing a fully oxygenated phosphorus atom is considerably more strained than the corresponding phosphorus(III) compound. It may lack sufficient stability to permit ready preparation at ordinary temperatures.

## Conclusions

A variety of approaches may be utilized to generate compounds containing both bromoaromatic and phosphonate units. Depending on the structure, such compounds may be utilized as additive or reactive dual functional flame retardants. The presence of bromine in these compounds imparts gas-phase flame retardant activity while the presence of phosphorus promotes solid-phase activity. Using reactive phospholanes containing bromoaromatic groups to initiate vinyl polymerization would lead to the formation of polymers containing flame-retarding units in the mainchain. Alternatively, oligomers containing these units may be used as flame-retarding additives for a variety of polymers.

## References


1. European Flame Retardants Association (www.flameretardants.eu) annual survey.
2. Weil, E.D. "An Attempt at a Balanced View of the Halogen Controversy", *Proceedings, 10th International Conference on Recent Advances in Flame Retardancy of Polymeric Materials*, Business Communications Company, Norwalk, CT, **1999**.
3. Georlette, P.; Simons J.; Costa, L. "Halogen-Containing Fire-Retardant Compounds" in Grand, A.F.; Wilkie, C.W. Eds., *Fire Retardancy of Polymeric Materials*, Marcel Dekker, Inc., New York, NY, **2000**; Ch. **8**, p 245-284.
4. Pettigrew, A. "Halogenated Flame Retardants", *Kirk-Othmer Encyclopedia of Chemical Technology*, 4th Ed., Wiley-Interscience, New York, NY, **1993**, Vol. 10.
5. Weil, E.D.; Mark, H.; Bikales, N.; Overberger, C.C.; Menges, G. Eds., *Encyclopedia of Polymer Science and Engineering*, 2nd Ed., Wiley-Interscience, New York, NY, **1990**, Vol. 11.
6. Avondo, G.; Vovell, G.C.; Delbourge, R. *Comb. and Flame*, **1987**, *21*, 7.
7. Green, J. *J. Fire Sci.*, **1994**, *12*, 257.
8. Green, J. *J. Fire Sci.*, **1994**, *12*, 338.
9. Rajca, A.; Padmakumarm, R.; Smithhisler, D.J.; Desai, S.; Ross, C.R.; Stezowski, J.J. *J. Org. Chem.*, **1994**, *59*, 7701.

10. Morgan, C.R.; Hoffman, H.A.; Cranchelli, *J. Amer. Chem. Soc.*, **1953**, 75, 4602
11. Van Der Made, A.W.; Van Der Made, R.H. *J. Org. Chem.*, **1993**, 58, 1262.
12. Berry, D.J.; Wakefield, B.J. *J. Chem. Soc.*, **1969**, 2342.
13. Koopes, W.M.; Adolph, H.G. *J. Org. Chem.*, **1981**,46, 406.
14. Ninagawa, A.; Matsuda, H. *Macromol. Chem.*, **1979**, 180, 2123.
15. Howell, B.A.; Cho, Y.-J. *Polym. Mater. Sci. Eng.,* **2005**, 93, 963.
16. Howell, B.A.; Cho, Y.-J. *J. Therm. Anal. Cal.*, **2006**, 85, 73.
17. Grand, A.F.; Wilkie, C.A., Eds., *The Fire Retardancy of Polymers*, CRC Press, Boca Ration, FL, **2000**, pp.122-3.
18. Uzibor, J. M.S. thesis, Central Michigan University, **2005**.
19. Howell, B.A.; Uzibor, J. *J. Vinyl Addit. Technol.*, **2006**, 12, 198.
20. Howell, B.A.; Uzibor, J. *J. Therm. Anal. Cal.*, **2006**, 85, 45.
21. Rajca, A.; Padmakumarm, R.; Smithhisler, D.J.; Desai, S.; Ross, C.R.; Stezowski, J.J. *J. Org. Chem.*, **1994**, 59, 7701.
22. Tamaka, K.; Kishigami, S.; Toda, F. *J. Am. Chem. Soc.*, **1990**, 55, 2981.
23. Munoz, A.; Hubert C.; Luche, J.-L. *J. Org. Chem.*,**1996**, 61, 6015.

## Chapter 16

# Development of Multifunctional Flame Retardants for Polymeric Materials

**Bob A. Howell**

**Center for Applications in Polymer Science, Central Michigan University, Mt. Pleasant, MI 48859-0001**
**(bob.a.howell@cmich.edu)**

Effective additives are required to impart a measure of fire retardancy to polymeric materials used in a variety of applications. Traditionally, these have been gas-phase active additives, most commonly organohalogen compounds, or solid-phase active agents, often organophosphorus compounds. Organosphosphorus flame retardants are often very effective but may suffer from a cost disadvantage when compared with their organobromine counterparts. Organohalogen flame retardants are usually quite effective but their use is subject to several environmental concerns. The development of additives that could simultaneously promote both types of fire retardant action could make available flame retardants that are both more cost effective and more environmentally friendly than those currently in use. Several sets of compounds with the potential to display both solid-phase and gas-phase flame retardant activity have been prepared and evaluated.

 

## Introduction

Currently, there are three forces driving the development of new flame retardant agents for polymers. The first is the ever-increasing need to expand the use range of these materials, i.e., to utilize them in applications at the limits permitted by the properties of the materials—at high temperatures or high stress loads, for example. Flame retardants must be found which are effective under these conditions.

Secondly, formulators wish to use as little flame retardant as possible—largely, to reduce the cost of the finished article. Thirdly, there is an ever-increasing demand for more environmentally benign (i.e. "greener") flame retardants. Not withstanding their superior effectiveness, limitations on the use of organohalogen flame retardants are now being sought in many parts of the world (*1-3*). Effective flame retardants that contain lower levels of halogen (or no halogen at all) are needed to be responsive to this concern.

All these considerations require that more effective flame retardants be found. In particular, if compounds which would simultaneously promote char formation at the surface of the polymer *and at the same time* display effective gas-phase disruption of the flame propagating reactions could be found a significant advance in flame retardant technology might be possible. In fact, there is some evidence that such agents might display a synergy of action (i.e., that the effectiveness of either mode of action might be enhanced in the presence of the other) (*4-6*). This would mean that smaller amounts of total additive would be required to impart the desired level of flame retardancy. This should both lower the cost required to achieve an acceptable level of flammability and reduce the potential environmental impact that might accompany decomposition or release of the additive.

A way of achieving enhanced flame retardant activity is to construct compounds which display more than a single mode of action or that are capable of a synergy of flame suppressant properties. Compounds containing both a high level of halogen, in particular bromine, and another element, which may be converted to a cross-linking, char-promoting agent during the combustion process, might display these characteristics. Organic compounds, which promote char formation, are those containing phosphorus, nitrogen, sulfur, phenoxy oxygen, and a few other heteroatoms. Many organophosphorus compounds are very effective solid phase flame retardants particularly when used with oxygen-containing polymers (*7*). Nitrogen compounds such melamine, melamine phosphate, melamine polyphosphate, ethylenediamine phosphate are frequently used as synergists to reinforce the effect of phosphorus compounds (*8-11*). Currently nitrogen-containing flame retardants are mainly used in the textile industry.

Organohaogen compounds, particularly brominated aromatics, continue to be among the most widely used flame retardants (*2,12*). The widespread use of these compounds reflect both their effectiveness and their relatively low cost.

They function by liberating, upon thermal decomposition, hydrogen halide or halogen atoms, which interrupt gas phase flame propagating reactions. The incorporation of bromine into a compound containing phosphorous, nitrogen, or other element known to enhance char formation at the surface of burning polymers might generate flame retardant agents which while retaining the good gas phase activity characteristic of organohalogen compounds would also display good solid phase activity.

## Experimental

### Methods and Instrumentation

All new materials were fully characterized by spectroscopic, thermal and Nuclear magnetic resonance spectra ($^{1}$H and $^{13}$C) were obtained using solutions in deuterochloroform, dimethyl sulfoxide-$d_6$ or toluene-$d_8$ containing tetramethylsilane (TMS) as an internal reference and a Varian Mercury 300 spectrometer. Infrared spectra were obtained using solid solutions (1 %) in anhydrous potassium bromide (as pellets) or thin films between sodium chloride discs using a model 560 Nicolet MAGN-IR spectrophotometer. Melting points were determined by differential scanning calorimetry (DSC) using a TA Instruments Thermal Analyst model 2100 system equipped with a model 2910 MDSC cell. The sample compartment chromatographic methods. was subject to a constant purge of dry nitrogen at 50 mL/min. Thermal degradation characteristics were examined by thermogravimetry using a TA Instruments model 2950 TGA unit interfaced with the Thermal Analyst Control module. The TGA cell was swept with nitrogen at 50 mL/min during degradation runs. The sample size was 5-10 mg in a platinum sample pan. The temperature was ramped at a rate of 5 °C/min.

## Results and Discussion

Several series of compounds with the potential to display unique flame retardant characteristics have been prepared and examined by thermogravimetry.

### Brominated Nitrogen and Phosphorus-Containing Compounds with High Thermal Stability

Phosphonium salts were prepared and examined by treating 1,3,5-tri(bromomethyl)benzene with phosphines. Examination of the decomposition of these materials by themogravimety revealed that several of them have decomposition temperatures of greater than 300 °C (Figure 1) (*13*). These may be useful for flame retarding polymeric materials that undergo combustion at relatively high temperatures. The symmetrical tri(bromomethyl) benzene has also been utilized as a substrate for the preparation of a series of brominated aryl ethers (Scheme 1) (*14*).

The thermal properties of these compounds may be found in Table I. Three of these are solids with good potential as flame retardant agents.

**Table I. Thermal Characteristics of Variously Brominated 1,3,5-(Phenoxymethyl)benzenes.**

| *Compound (Bromophenoxy Substituent)* | *Melting Point (°C, DSC)* | *Decomposition Temperature (°C, TGA)* |
|---|---|---|
| 4-Bromophenoxy | | 255 |
| 2,4-Dibromophenoxy | 152 | 307 |
| 2,4,6-Tribromophenoxy | 277 | 295 |
| Pentabromophenoxy | 290 | 299 |

Two series of compounds which contain high levels of nitrogen as well as halogen have been generated by treating 2,4,6-trichloro-1,3,5-triazine with an appropriate nucleophilic reagent. The first was a series of bromophenoxy derivaties obtained by treating the trichlorotriazine with the corresponding phenoxide (*15*). This is illustrated in Scheme 2 for the preparation of 2,4,6-tri(4-bromophenoxy)-1,3,5-triazine. The thermal characteristics of these compounds are displayed in Table II.

**Table II. Thermal Characteristics of 2,4,6-Tri(bromo$_x$phenoxy)-1,3,5-triazines**

| *Compound (Bromophenoxy Substituent)* | *Melting Point (°C, DSC)* | *Decomposition Temperature (°C, TGA)* |
|---|---|---|
| 4-Bromophenoxy | 235 | 372 |
| 2,4-Dibromophenoxy | 179 | 385 |
| 2,4,6-Tribromophenoxy | 229 | 398 |
| Pentabromophenoxy | 371 | 408 |

$CH_2Br$ $BrCH_2$ $CH_2Br$ + $R_1$ P $R_2$ $R_3$ —DMF, Δ→ $R_1R_2R_3P^+CH_2$ (Br$^-$) trisubstituted benzene phosphonium salt

| $R_1$,$R_2$, $R_3$ | Yield (%) | Decomposition Temperature (°C) |
|---|---|---|
| Triphenyl | 97.3 | 364 |
| *n*-Butyldiphenyl | 91.9 | 297 |
| *n*-Propyldiphenyl | 97.6 | 293 |
| Di-*t*-butylphenyl | 87.0 | 244 |
| Benzyldiphenyl | 96.1 | 344 |
| Methyldiphenyl | 35.0 | 324 |
| Tri-*p*-chlorophenyl | 87.7 | 330 |
| Tri-*t*-butyl | 70.2 | 250 |

*Figure 1. Thermal Decomposition Characteristics of Phosphonium Salts Derived from 1,3,5-Tri(bromomethyl)benzene.*

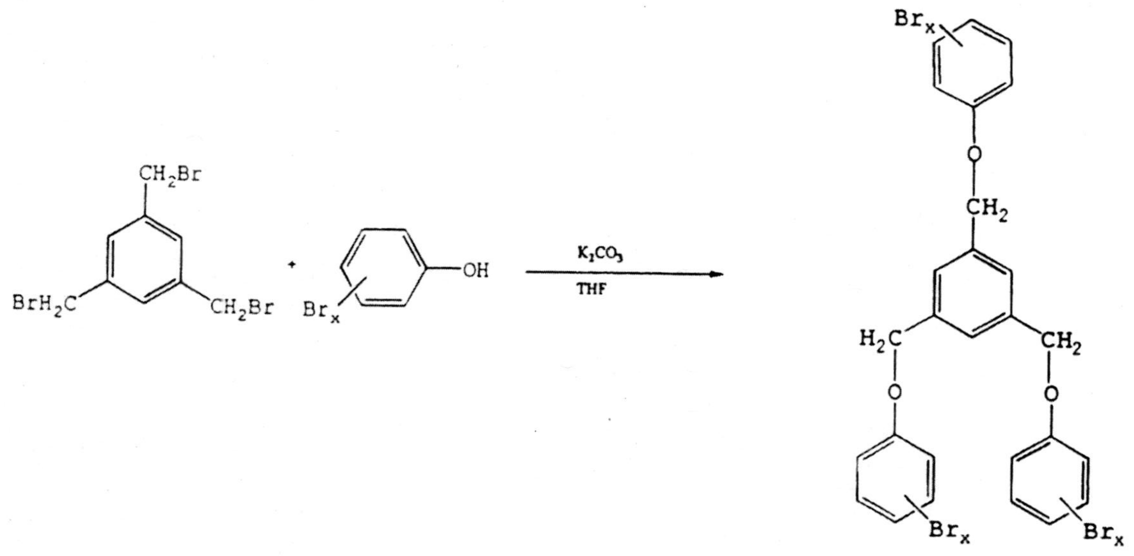

*Scheme 1. Preparation of 1,3,5-Tri[(bromo)$_x$phenoxymeihly]benzene.*

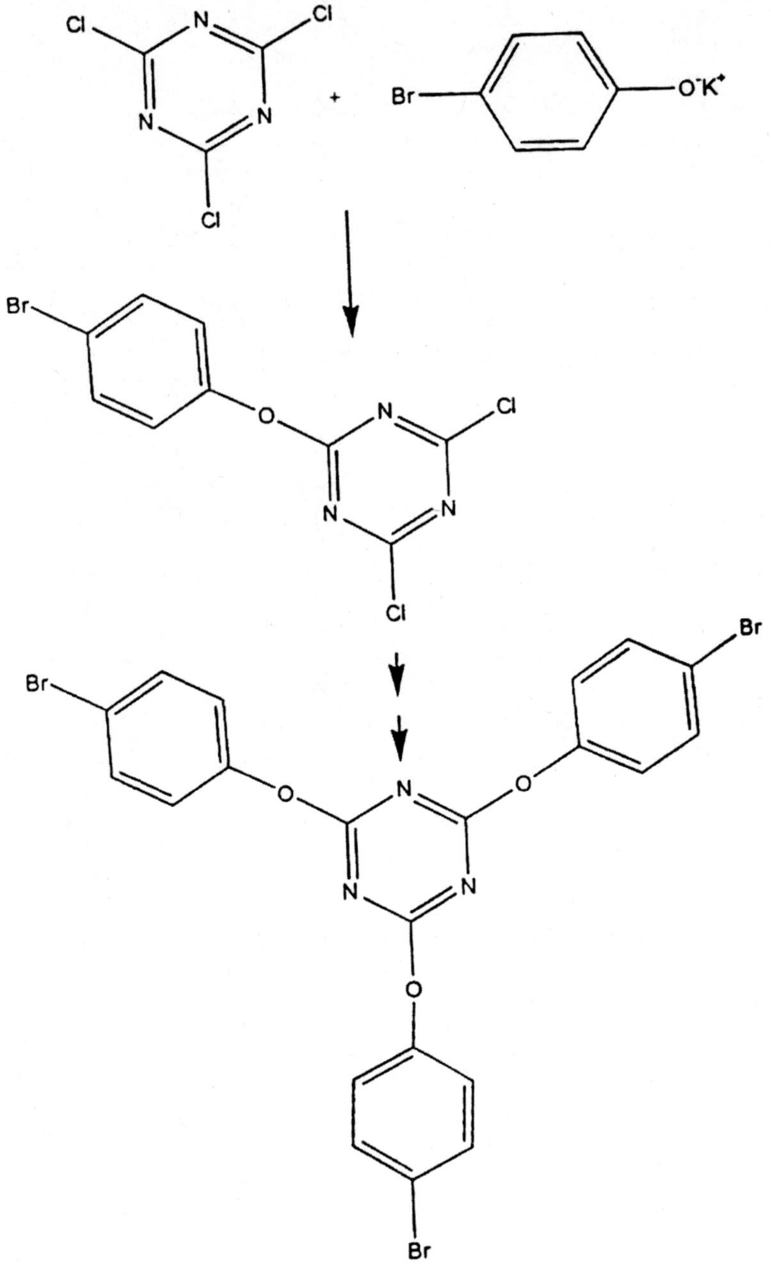

*Scheme 2. Synthesis of 2,4,6-Tri(4-bromophenoxy)-1,3,5-triazine.*

A second series, containing a higher level of nitrogen, was obtained using brominated anilines as the nucleophile. This is illustrated in Scheme 3 (16,17). All these compounds are stable solids and contain high levels of both nitrogen (six nitrogen atoms per molecule) and bromine (three to fifteen bromine atoms per molecule). The thermal degradation characteristics for these compounds were determined by thermogravimetry. A thermogram for the decomposition of 2,4,6-tri(4-bromoanilino)-1,3,5-triazine is shown in Figure 2. The decomposition process consists of three stages. The first centered at 406 °C represents a loss of 39.7% of the initial mass and probably reflects the loss of a 4-bromoanilino group.

The second stage is centered at 545 °C and corresponds to an additional loss of mass (23 % of the initial mass present) and probably reflects the loss of a second bromoanilino group. The final stage of decomposition is centered at 690 °C and corresponds to the loss of the final 4-bromoanilino group. The residue, 24% of the initial mass of the sample, corresponds to the triazine core of the compound. Other (bromoanilino)triazines degrade similarly. The degradation data are collected in Table III.

All these compounds undergo initial decomposition at temperatures greater than 400 °C and degradation occurs in a stepwise manner reflecting, in the main, sequential loss of bromoanilinino groups. As the level of bromine present increases the initial decomposition temperature increases somewhat. These observations suggest that the compounds may be good candidates for high temperature flame retardants and further that both gas-phase action (volatilization of bromine-containing fragments) and solid phase action (formation of a nitrogen-rich residue) will be present.

A series of bromoarylphosphates which could simultaneously promote char formation at the surface of a burning polymer and at the same time display effective gas-phase disruption of flame propagating reactions has been prepared from bromophenols and diethylphosphite (*18,19*).

For example, (4-bromophenyl)diethylphosphate was obtained as a yellowish oil (78.6% yield) from reaction of 4-bromophenol with diethylphosphite. This material was characterized by spectroscopic and thermal methods. The proton NMR spectrum of the phosphate is displayed below in Figure 3. The aromatic region of the spectrum contains an AB pattern characteristic of a *p*-disubstituted phenyl group. Coupling to phosphorus is clearly evident in the quartet at δ 4.2 corresponding to the methylene of the ethyl group. Spectra for other members of the series were equally definite for the structure.

The carbon-13 NMR spectral data are displayed in Table IV. The infrared spectrum contains characteristic absorptions at 3094 $cm^{-1}$ (w) for aromatic $C_{sp2}$-H stretching, 2985 $cm^{-1}$ (s) for aliphatic $C_{sp3}$-H stretching, 1584 $cm^{-1}$ (w) and 1485 $cm^{-1}$ (s) for the phenyl nucleus, 1281 $cm^{-1}$ (s) and 1220 $cm^{-1}$ (s) for aromatic P-O-C stretching, 1098 $cm^{-1}$ (s) and 781 $cm^{-1}$ (s) for O-P-O stretching, 1033 $cm^{-1}$ (s) and 960 $cm^{-1}$ (s) for P-O-C aliphatic stretching vibrations.

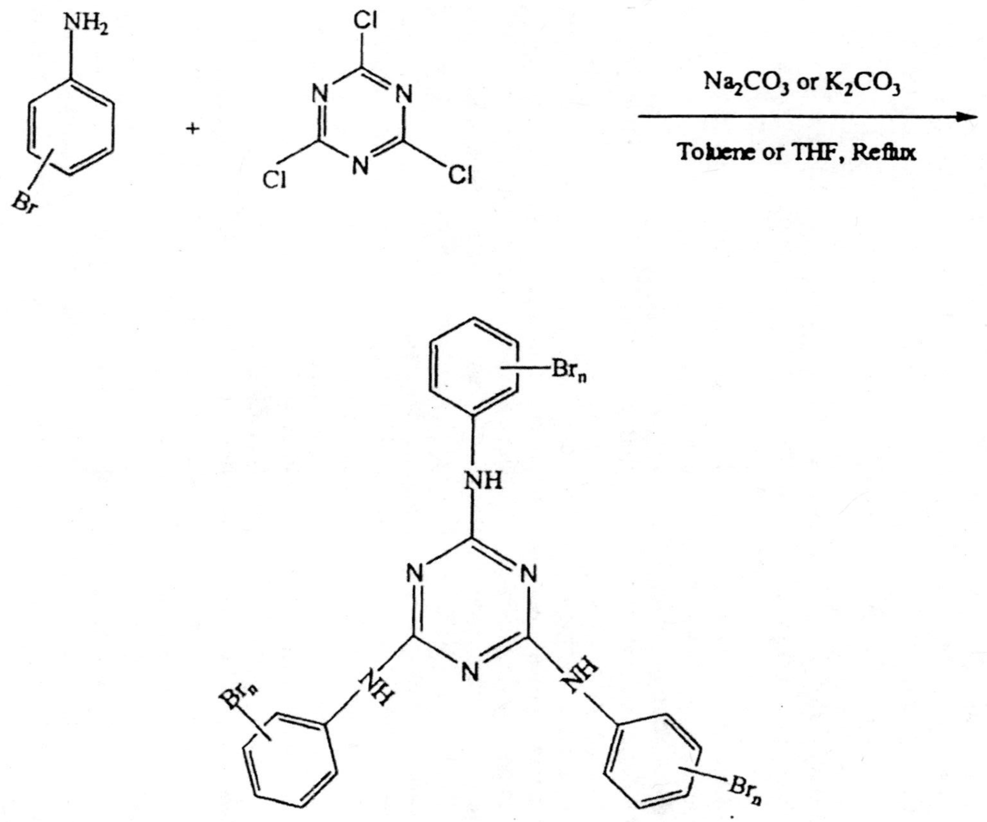

$NH_2$ Br + Cl N N Cl N Cl

$Na_2CO_3$ or $K_2CO_3$

Toluene or THF, Reflux

$Br_n$ NH N N N NH NH $Br_n$ $Br_n$

*Scheme 3. Synthesis of 2,4,6-Tri[(bromo)$_x$anilino]-1,3,5-triazines.*

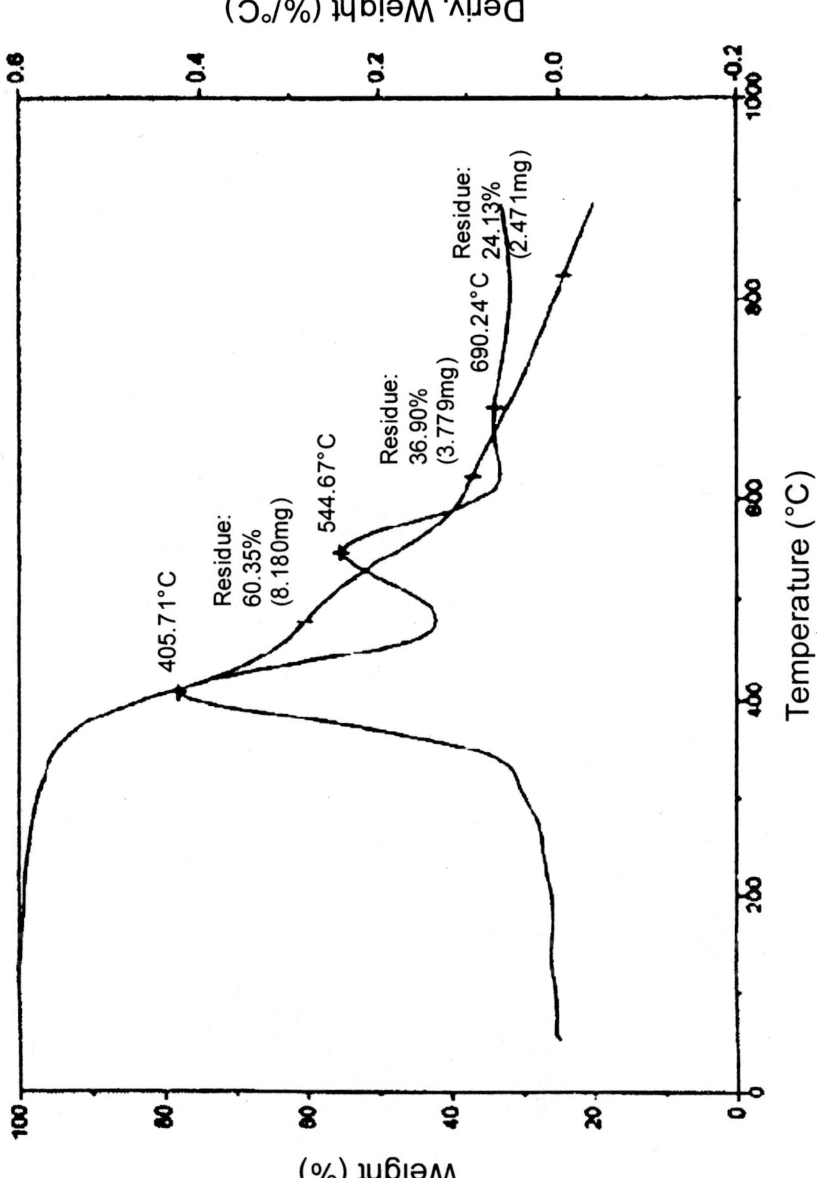


*Figure 2. Thermal Decomposition of 2,4,6- Tri( 4-bromoanilino )-1 ,3,5-triazine*

**Table III. Decomposition Data for 2,4,6- Tri[(bromo )$_x$anilino ]-1,3,5-triazines**

| Compound (number of bromine atoms) | First Stage | | Second Stage | | Third Stage | | |
|---|---|---|---|---|---|---|---|
| | dp[a] (°C,TGA) | Mass Loss[b] (%) | dp[a] (°C,TGA) | Mass Loss[b] (%) | dp[a] (°C,TGA) | Mass Loss[b] (%) | Residue (%) |
| 1 | 405.7 | 37 | 544.7 | 23 | 690.2 | 13 | 24.13 |
| 2 | 418.9 | 42 | 551.3 | 25 | 865.6 | 29 | 3.68 |
| 3 | 447.1 | 38 | 534.7 | 37 | 758.1 | 23 | 2.35 |
| 5 | 430.5 | 42 | 529.8 | 40 | 691.9 | 15 | 2.35 |

[a] Decomposition point taken as the maximum in derivative plot of weight loss versus temperature.

[b] Mass loss as a percentage of the initial sample mass.

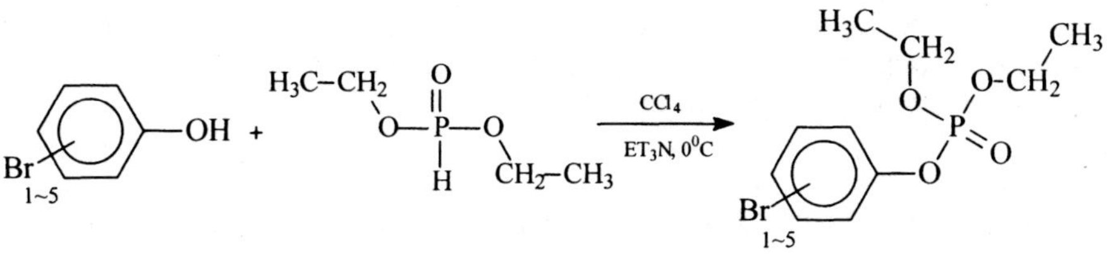

*Scheme 4. Synthesis of [(Bromo)$_x$phenyl]diethylphosphate.*

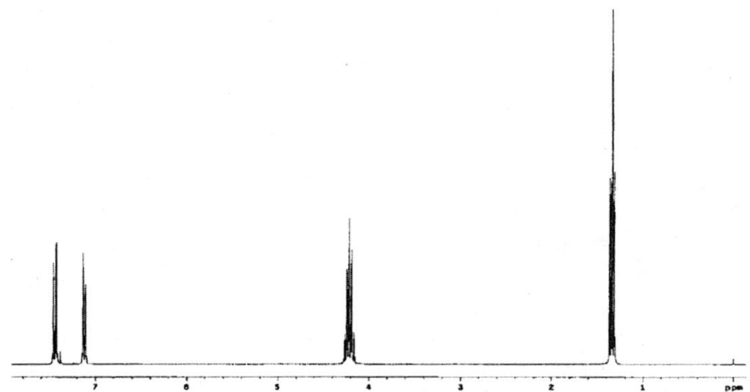


*Figure 3. Proton NMR Spectrum of (4-Bromophenyl)diethylphosphate*

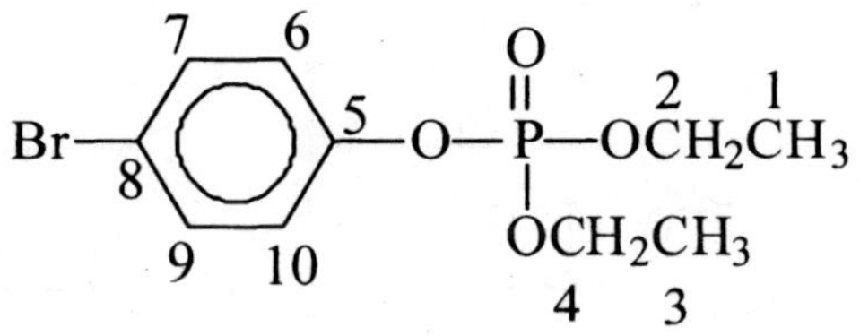

**Table IV. $^{13}$C-NMR Spectral Data for (4-Bromophenyl)diethylphosphate**

| Chemical Shift (δ) | Carbon Atoms |
|---|---|
| 15.9 | $C_1$ and $C_3$ |
| 64.6 | $C_2$ and $C_4$ |
| 117.7 | $C_8$ |
| 121.6 | $C_6$ and $C_{10}$ |
| 132.5 | $C_7$ and $C_9$ |
| 149.5 | $C_5$ |

The molecular weight of this material is 308 g/mol as determined by mass spectrometry and its decomposition onset temperature (TGA) is 231 °C. The mass spectrum of this compound is displayed in Figure 4. From the intensities of the $M^+$ and $[M+2]^+$ peaks, m/z 308 and 310, it is clear that the compound contains a single bromine atom. Other [(bomo)$_n$phenyl]diethylphosphates, (2,4-dibromophenyl)diethylphos-phate, (2,4,6-dribromophenyl)diethylphosphate, (pentabromophenyl)diethyl-phosphate] were characterized in an analogous manner.

The infrared spectra of all the (bromophenyl)diethylphosphates synthesized contain prominent bands for phosphorus-oxygen stretching. These bands represent one of the most widely used characterization features of organophosphorus compounds. Compounds containing a P-O-C unit are characterized by spectra containing intense absorption bonds near 1000 $cm^{-1}$. This absorption appears at somewhat higher frequency (905 – 1060 $cm^{-1}$) for pentavalent phosphorus compounds than for trivalent (850 -1034 $cm^{-1}$) (20). In addition, P-O-C (aromatic) stretching vibrations appear at higher frequency than do those of P-O-C (aliphatic). Both of these absorptions are prominent features in spectra of the compounds described here. These are collected in Table V.

**Table V. Infrared Absorptions for the P-O-C unit in [(Bromo)$_n$phenyl]diethylphosphates**

| Stretching Vibration / Flame Retardant | P-O-C (aromatic) ($cm^{-1}$) | | P-O-C (aliphatic) ($cm^{-1}$) | |
|---|---|---|---|---|
| (4-Bromophenyl)diethylphosphate | 1281 | 1220 | 1033 | 960 |
| (2,4-Dibromophenyl)diethylphosphate | 1289 | 1236 | 1038 | 956 |
| (2,4,6-Tribromophenyl)diethylphosphate | 1290 | 1248 | 1045 | 983 |
| (Pentabromophenyl)diethylphosphate | 1319 | 1292 | 1040 | 970 |

A prominent characteristic in the infrared spectra of organophosphorus compounds are bands arising from O-P-O stretching vibrations. These bands usually appear between 1101 and 780 $cm^{-1}$. These bands for the (bromo-phenyl)diethylphosphates described here are listed in Table VI.

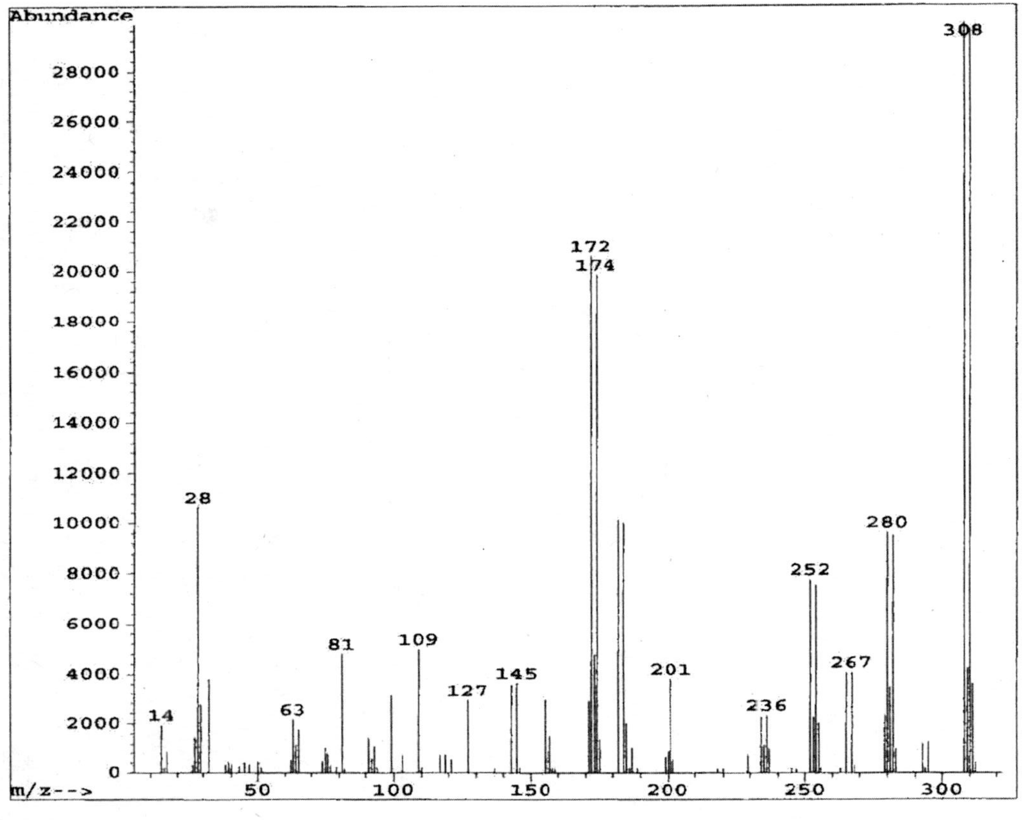


*Figure 4. Mass Spectrum of (4-Bromophenyl)diethylphosphate*

**Table VI. Infrared Absorptions for the O-P-O unit in [(Bromo)$_n$phenyl]diethylphosphates**

| Flame Retardant \ Stretching Vibration | P-O-C (aromatic) ($cm^{-1}$) | P-O-C (aliphatic) ($cm^{-1}$) |
|---|---|---|
| (4-Bromophenyl)diethylphosphate | 1098 | 781 |
| (2,4-Dibromophenyl)diethylphosphate | 1098 | 794 |
| (2,4,6-Tribromophenyl)diethylphosphate | 1102 | 794 |
| (Pentabromophenyl)diethylphosphate | 1088 | 818 |

The thermal degradation characteristics of these compounds are shown in Table VII. It might be noted that all undergo decomposition at approximately 250 °C and that two are crystalline solids.

**Table VII. Thermal Characteristics of [(Bromo)$_x$phenyl]diethylphosphates.**

| *Compound (Bromophenyl Substituent)* | *Melting Point ( °C, DSC)* | *Decomposition Temperature ( °C, TGA)* |
|---|---|---|
| 4-Bromophenyl | | 231 |
| 2,4-Dibromophenyl | | 265 |
| 2,4,6-Tribromophenyl | 73 | 259 |
| Pentabromophenyl | 137 | 242 |

The effectiveness of these compounds as flame retardants for poly(styrene) was evaluated using the UL-94 vertical burn test (ASTM D3801). Plaques for testing were prepared by dissolving the flame retardant in styrene monomer at the appropriate concentration, carrying out partial polymerization (AIBN-initiated) to generate a syrup, and curing the syrup in a mold of dimensions 2x13x127 mm. The results of the flame test are collected in Table VIII. As can be seen all the additives are quite effective as flame retardants for poly(styrene) at the 5% level. Even at a low loading of 3% an appreciable flame retarding influence is observed.

**Table VIII. Impact of the Pressence of Brominated Aryl Phosphates on the Flammability of Poly(styrene).**

| *Flame Retardant* | *Loading (wt.%)* | *UL-94 Rating*[a] |
|---|---|---|
| (4-Bromophenyl)diethylphosphate | 3 | V-1 |
| (2,4-Dibromophenyl)diethylphosphate | | V-1 |
| (2,4,6-Tribromophenyl)diethylphosphate | | V-1 |
| (Pentabromophenyl)diethylphosphate | | V-0 |
| (4-Bromophenyl)diethylphosphate | 5 | V-0 |
| (2,4-Dibromophenyl)diethylphosphate | | V-0 |
| (2,4,6-Tribromophenyl)diethylphosphate | | V-0 |
| (Pentabromophenyl)diethylphosphate | | V-0 |

[a] A UL-94 rating of V-1 indicates that buring stops (without flaming drips) within 60 seconds after two applications of ten seconds each of a flame to a test bar; V-0 indicates that buring stops (without flaming drips) within 10 seconds after two applications of ten seconds each of a flame to a test bar and is the most desirable performance rating.

## Reactive Phosphacyclic Flame Retardants for Poly(styrene)

Poly(styrene)s containing phosphonate moieties in the mainchain have been prepared using dioxaphospholanes as polymerization initiators. In principle, several suitable reactive dioxaphospholanes may be prepared from benzpinacol (21). Some of these are shown below. The synthesis of one of these, 2,4,5,5-pentaphenyl-1,3,2-dioxaphospholane, is outlined in Scheme 5.

The cyclic phospholane can be obtained in good yield as a white crystalline solid with a melting point of 72 °C (DSC) (*22,23*). The proton NMR spectrum contains absorption for aromatic protons and the phosphorus spectrum consists of a singlet at $\delta$ 22.1 ppm, relative to the absorption for 85% aqueous phosphoric acid as external reference. The infrared spectrum of this contains an aromatic absorption band at 1601 $cm^{-1}$ as well as strong bands for O-P-O absorption at 1059 and 759 $cm^{-1}$.

This five-membered ring structure contains a sterically strained carbon-carbon bond that might be expected to undergo thermally induced homolysis at modest temperatures. This may be readily demonstrated using thermogravimetry. The compound readily undergoes thermal decomposition with a maximum rate of degradation at 120 °C and a degradation onset temperature of 70 °C. Because of the ease with which the central carbon-carbon bond of this bond undergoes homolysis, it should function as a polymerization initiator. This is illustrated in Scheme 6 for the polymerization of styrene.

If the compound functions as initiator, each polymer chain generated should contain one phosphorus moiety. Alternatively, it is possible that the phosphorus

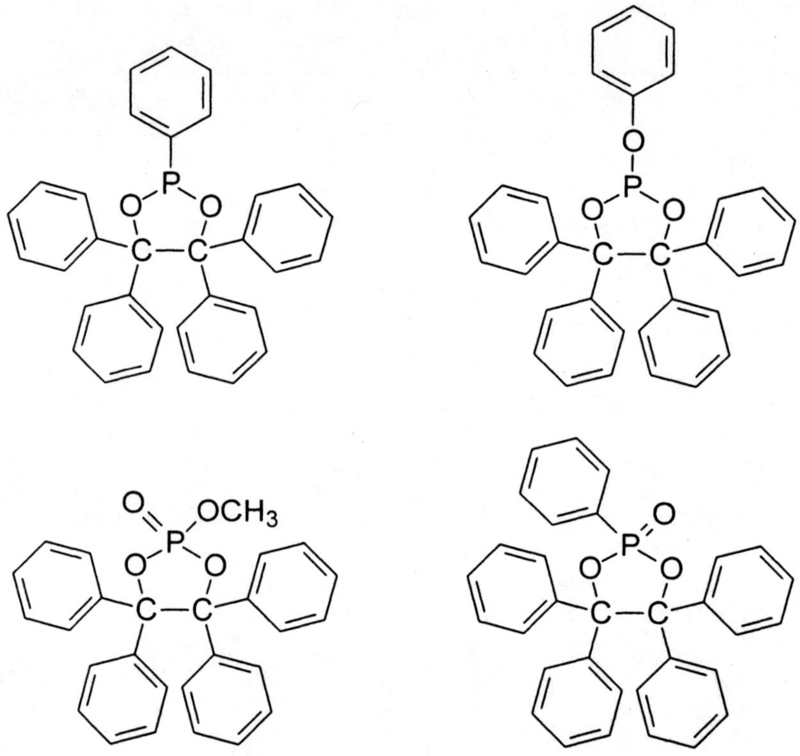

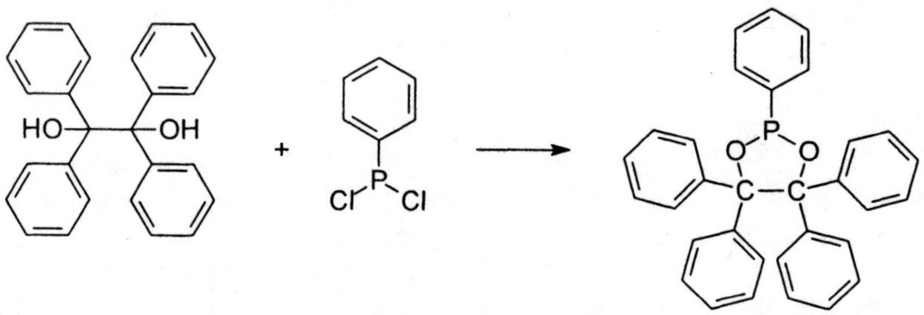

Scheme 5. Synthesis of 2,4,4,5,5-Pentaphenyl-1,3,2-dioxaphospholane.

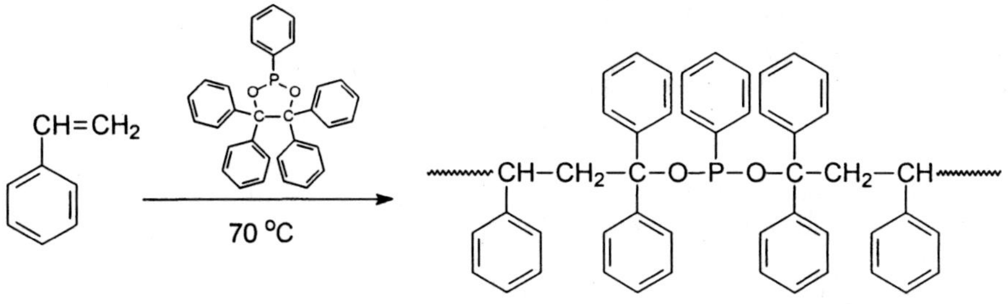

*Scheme 6. Styrene Polymerization Initiated with 2,4,4,5,5-Pentaphenyl-1,3,2-dioxaphospholane.*

compound could also function as a comonomer in the polymerization, i.e., the propagating poly(styryl) radical could add to the strained carbon-carbon bond.

A series of mixtures was prepared in standard polymerization tubes. The first tube contained pure styrene monomer and was used as a control. The remaining tubes contained solutions of 1, 5 and 10% by mass of 2,4,4,5,5-pentaphenyl-1,3,2-dioxaphospholane in styrene monomer, respectively. Tubes were placed in an oil bath maintained at 70 °C. Progress of polymerization was followed by removing aliquots of the mixture as a function of time for viscosity measurement. The aliquots were diluted with benzene (1g of mixture/100 ml of benzene). The relative viscosity increased as a function of time reflecting the extent of polymerization. Size exclusion chromatographic data for the polymers formed are contained in Table IX.

**Table IX. Size Exclusion Chromatographic Characterization of Poly(styrene) Produced Using 2,4,4,5,5-Pentaphenyl-1,3,2-dioxaphospholane as Initiator at 70° C.**

| Initiator Present (Wgt. %) | $M_n$ | $M_w$ | Polydispersity |
|---|---|---|---|
| 0 | N[1] | N[1] | N[1] |
| 1 | 180000 | 365400 | 2.03 |
| 5 | 120000 | 302400 | 2.52 |
| 10 | 90000 | 273600 | 3.04 |

N[1]: no polymer formed.

From the chromatographic data, it is apparent that thermal decomposition of 2,4,4,5,5-pentaphenyl-1,3,2-dioxaphospholane is effective in initiating styrene polymerization. In fact, when no phospholane was present in the monomer, no polymer was produced at 70 °C. In the presence of phospholane, polymer is readily produced under the same conditions. Further, the molecular mass of the polymer formed decreases regularly, as expected, as the concentration of initiator (phospholane) is increased. Analysis of the of polymers by both infrared and phosphorus-31 NMR spectroscopy indicated that phosphorus units were incorporated into the polymer mainchain at a much higher level than can be accounted for by initiation, i.e., the phosphorus moiety probably functions as both initiator and comonomer.

## Thermal Properties of Styrene Polymers Containing Phosphorus Units

The thermal stability of the styrene polymers generated using one, five and ten mass percent 2,4,4,5,5-pentaphenyl-1,3,2-dioxaphospholane as initiator was examined using thermogravimetry. The relevant decomposition data are displayed in Table X.

**Table X. Comparison of the Extrapolated Onset and Maximum Decomposition Temperature of Styrene Polymers Produced in the Presence of One, Five, and Ten Percent 2,4,4,5,5-Pentaphenyl-1,3,2-dioxaphospholane at 70° C.**

| *Phospholane in Polymerization Mixture (Wgt. %)* | *Extrapolated Onset Temperature for Decomposition (°C)* | *Maximum Decomposition Rate Temperature (°C)* |
|---|---|---|
| 0 | 418.1 | 438.9 |
| 1 | 424.2 | 443.0 |
| 5 | 429.3 | 448.3 |
| 10 | 433.2 | 454.1 |

From these results, it would appear that the thermal stability of the polymers containing a phospholane unit is similar to that of poly(styrene), i.e., incorporation of the phospholane into the polymer mainchain does not diminish the thermal stability of poly(styrene).

### Evaluation of Flammability

The flammability of these polymers was evaluated using the UL 94 vertical burn test. Results are presented in Table XI.

The results presented in Table XI suggest that, at the level of incorporation, the phospholane imparts a modest flame retardancy to poly(styrene).

A more recent focus has been the development of dioxaphospholanes or similar compounds which contain bromine in the aromatic groups (*24*). Some examples are shown in Scheme 7.

These compounds retain all the useful reactivity inherent in the phospholane structure but also contain aromatic bromine which should impart some gas-phase flame retardant activity.

**Table XI. Flammability Behavior of Styrene Polymers Generated by Initiation with 2,4,4,5,5-Pentaphenyl-1,3,2-dioxaphospholane at 70 °C.**

| *Observation* | *Level of Phospholane in Polymerization Mixture (Wgt Percent)* | | | |
|---|---|---|---|---|
| | 0% | 1% | 5% | 10% |
| Total flaming combustion for each specimen | 30s | 30s | 30s | 30s |
| Total flaming combustion for all 5 specimens of any set | 250s | 250s | 250s | 80s |
| Flaming and glowing combustion for each specimen after second burner | 60s | 60s | 60s | 60s |
| Cotton ignited by flaming drips from any specimen | YES | YES | YES | NO |
| Glowing or flaming combustion of any specimen to holding clamp | YES | YES | NO | NO |
| **Classification** | | | **V-1** | **<V-1** |

## Conclusions

A variety of approaches have been utilized to generate compounds containing both halogen (bromine) and either phosphorous or nitrogen. These compounds range from phosphonuim bromide salts which display relatively high decomposition temperatures to (bromoaryl)phosphates which undergo decomposition at much more modest temperatures to bromoaryldioxaphospholanes which are sufficiently reactive to be covalently incorporated into vinyl

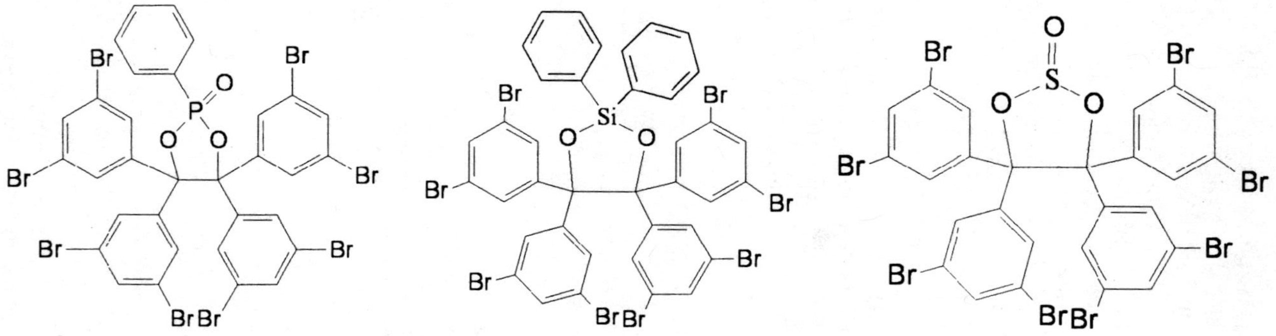

*Scheme 7.*

polymers. These compounds offer the potential to display dual functional behavior as flame retardants, i.e., to maintain the excellent gas phase activity associated with organohalogen compounds while, at the same time, promoting the development of protective char at the solid phase.

## References


1. Weil, E.D. "An Attempt at a Balanced View of the Halogen Controversy", *Proceedings, 10th International Conference on Recent Advances in Flame Retardancy of Polymeric Materials*, Business Communications Company, Norwalk, CT, 1999.
2. Georlette, P.; Simons J.; Costa, L. "Halogen-Containing Fire-Retardant Compounds", in A.F. Grand and C.W. Wilkie, Eds., *Fire Retardancy of Polymeric Materials*, Marcel Dekker, Inc., New York, NY, **2000**, Ch. 8, pp. 245-284.
3. Weil, E.D. *Polym. Degr. Stab.*, **1996**, *54*, 125.
4. Avondo, G.; Vovell, G.C.; Delbourge, R. *Comb. and Flame,* **1987**, *31*, 7.
5. Green, J. *J. Fire Sci.*, **1994**, *12*, 257.
6. Green, J. *J. Fire Sci.*, **1994**, *12*, 338.
7. Weil, E.D.; Mark, H.; Bikales, N.; Overberger, C.C.; Menges, G.; Eds., *Encyclopedia of Polymer Science and Engineering*, 2nd Ed., Wiley-Interscience, New York, NY, **1990**, Vol. 11.
8. Van Der Made, A.W.; Van Der Made, R.H. *J. Org. Chem.,* **1993**, *58*, 1262.
9. Berry, D.J.; Wakefield, B.J. *J. Chem* Soc. C, **1969**, 2342.
10. Koopes, W.M.; Adolph, H.G. *J. Org. Chem.,* **1981**, *46*, 406.
11. Ninagawa, A.; Matsuda, H. *Macromol. Chem.*, **1979**, *180*, 2123.
12. Pettigrew, A. *Halogenated Flame Retardants, Kirk-Othmer Encyclopedia of Chemical Technology,* 41h Edition, **1993**, Vol 10.
13. Howell, B.A.; Johnston, K.F.; Liu, C.; *Phosphorus, Sulfur* and *Silicon,* **1999**, *147*, 100.
14. Howell, B.A.; Zeng, W.; Uhl, F.M. *Polym. Mater. Sci. Eng.*, **2000**, *83*, 100.
15. Howell, B.A.; Uhl, F.M. *Polym. Mater. Sci. Eng.*, **1998**, *77*, 101.
16. Howell, B.A.; Wu, H. *Polym. Mater. Sci. Eng.*, **2000**, *83*, 72.
17. Howell, B.A.; Wu, H. *J. Therm. Anal. Cal.*, **2006**, *83*, 79
18. Howell, B.A.; Cho, Y.-J. *Polym. Mater. Sci. Eng.*, **2005**, *93*, 963.
19. Howell, B.A.; Cho, Y.-J. *J. Therm. Anal. Cal.*, **2006**, *85*, 73.
20. Thomas, L.C. Interpretation of the Infrared Spectra of Organophosphorus Compounds, Adacemic Press, London, 1974, p. 46.
21. Uzibor, J. M.S. Thesis, Central Michigan University, **2005**.
22. Howell, B.A.; Uzibor, J. *J. Vinyl Addit. Technol.*, **2006**, *12*, 198.
23. Howell, B.A.; Uzibor, J. *J. Therm. Anal. Cal.*, **2006**, *85*, 45.
24. Howell, B.A.; Cho, Y.-J. *Polym. Mater. Sci. Eng.*, **2008**, *98*, 361.

# Chapter 17

# Flammability of Fluoropolymers

**Shiow-Ching Lin and Bradley Kent**

**Solvay Solexis Inc., 10 Leonard Lane, West Deptford, NJ 08086**

Fluoropolymers are an important class of polymeric materials useful in many industrial applications, such as wire and cable, semiconductor tools and high performance industrial coatings. The common properties of fluoropolymers include low dielectric constant, good chemical resistance, low surface tension, good mechanical properties and low flammability. Flame resistance of fluoropolymers has been a unique advantage leading to their usefulness in various industrial applications. This article intends to discuss the flammability of commercially available linear fluoropolymers. The modified fluoropolymer resins, such as fluorine containing aromatic polymers, are outside the scope of this discussion.

## Major Thermal Degradation Mechanisms of Fluoropolymers

This chapter deals with the homopolymers and copolymers of fluorinated ethylene monomers due to their long industrial application history. They are linear thermoplastics widely used in cookware, chemical process industry (CPI) and semiconductor manufacturing. They are basically divided into two groups, perfluorinated and partially fluorinated polymers. Perfluorinated linear polymers include polychlorotrifluoroethylene (PCTFE), homo- (PTFE) and co-polymers of tetrafluoroethylene (TFE), such as FEP, PFA and MFA which are copolymers of TFE with hexafluoropropylene, perfluoropropyl vinyl ether and perfluoromethyl vinyl ether, respectively. Partially fluorinated polymers are polyvinyl fluoride (PVF), polyvinylidene fluoride (PVDF), alternating copolymer of ethylene and TFE (ETFE), and alternating copolymer of ethylene and chlorotrifluoroethylene (ECTFE).

 

Figure 1 shows the thermal degradation behavior of fluoropolymers in air characterized by dynamic thermogravimetry (TGA). PTFE has the highest decomposition temperature and degrades quickly to zero char residue. ETFE has a lower decomposition temperature and also produces no char. ECTFE thermally degrades at the earliest stage and, however, produces the most significant intermediate char residue among these fluoropolymers. Similarly, PVDF yields about the same amount of intermediate char as ECTFE. In air, the formed char undergoes slow ablative oxidation at high temperatures. Table I summarizes the experimental results comparing to the theoretically calculated char residue based on the assumption that dehydrohalogenation is the sole degradation mechanism.

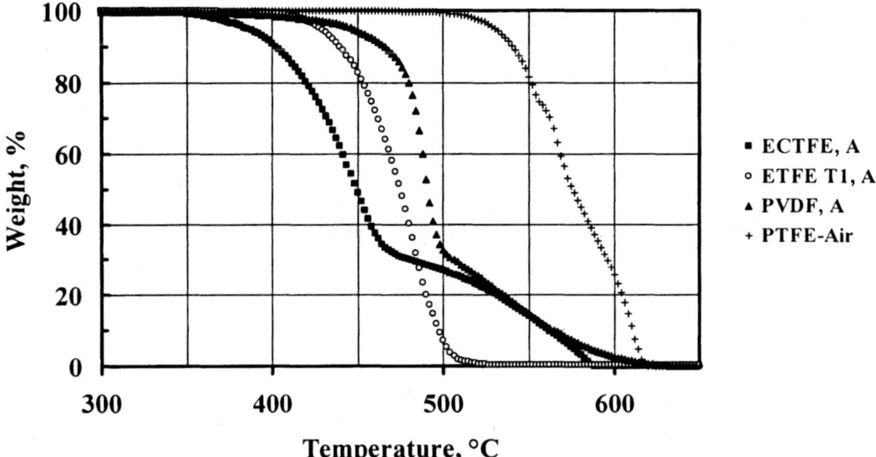


*Figure 1. Dynamic TGA results of fluoropolymers in Air*

It is known that the C-F bond (105.4 kcal/mole) has a much higher bond dissociation energy than the C-Cl bond (78.5 kcal/mole) (*1*) to resist thermal and thermo-oxidative degradations. ECTFE is expected to initially undergo dehydrochlorination catalyzed by oxygen in air attributed to the low bond dissociation energy of C-Cl. The thus formed carbon-carbon unsaturation sequentially accelerates dehydrofluorination and forms conjugated fluorinated polyene or char. Since the dissociation energy of the C-F bond is much higher than the C-C bond (83.1 kcal/mole), the thermal degradation temperature of a perfluorinated polyolefin such as PTFE is high and the degradation mode is therefore in favor of random C-C chain scissions along the polymer backbone. The data in Table I suggests that ECTFE undergoes almost completely thermal dehydrohalogenation. For PTFE, it has been concluded that the depolymerization to TFE monomer is the main thermal decomposition reaction (*2*). Other perfluorinated polymers such as PCTFE, FEP, PFA and MFA have also been found to undergo random scission degradation thermally and thermo-oxidatively (*3*). Similarly ETFE has been confirmed to thermally undergo random chain scission (*4*). As suggested by char yield data, PVDF is also thermally dehydrofluorinated to fluorinated polyene with a small fraction of random chain scission. PVF dehydrofluorinates at a low degradation temperature, followed with backbone cleavage by chain scission around 450 °C *(3)*.

**Table I. TGA parameters of fluoropolymers**

| | ECTFE | PVDF | ETFE | PTFE |
|---|---|---|---|---|
| Char residue after -HX | | | | |
| Theoretical, % | 33 | 41 | 41 | 100 |
| Actual, % | 32~34 | ~32 | ~0 | 0 |
| Decomposition T, °C | | | | |
| In air | | | | |
| Onset | 355 | 415 | 410 | 500 |
| Extrapolation | 415 | 480 | 450 | 535 |
| In N2 | | | | |
| Onset | 425 | 450 | 440 | 500 |
| Extrapolation | 450 | 495 | 495 | 550 |

Onset decomposition temperature is taken by extrapolating weight loss curve to zero % of weight loss.

## Combustion of Fluoropolymers

Polymer combustion is the extreme thermal oxidation of the thermally degraded combustible volatiles and involves vigorous flaming degradation. Three basic elements, heat, fuel and oxygen, are involved in the combustion process. Figure 2 represents the simplified scheme of polymer combustion (*5*). The material is heated initially in air from an external source and, eventually, thermally and thermo-oxidatively degraded to give volatile flammable and non-flammable pyrolytic gases and char. The flammable pyrolytic gas mixes with oxygen in air, and undergoes cracking and vigorous free radical chain reactions. The heat flux generated from this flaming process feeds back to solid polymer to provide a continuous burning cycle. When the flame heat flux is lower than the critical heat flux, the material becomes self-extinguishable.

Burning is the result of the extreme oxidation of combustible pyrolytic volatiles after mixing with oxygen. Char yield at high temperature has been commonly used to interpret the flame resistance of a polymer. However, the amount and the speed of combustible gas release at pyrolytic conditions are actually more meaningful indicators. The release of non-combustible pyrolytic gases during degradation reactions may actually inhibit the flaming process due to the quenching effect of the free radical chain reaction. Hydrogen halides have been concluded to serve as free radical scavengers *(5)*. The elimination of hydrogen halides from ECTFE and PVDF is quite helpful to reduce the flammability of a polymer in the gas phase in addition to the formation of polyene char. The eliminated HCl from thermal degradation of ECTFE provides a higher radical quenching effect than HF.

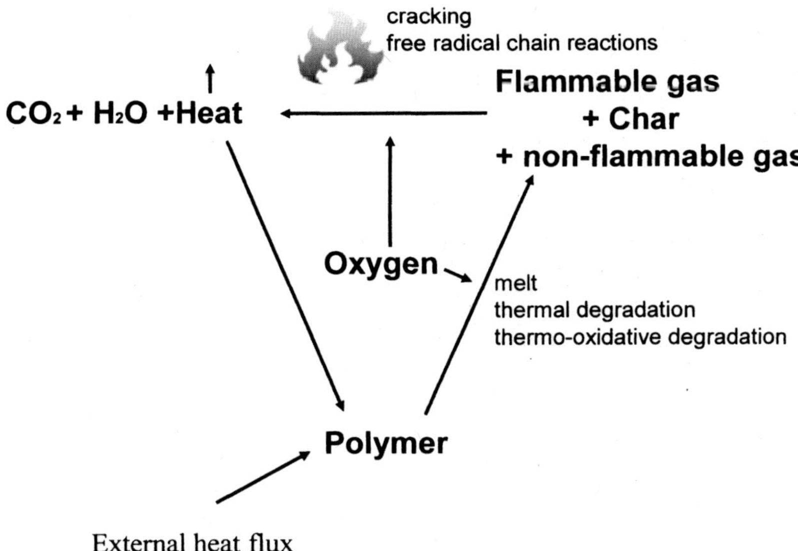


*Figure 2. Burning cycle of a polymer*

## Flammability Characteristics

Perfluorinated polymers undergo almost only chain scission to generate volatile combustible gases. Due to the high content of fluorine atoms, the combustion gases give a low heat release rate, which may be too low to continue the burning cycle. Therefore, these polymers normally have greater than 95% for the limiting oxygen index (LOI) and are flame resistant. Partially fluorinated polymers, as discussed earlier, may undergo chain scission and/or dehydrohalogenation depending on segment structure, and halogen species and their contents. Table II exhibits LOI of partially fluorinated polymers.

PVF undergoes thermal dehydrofluorination at high temperature. Because of high hydrocarbon content, it also produces combustible hydrocarbon pyrolytic gases to give a high heat release rate to yield a low LOI. ETFE, PVDF and ECTFE contain identical mole fractions of $-CH_2-$ groups and have significant LOI differences. As shown in Figure 1, ETFE undergoes thermal random chain scission without char formation different from the dehydrohalogenation possessed by ECTFE. PVDF thermally pyrolyzes by a small fraction of random chain scission and a major dehydrofluorination as discussed earlier.

**Table II. LOI of partially fluorinated polymers *(3)***

| Polymer | LOI |
|---|---|
| Polyvinyl fluoride (PVF) | 22.6 |
| Copolymer of ethylene and tetrafluoroethylene (ETFE) | 30 |
| Polyvinylidene fluoride (PVDF) | 44 |
| Copolymer of ethylene and chlorotrifluoroethylene (ECTFE) | 52 minimum |

To clarify how the degradation mechanism affects the burning behavior of partially fluorinated polymers having similar empirical formulas, cone calorimetry was used to characterize ECTFE and ETFE, including ignition, heat release rate and critical heat flux. Ignition time is the time between the application of a heat source to a material and the onset of self-sustained combustion, with flame observed over the specimen surface for at least 4 seconds. In order for a polymer to be ignitable under a given external heat flux, the polymer has to degrade at a rate to produce enough flammable gas. A higher production rate of flammable gas reflects a lower ignition time. Therefore the ignitability of a polymer depends on the thermo-oxidative degradation mechanism and degradation rate of a polymer at a given heat flux. Figure 3 shows typical heat release diagrams of ECTFE and ETFE plaques at a constant irradiation heat flux of 50 $kW/m^2$. ETFE shows an ignition time around 70 seconds and a high flame heat release in a short period at this particular heat flux. In slightly more than 400 seconds, ETFE is totally consumed by fire. ECTFE has a completely different response to this external heat flux. It shows no heat release up to 220 seconds and exhibits no ignition at the end of 900 seconds. Heat release rate increases slightly after 220 seconds possibly attributed to surface oxidative ablation of the produced carbon char.

The higher heat release rate of ETFE can be used to interpret its lower LOI in comparison to ECTFE. Because of similar empirical formula, ETFE, PVDF and ECTFE are expected to have quite similar heat of combustion. Due to their difference in thermal degradation mechanisms, their heat release rates are drastically different to provide different heat flux feedbacks to support burning process. ETFE quickly degrades to combustible gases to give a fast rate of heat release. ECTFE only undergoes slow ablative oxidation of char after dehydrohalogenation. A small amount of PVDF thermally degrades to combustible gases which contribute to a moderate LOI.

Table III provides additional ignition parameters of fluoropolymers as additional reference. Pilot- or auto-ignition temperature is the temperature at which a material ignites in a normal atmosphere with or without an external source of ignition, respectively. Both parameters are also used to indicate the ignitability of a combustible material. Based on the results, ECTFE exhibits its high ignition resistance similar to the conclusion from cone calorimetric results.

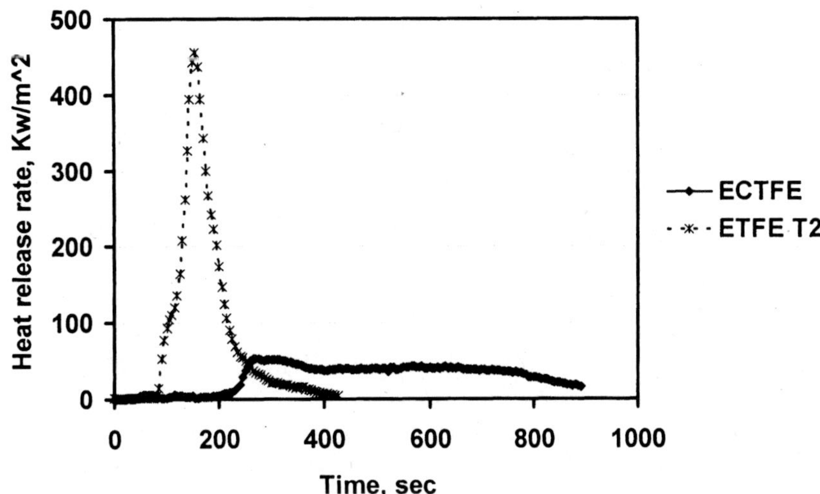


*Figure 3. Typical cone calorimetric diagram of ECTFE and ETFE plaques at an irradiation heat flux of 50 kW/m².*

**Table III. Ignition temperatures of fluoropolymers**

| Polymer | Pilot-Ignition T., °C | Auto-Ignition T, °C |
|---|---|---|
| PTFE | | 580[6] |
| | 560[7] | 565[8] |
| | | 620~675[9] |
| ETFE | 470[7] | 555[8] |
| | | 550[9] |
| PVDF | | 630[8] |
| ECTFE | | 655[8] |

Even though ECTFE undergoes dehydrohalogenation as the preferred degradation mechanism to contribute ignition resistance and a low LOI, the selectivity of degradation mechanism may be changed or the pyrolytic char may catch fire when the external heat flux is intensified to exceed a certain extent. To further analyze the ignition resistance of fluoropolymers, the critical heat fluxes of ECTFE and ETFE have been characterized. The critical heat flux is the minimum external heat flux to cause a material to be ignited by a pilot. Table IV shows the critical heat fluxes of ETFE and ECTFE along with the results of PVDF and PTFE obtained from literature (*10*). Due to high char formation and early decomposition, a high critical heat flux of ECTFE is expected to lead to its high ignition resistance.

**Table IV. Critical heat fluxes of fluoropolymers**

| Powder Coating | Critical Heat Flux, $kW/m^2$ |
|---|---|
| ECTFE | 74 |
| ETFE | 14.5~16.5 |
| PVDF | 35 |
| PTFE | 43 |

## Coating Applications

Other than the intrinsic fire retardancy of fluoropolymers, excellent chemical resistance is an additional advantage over other polymeric products. Perfluorinated polymers have long been used as non-stick coatings for cookware as a result of their release properties. Taking advantages of chemical resistance, corrosion protection, easy fabrication and fire resistance, ECTFE, ETFE and PVDF have been the most suitable fluoropolymers for coating applications. One particular coating usage with high application volume is exhaust duct coating for semiconductor fabs. The duct is made of stainless steel to provide mechanical strength. Its inner surface is protected from corrosive exhaust gas such as HF by an organic coating such as ECTFE and ETFE. Both coatings can be easily applied by electrostatic powder spray process.

It is understandable that corrosion resistance of a coating can be further enhanced by building up its thickness to prevent chemicals from penetration. However, it is also very important to have the coating material resist fire and its propagation, when coating thickness is increased. Factory Mutual has established a testing standard (*11*) known as FM 4922 as a guideline. Judged from the thermal pyrolytic mechanism, ECTFE coating forms char early from dehydrohalogenation and is anticipated to resist fire propagation even if the coating thickness is increased. ETFE undergoes random chain scission to produce volatile flammable fragments. The concern of fire resistance and propagation of ETFE coating may increase with coating thickness.

Figure 4 shows the Cone Calorimetric testing results of ECTFE and ETFE coatings with varied coating thickness. The testing was carried out at three different heat fluxes, 30, 50 and 70 KW/m$^2$. As anticipated, the peak heat release rate of ECTFE at a given external heat flux is much less than that of ETFE, when the coating thickness is kept the same. As well, the fire resistance of ECTFE coating is much independent of coating thickness when the same external heat flux is applied. The results imply that ECTFE coating thickness can be increased to further enhance corrosion protection to metal substrate without the concern of fire protection reduction. Practically, this has been demonstrated through FM 4922 tests.

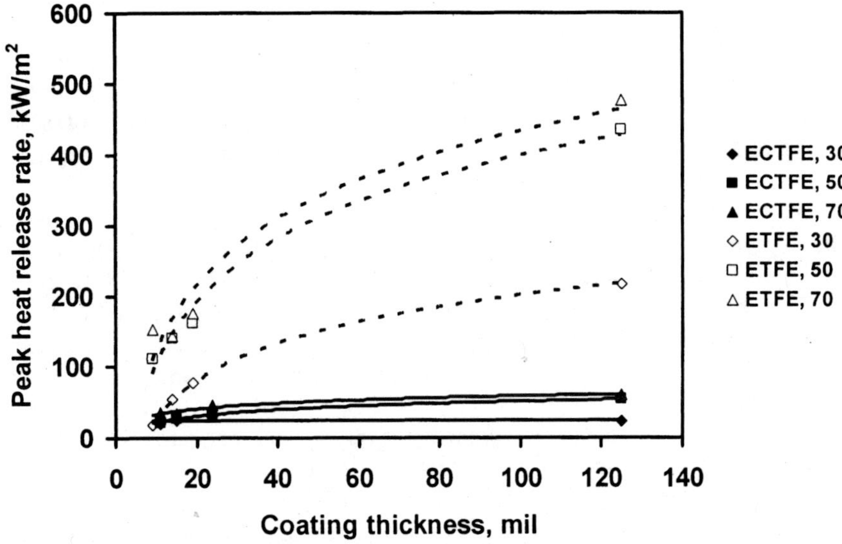


*Figure 4. Coating thickness effect on fire behavior of ECTFE and ETFE coatings.*

## Melt Processed Applications

Fluoropolymers also find use in melt processed applications, based on extrusion or molding processes. The largest use is in the wire and cable industry, which includes coated copper wires, jacketed fiber optic cables, and plastic conduits and pipes for use in the plenum space of buildings. Building codes in the United States have called for materials with low smoke generation and flame propagation characteristics for many years and similar approaches are being adoped in other countries.

Fluoropolymers are particularly used for the desirable combination of excellent electrical properties, as measured by the dielectric constant and dissipation factor, and excellent flame and smoke characteristics. Test protocols have been established, such as the NFPA 262 tunnel test in which coated cables are burned to determine peak and average smoke release along with flame propagation distance (*12*). Cables constructed with fluoropolymer insulation on the primary conductors and/or the outer jackets are more likely to pass this test than other non-fluorinated polymers. For certain cable constructions, especially those containing fillers, spacers, and other internal elements, additives may be used to reduce smoke generation. For example, in fiber optic cable constructions, natural PVDF is often not able to provide sufficiently low smoke generation, so additives such as molybdates (*13*) or tungstates (*14*) can be used to increase the effective LOI from about 44% up to a range of 65 to 100%.

Besides copper and fiber optic communication cable insulation and coatings, fluoropolymers are also used for plenum rated conduits and piping to provide either additional protection for cables in the plenum space or a fluid transportation system for cooling fluids or waste streams. Due to mechanical property requirements, PVDF formulated for low smoke generation is often used for these applications. Plenum rated conduits must pass the flammability and other test requirements specified in UL 2024 (*15*). Fluoropolymers used in other building components must pass the flame and smoke tests specified in UL 723 (*16*), which have been passed by PVDF formulated for low smoke generation.

Melt-processed fluoropolymers are useful in semiconductor manufacturing because of their high purity of certain types (low extractibles into the fluids used for semiconductor processes). Flammability properties are also important in applications such as plastic wet benches used in semiconductor fabs owing to large financial losses from fires sustained in certain facilities. Factory Mutual developed a test protocol known as FM 4910 (*17*) to validate materials for wet bench construction. This test comprises a vertical burn of a large sheet of plastic in which smoke generation and flame propagation distance are measured. Both PVDF and ECTFE have passed this test and various equipment fabricators use these polymers to produce wet benches, sinks, and other system components to meet the strict flammability requirements of the industry. Because of the high purity requirement for low extractibles, only natural PVDF and ECTFE can be used, while natural ETFE cannot pass the FM 4910 test due to its flammability characteristics described earlier.

## Acknowledgement

The authors thank Solvay Solexis to support the effort to complete all tasks to arrive at meaningful conclusions.

## References

1. L. Pauling, *Nature of the Chemical Bond, 3rd Ed.*, Cornell University Press, Ithaca, New York, **1960**.
2. Kerbow, D. L., "*Ethylene/Tetrafluoroethylene Copolymer Resins*" in Modern Fluoropolymers, Edited by J. Scheirs, John Wiley & Son, **1997**, p.301.
3. Scheirs, J., "*Structure/Property Considerations for Fluoropolymers and Fluoroelastomers to Avoid In-Service Failure*", In Modern Fluoropolymers, Edited by J. Scheirs, John Wiley & Son, **1977**, P.1.
4. Wall, L. A., "Thermal Decomposition of Fluoropolymers" in Fluoropolymers, Edited by Wall, L. A., Wiley, **1972**, 381
5. P. C. Warren, "Stabilization against Burning" in Polymer Stabilization, Edited by W. L. Hawkins, Wiley-Interscience, **1972**, p.313.
6. J. Troitzsch, *International Plastics Flammability Handbook*, Hanser-Gardener, Munch, 1990.
7. L. Shea, Semiconductor Fabtech, **1999**, *9*, 155.
8. SwRI Contract Report. Unpublished results.
9. DuPont brochure. DuPont website.
10. Appendix I, "*Fire Characteristics of Hall B Fuels*", NFPA Fire Protection Handbook, **1997**.
11. FM Global Technology LLC, "*Approval Standard for Fume Exhaust Ducts or Fume and Smoke Exhaust Ducts – Class Number 4922*", **2002**.
12. NFPA 262 "*Standard Method of Test for Flame Travel and Smoke of Wires and Cables for Use in Air-Handling Spaces*". National Fire Protection Association, Quincy, MA. **2007**.
13. US 4,898,906, "*Compositions based on vinylidene fluoride polymers which have a reduced smoke-generating capacity*" to Solvay & Cie. February 6, **1990**.
14. US 5,919,852, "*Low smoke, flame retardant, VDF polymers*" to Elf Atochem. July 6, **1999**.
15. UL 2024 "*Optical Fiber Cable Raceway*", Underwriters Laboratories, Northbrook, IL. **2004**
16. UL 723 "*Surface Burning Characteristics of Building Materials*", Underwriters Laboratories, Northbrook, IL. **1998**.
17. FM Global Technology LLC, "*Test Standard for FM Approvals Cleanroom Materials Flammability Test Protocol – Class Number 4910*", **2006**.

## Chapter 18

# Up-Cycling of Rubber-Plastic Waste to Fire-Retardant Systems

**P. Anna, Sz. Matkó, I. Répási, Gy. Bertalan, and Gy. Marosi***

**Department of Organic Chemistry and Technology, Budapest University of Technology and Economics, H–1111 Budapest, Műegyetem rkp. 3. Hungary**
***Corresponding author: gmarosi@mail.bme.hu**

Various flame retardant, recycled ground rubber containing compounds were prepared in a polyethylene-ethylene vinyl acetate blend matrix with the help of a reactive surfactant. The beneficial effect of ground rubber was demonstrated on the flame retardancy, noise damping, and environmental impact of the compounds.

"Waste management will be the greatest challenge for the global economy in the 21$^{st}$ Century."(*1*) The steadily increasing amount of waste are represented in large percentage by polymers, mainly by polyolefins (*2, 3*) and tire rubber (*4*), which do not decompose easily. EU directives prescribe the compulsory proportion of polymers to be recycled in form of products (*5*). It seems to be a practical solution to recycle these materials in composite form where the matrices are thermoplastic polyolefins and the filler is cross-linked ground tire rubber. The mechanical properties of these compounds, however, are moderate without special interfacial treatments, hence their application is limited (*6, 7*). Up-cycling of these type of compounds providing them with one or more special advantageous characteristics (such as flame retardancy, noise dumping, heat insulation etc.) could offer an acceptable solution. Special characteristics allow special applications, where the high mechanical strength is not a critical requirement (*8*). In favorable composites the special characteristics should be promoted by the recycled polymer, e.g. the flame retardancy by recycled polyurethane or noise damping by recycled ground tire rubber. The application of virgin thermoplastic polyurethane in intumescent compounds as charring components is known (*9,10,11*), however, the effect of a recycled, cross-linked polyurethane has not yet been investigated. The effect of ground tire rubber on intumescent flame retardancy and on noise damping has not yet been studied either.

Considering the described aspects, flame retarded composites were prepared consisting of thermoplastic matrices and cross-linked waste polymers, such as ground rubber and polyurethane. Intumescent system and metal hydroxide were applied as flame retardant additives. The beneficial effect of the recycled rubber was demonstrated on flame retardancy, on vibration damping and on environmental impact during forced burning of the compounds.

## Experimental

### Materials

Low density polyethylene (PE): Tipolen AE 2016 (TVK Rt. Hungary). Ethylene vinyl acetate copolymer (EVA): Ibucell K-100 (H. B. Fuller, Germany). Recycled ground rubber (R): 200-600 μm fraction of rubber powder prepared by an ultra high pressure water jet cutting process (Regum Ltd. Hungary). Ammonium polyphosphate (APP): Exolit AP 422 (Clariant Ltd. Germany). Aluminium trihydrate (ATH): Alolt DLS 60 (MAL ZRt., Hungary), a precipitated product with d50 = 1.6 μm. Recycled polyurethane foam (PU): Ground of Freon freed waste polyurethane foam recovered from fridges (Amatech-Polycel Inc. Germany). Reactive surfactant (RS): prepared from maleic anhydride (MA) and conjugated linoleic acid (CLA) (Reaction 1) in an in situ Diels-Alder addition reaction during the compounding process as described in the literature (*12*). Foaming agent: Expancel 098 MB 120 (Akzo Nobel) a foaming master batch. Cross-linking agent: Luperox F90P (Arkema Ink.), a dilkil peroxide. Precipitated barium sulfate (B) (Biotech Hungaria Ltd.).

COOH + OC O CO ⟶

⟶ COOH OC O CO

*Reaction 1. Formation of RS from MA and CLA in a Diels-Alder reaction*

### Preparation

The components were homogenized in a Brabender Plasti Corder PL 2000 apparatus, at 150 °C, with a rotor speed of 50 rpm, in 10 min. The blends were compressed in a Collin P 200 E laboratory compression moulding machine at 150 °C, with 10 bar, for 10 min. The foamed structure was prepared in three steps. 1) Mixing the polymer ingredients and 5% foaming master batch and 2% cross linking peroxide at 110 °C. 2) Sheet formation in a press at 110 °C. 3) Foaming between open lids of press at 150 °C. Fracture surface was prepared by breaking the sample previously cooled in liquid nitrogen.

### Characterization

Differential Scanning Calorimetry (DSC): using SETARAM DSC 92 equipment, 10 mg sample in Al crucible, 10 °C/min heating rate, air atmosphere. Scanning Electron Microsopy (SEM): JEOL JSM-6380LA at high vacuum, accelerant voltage 20 kV. UL 94 burning test: performed according to ASTM D 635/77 standard. Oxygen Index (LOI) test: performed according to the ASTM D2863-06a standard. Cone calorimetry: performed in Mass Loss Cone Calorimeter (Fire Testing Technology Ltd.) according to the ISO 13927:2001 standard (heat release detection by thermocouples instead of oxygen consumption) with radiation of 50 $kW/m^2$, without direct ignition. Dynamic Mechanical Analysis (DMA): using AR 2000 Rheometer (TA Instruments Inc.).Test-pice 2x10x60 $mm^3$, clamping length 40 mm, oscillation stress 10 exp 5 Pa, normal force 5 N. Fourier Transform Infrared spectroscopic (FT-IR) gas analysis: the gas evolving from the samples being exposed to 50 $kW/m^2$ radiation in the mass loss calorimeter was analysed by a Bruker Tensor 37 FTIR as described earlier (*13*).

## Results and Discussion

### Structure and Flammability of Ground Rubber Containing Compound

Compositions of reference, flame retardant and noise damping compounds are given in the Table I.

The SEM microscopic images on Figure 1 and 2 show the fracture surface of PE/R and PE/E/R compounds. On the image of PE/R compound an evident detachment of the components can be observed occurring during the sample preparation. This detachment indicates a week interaction between the rubber and the matrix. In the presence of the elastomer (E) component consisting of

**Table I. Composition of compounds**

| *Symbols of compounds* | *Components* | | | | | | |
|---|---|---|---|---|---|---|---|
| | PE | E EVA+RS | R | PU | APP | ATH | B |
| PE/E | 70 | 30+0 | | | | | |
| PE/R | 85 | | 15 | | | | |
| PE/E/R | 59 | 25+1 | 15 | | | | |
| PE/E/PU-APP | 13.1 | 35.9+1 | 25 | 25 | | | |
| PE/E/R/PU-APP | 8.3 | 27.7+1 | 15 | 20.6 | 32.4 | | |
| PE/E/B | 13.7 | 18.3+1 | | | | | 64.3 |
| PE/E/ATH/B | 11.7 | 18.3+1 | | | | 43.3 | 25.7 |
| PE/E/R/ATH/B | 11.7 | 18.3+1 | 5 | | | 38.3 | 25.7 |

EVA and maleated linoleic acid, the detachment takes place only partly indicating better interfacial adhesion (Figure 2). This matrix was used in further studies.

Before the preparation of flame retardant compositions, the thermal behavior of rubber-containing compound was investigated by DSC measurements. The DSC curve of the PE/E/R compound is given in Figure 3, where the curve of rubber as reference can be seen as well. The rubber in itself shows an early decomposition, starting at about 150 °C. If rubber is incorporated in the complex matrix its thermal stability is increased, as it is shown on the curve of the PE/E/R compound.

The presence of rubber was expected to reduce the burning rate of its compounds. However, an opposite effect can be seen in Figure 4, where the horizontal burning rates of PE matrix, PE/R and PR/E/R compounds are given. The ground rubber increase the burning rate considerably.

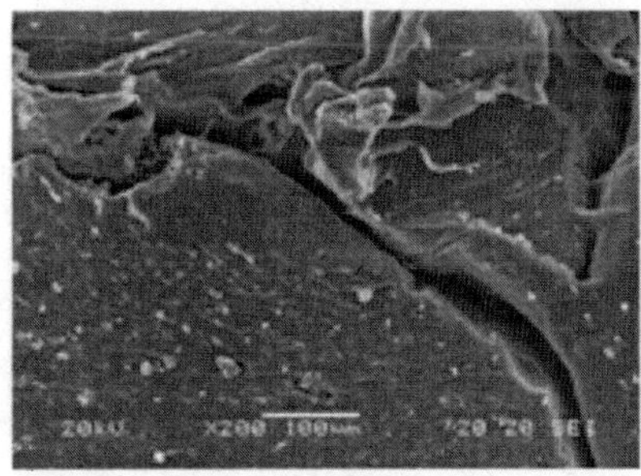

*Figure 1. Fracture surface of PE/R compound*

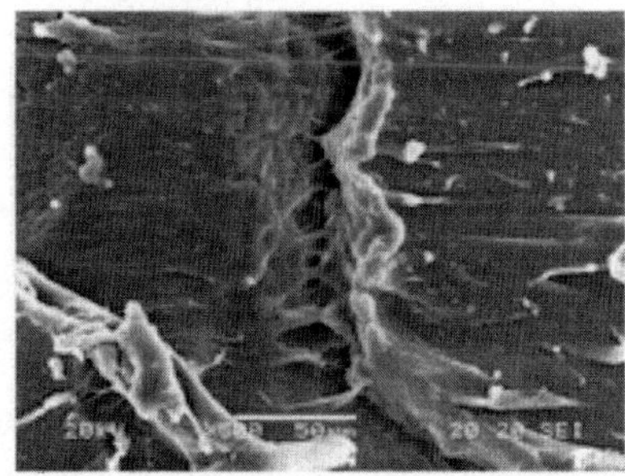

*Figure 2. Fracture surface of PE/E/R compound*

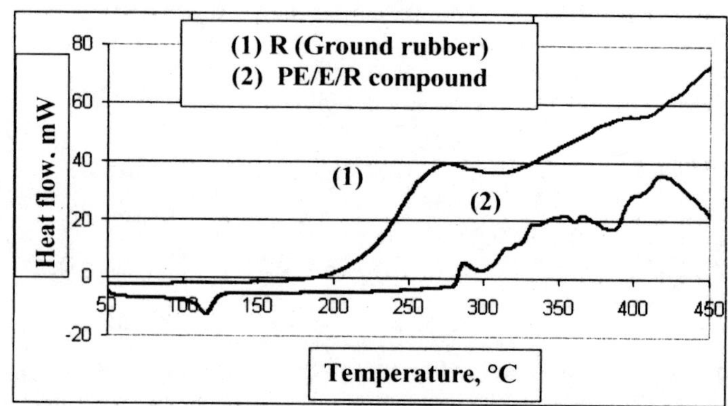


*Figure 3. DSC curves of ground rubber and compound containing it*

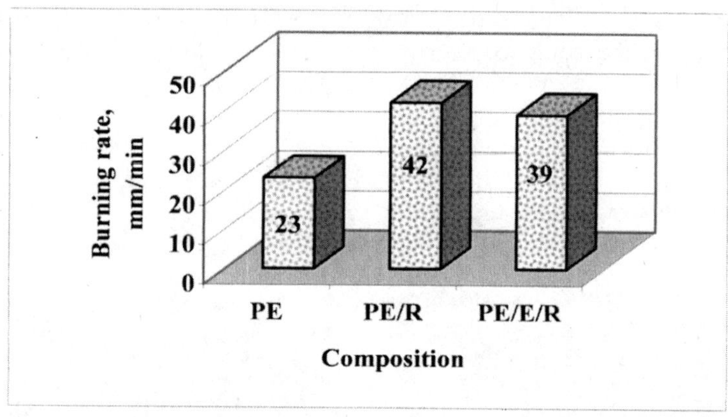


*Figure 4. Horizontal burning rate of PE matrix and PE/R, PE/E/R compounds*

## Influence of Ground Rubber on the Flammability of Intumescent Flame-Retardant Compound

Two types of rubber containing flame-retardant compounds were formulated. One contains an intumescent system consisting of APP as the acid source and recycled PU as the charring and blowing component, while the other contains a metal hydroxide, aluminium trihydrate (ATH). The flammability characteristics of the compounds are shown in Table II. The intumescent additive system provides excellent flame retardancy with high LOI value and V-0 UL 94 grade compared to the flammability of the PE/E matrix component. The incorporation of rubber decreased the LOI value, but did not affect the UL 94 grade.

**Table II. Flammability characteristics of compounds**

| *Composition* | *LOI (vol%)* | *UL 94* | *Comment* |
|---|---|---|---|
| PE/E | 18 | HB | - |
| PE/E/R | 17 | HB | Flexible, rubber smell |
| PE/E /PU-APP | 32 | V-0 | Flexibile |
| PE/E/R/PU-APP | 30 | V-0 | No smell, flexible |
| PE/E/B | 19 | HB | Flexibile |
| PE/E/ /ATH/B | 31 | V-1 | Flexibile |
| PE/E/R/ATH/B | 30 | V-0 | Flexible |

These results do not show any positive effect of the rubber in respect of flammability, but in fact such effect exists. If we characterize the flammability with the more sensitive method of cone calorimetry, a considerable affect can be detected, as shown in Figure 5. The recycled PU/APP additive strongly reduces the heat release compared to the polymer matrix. The introduction of rubber reduces the peak of heat release further and shifts the ignition (TTI) to longer time. The positive effect of rubber on the flame retardancy can be explained by increased char formation, which could have been detected by cone calorimeter. The residual mass of the discussed compounds is shown in Figure 6. The rubber containing compound has the highest residual mass nearly in the whole time range of the test, consequently the rubber promotes the char formation.

A particular additional positive effect of the applied intumescent system is the reduction of rubbery smell of the compound. From the PE/E/R/PU-APP compound a foamed structure could be prepared by simultaneous addition of foaming and cross-linking agents during a low temperature compounding process followed by a higher temperature pressing. The foamed structure is shown in Figure 7.

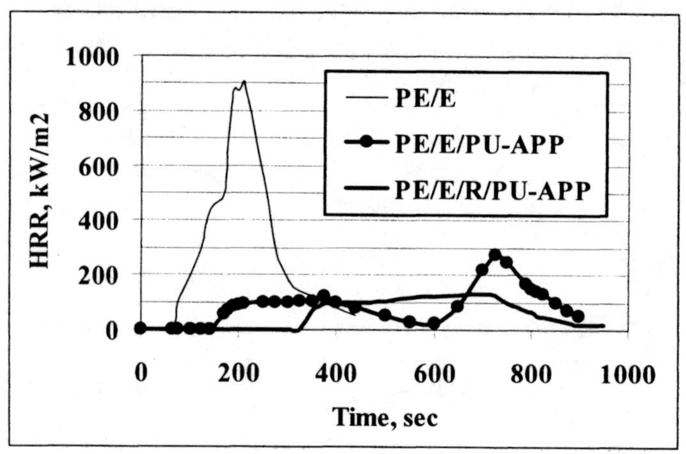


*Figure 5. Mass loss calorimetric Heat Release Rate (HRR) of compounds*

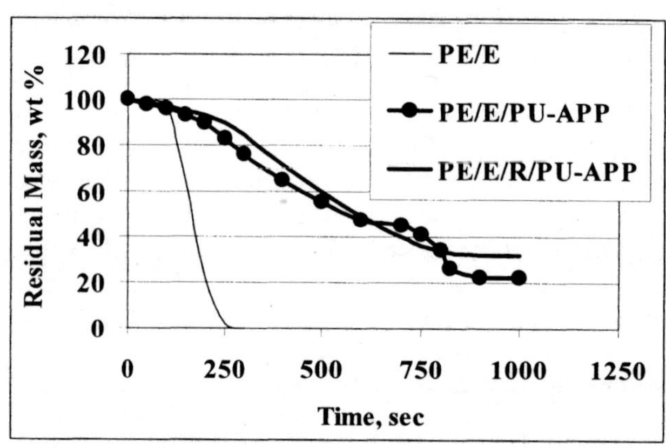


*Figure 6. Residual mass versus timeime in mass loss calorimeter*

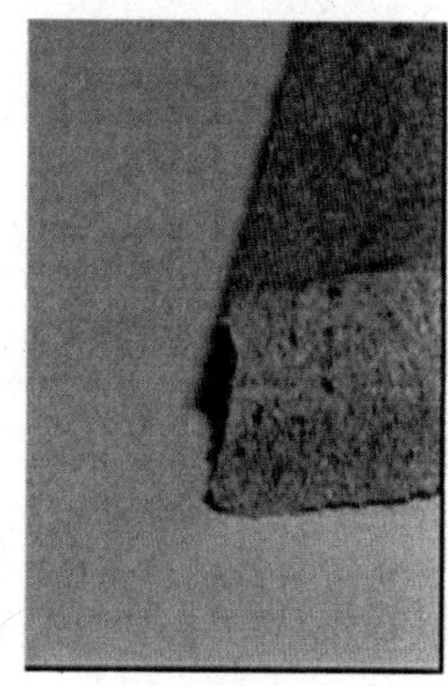

*Figure 7. Foamed PE/E/R/PU-APP system*

## Influence of Ground Rubber on Vibration Damping of Aluminium Trihydrate Containing Flame Retardant Compound

ATH containing flame retardant compounds prepared for vibration damping product contained barium sulfate additive to enhance the damping. The damping was characterized by dynamic mechanical analysis (DMA) where the complex torsion modulus (G*) was measured as a function of deformation frequency (ω), which are in correlation with the storage (G′) and loss (G′′) modulus as described in Scheme 1 (*14*),

Loss modulus G′′ is defined as proportional to the energy dissipated during one loading cycle. It represents, for example, energy lost as heat, and is a measure of vibration energy that has been converted during vibration and that cannot be recovered. Thus, the higher the loss modulus, the higher the vibration damping.

$$|G^*| = \sqrt{(G'(\omega))^2+(G''(\omega))^2}$$
$$G'(\omega) = |G^*| \cos\delta$$
$$G''(\omega) = |G^*| \sin\delta$$
$$tg\delta = G''(\omega)/G'(\omega)$$

*Scheme 1. Formulae for calculating modulus and loss factor tgδ (14)*

The G′′ of the PE/E/R/ATH/B compound is compared with compound without rubber and compound without rubber and ATH in Figure 8. As shown the G′′ below 60 Hz has a very low value for all three compounds. At higher frequencies, however, the G′′ begins to increase, but not with uniform rate. The increase of G′′ of the R containing compound is much higher so it reaches a value nearly two times higher than that of the two other compounds at vibration values of 100 Hz. It can be concluded that the presence of R positively influences the vibration damping character of polymeric compounds.

## Influence of Ground Rubber on the Smoke Released from Intumescent Flame Retardant Compounds

Flame retardant materials decompose under forced burning conditions, therefore, the knowledge of combustion products, and particularly the gases emitted by the burning materials, is important especially if the materials are applied in the interior of buildings. The environmental effect can be estimated based on the quality and quantity of the evolved gases. These were measured by a gas analyzer system coupled to cone calorimeter through an interface leading the evolved gases in a long path gas cell of FT-IR spectrometer equipped with a gas flow control and a data processing unit.

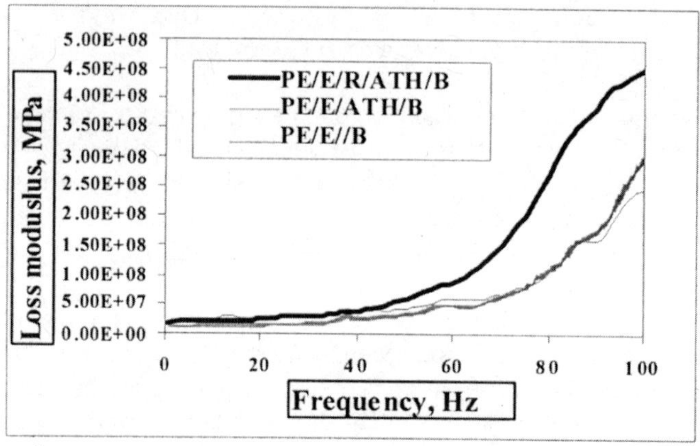


*Figure 8. Loss modulus of noise damping compounds*

Figure 9 shows the overview spectra of PE/E blend (***a***) and of the rubber containing flame retardant PE/E/R/PU-APP compound (***b***), where such components as water (in gase phase 1594, 3657; 3755 $cm^{-1}$), carbon dioxide (2355 $cm^{-1}$), carbon monoxide (2177 $cm^{-1}$), acetic acid (1180 $cm^{-1}$) and ammonia (855 cm-1), can be well identified.

The absorbance versus burning time curves of $CO_2$ are shown in Figure 10 for the PE/E/ blend (***a***), for PE/E/PU-APP compound without rubber (***b***) and for the complete PE/E/R/PU-APP compound containing rubber (***c***). The values of integrated $CO_2$ and CO curves (in Figure 10 a, b and c) were used for comparison of their quantities. The formed carbon dioxide from the PE/E/PU-APP compound (Figure 10/(***b***) is much lower than that of the polymer matrix (Figure 10/(***a***). The formation of carbon dioxide is further reduced by incorporation of rubber (Figure 10/(***c***). The reduced formation of carbon dioxide can be explained by the increased char formation in the presence of rubber (Figure 7). Consequently, the addition of recycled ground rubber to intumescent flame retarded compound reduces the emission of carbon dioxide during combustion.

## Conclusions

Compounds containing recycled ground rubber were prepared in a polyethylene-ethylene/vinyl acetate copolymer blend matrix. The interaction between components increased by incorporation of a reactive surfactant prepared by Diels-Alder addition of maleic anhydride on isomerised linoleic acid during the compounding. The influence of ground rubber additive on various properties

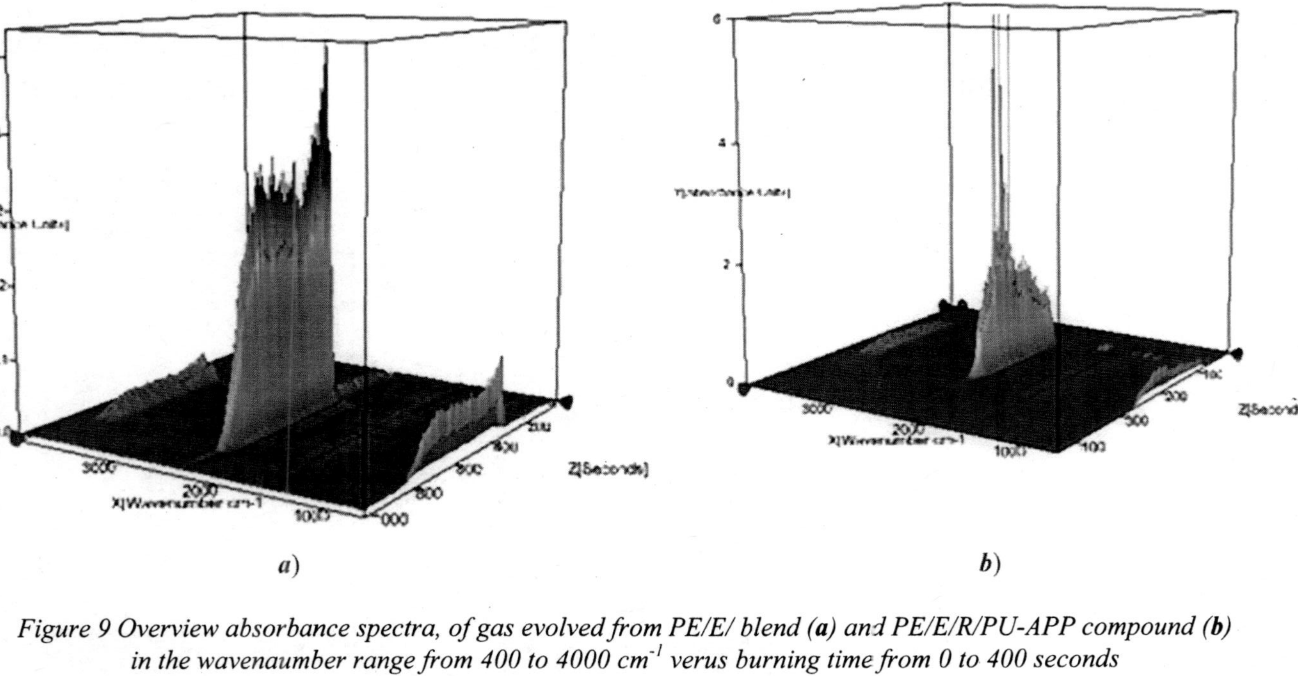


*Figure 9 Overview absorbance spectra, of gas evolved from PE/E/ blend (**a**) and PE/E/R/PU-APP compound (**b**) in the wavenaumber range from 400 to 4000 cm$^{-1}$ verus burning time from 0 to 400 seconds*

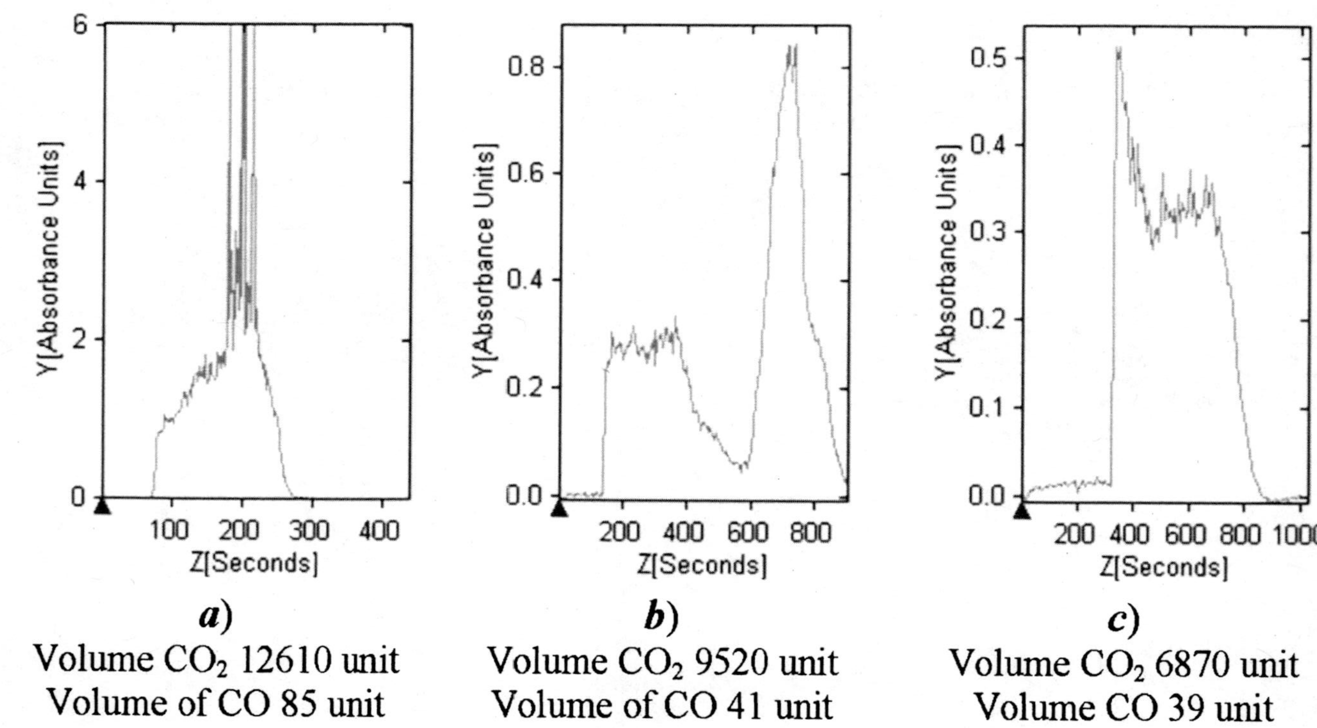


*Figure 10. IR absorbance versus treating time curves of $CO_2$ evolved from PE/E (**a**), PE/E/PU-APP (**b**), PE/E/R/PU-APP (**c**) compounds in cone calorimeter and the values calculated by the integration of the curves of $CO_2$ and CO gases*

such as burning rate, OI, UL 94, peak of heat release rate (PHRR), the integral value of curves of $CO_2$ and CO evolved during the cone calorimetric measurement and loss modulus of vibration damping compounds are summarised in Table III.

**Table III. Influence of ground rubber additive on propertie of various flame retardant compounds**

| *Compositins* | *Burning Rate* | *OI* | *UL 94* | *PHRR* | *$CO_2$ evolved* | *CO evolved* | *Loss Modulus 80 Hz* |
|---|---|---|---|---|---|---|---|
| | *mm/min* | *(%)* | | *(kW/m2)* | | | *(MPa)* |
| PE | 23 | | | | | | |
| PE/E | | 18 | HB | 900 | 12610 | 85 | |
| PE/R | 42 | | | | | | |
| PE/E/R | 39 | 17 | BH | | | | |
| PE/E/PU-APP | | 32 | V0 | 300 | 9520 | 41 | |
| PE/E/R/ PU-APP | | 30 | V0 | 120 | 6870 | 39 | |
| PE/E/B | | 19 | HB | | | | 100 |
| PE/E/ATH/B | | 31 | V1 | | | | 100 |
| PE/E/R/ATH/B | | 30 | V0 | | | | 300 |

The rubber in itself increase the flammability of the compounds. Incorporation of an intumescent additive system consisting of ammonium polyphosphate and recycled PU in the polyethylene-elastomer blend resulted in a flame retardant compound. Addition of recycled rubber improved the flame retardancy by increasing the ignition time and reducing the heat release rate in cone calorimeter, while the LOI value was moderately decreased. Addition of recycled ground rubber to intumescent flame retarded compound reduces the emission of carbon dioxide during a forced combustion. The improved flame retardant character and the reduced carbon dioxide emission can be explained by the increased char formation in the presence of ground rubber. The vibration damping character of a barium sulfate containing compound, flame retarded with alumina trihydrate, could also be improved by incorporation of recycled ground rubber.

## Acknowledgement

Our work was supported by the Hungarian Ministry of Education (Multirec Project, GVOP 3.1.1.-2004-05-0531/3.0), BUTE Natural- and Sport Scientific Assoc. and Hungarian Research Found OTKA T049121.

# References


1. *Focusing the 2020 (waste) vision*; Collins, R. , WME Environment Business Media 2003. 12.
2. Scaffaro, R.; Tzankova Dintcheva, N.;. La Mantia, F.P. *Polym. Degrad. Stab.y* **2006**, *91*, 3110-3116.
3. Baroulaki, I.; Karakasi, O.; Pappa, G.; Tarantili, P.A.; Economides, D.; Magoulas, K. *Composites Part A* **2006**, *37*, 1613-1625.
4. Adhikari, B.; De, D.; Maiti, S. *Prog. Polym. Sci.* **2000**, *25*, 909-948.
5. *EU Directive on Landfill of Waste* (Directive 99/31/EC), *End-of-life Vehicle Directive* (Directive 2000/53/EC).
6. Scaffaro, R.; Tzankova Dintcheva, N.; Nocilla, M.A.;. La Mantia, F.P. *Polym.r Degrad. Stab.y* **2005**, *90*, 281-287.
7. Sonnier, R.; Leroy, E.; Clerc, L.; Bergeret, A.; Lopez-Cuesta, J.M. *Polymer Testing* **2007**, *26*, 274-281.
8. Matkó Sz.; Marosi Gy.; Zubonyai F.; Garas S.; *Mûanyag és Gumi (Plastics and Rubber, Hungarian)* **2006**, *43*, 308-311.
9. Bugajny M.; Le Bras M.; Bourbigot S.; Delobel R. *Polym. Degrad. Stab.* **1999**, *64*, 157-163.
10. Bugajny M.; Le Bras M.; Bourbigot S.; Poutch F.; Lefebvre J-M. *J. Fire Sci.* **1999**, *17*, 494-513.
11. Bourbigot S.; Le Bras M.; Duquesne S.; Rochery M. *Macromol.Mater. Eng.* **2004**, *289*, 499-511.
12. Bertalan Gy.; Marois Gy.; Anna P.; Szenirmai K.; Bessenyei F.; Gyugos P.; Hung. Pat. 218,016. 2000.
13. Matkó Sz.; Répási I.; Szabó A.; Bodzay B.; Anna P.; Marosi Gy. *eXPRESS Polym. Lett.* **2008**, *Vol.2*, No.2, 126-132.
14. ISO 6721-1:2001, Plastics -- Determination of dynamic mechanical properties -- Part 1: General principles

# Metal-Containing Systems

## Chapter 19

# A Review of Transition Metal-Based Flame Retardants: Transition Metal Oxide/Salts, and Complexes

**Alexander B. Morgan**

**Advanced Polymers Group, Multiscale-Composites and Polymers Division, University of Dayton Research Institute, Dayton, OH 45469-0160**

This short review covers flame retardant chemistry other than the main classes of halogen, phosphorus, or metal hydroxide, and instead focuses on the transition metal element flame retardants that have shown some effectiveness in reducing polymer flammability. There appears to be a lot of promise in these flame retardant additives which utilize a wide range of chemistry and solid state structures to induce either char formation or highly effective vapor phase free radical inhibition for the burning plastic Not all of the materials reviewed in this paper are effective though by themselves, and this paper hopes to review some of the approaches and comment on why they may be promising new avenues for flame retardant research.

 

# The Periodic Table of Flame Retardants

Flame retardant chemistry today for polymeric materials focuses on just a few elements of the periodic table. Group VII (halogen) is still the most commonly used, with bromine and chlorine being the main active species in organohalogen compounds use to flame retard plastics (*1*). Fluorine and iodine are also effective in flame retardancy, but each of these elements has drawbacks which prevent their ready use in flame retardant additives. In the case of Fluorine, the C-F bond is too strong to release fluorine into the gas phase for flame retardant work in plastics, but if the C-F compound is a gas, it makes an excellent flame extinguishing agent (Halon extinguishers). Iodine tends not to be used since the C-I bond is quite weak and organoiodine compounds are not very stable. Stability is important to a flame retardant since it will need to be ready to activate when fire occurs, and for passive methods of fire protection, such as flame retardant additives in plastics, it may be years or more before such a compound is needed to protect against an ignition source. Organoiodine compounds do not have this level of stability, especially when exposed to light.

Phosphorus is another element commonly used to flame retard plastics, in both its red elemental form and in P(III)/P(V) oxidation states as both inorganic phosphate and organophosphorus compounds (*2*). Interesting, some other group V elements get used for flame retardancy, but only in specific structures. Nitrogen gets used in melamine (*3,4*) and a few other specialized compounds, as does antimony in the form of antimony oxide (*1,5*). Other elements such as Mg (*6*), Al, Si (*7*), B (*8*), and Sn (*9*) have shown some effectiveness as flame retardants, they are chemical structure specific and not universal to all most of the chemical compounds containing that atom. Carbon is a mixed category in regards to flame retardancy. Some chemical structures containing carbon will serve as flame retardant molecules, such as alkynes (*10*) or dehydroxybenzoin structures (*11*), but all carbon can combust, and again, it is not the carbon atom that does the chemistry of flame retardancy but the chemical structure of carbon atoms that does the work.

Certainly chemical structure plays a role in the effectiveness of Cl, Br, and P in flame retardancy, but it is the Cl/Br/P atoms that do most of the chemical work in providing flame retardant effects. With this observation, it bears asking if these elements are all there is to flame retardant chemistry, and the answer is definitely no. Certainly the ease of making varied chemical structures with Cl, Br, and P have led to their rapid adoption and use as flame retardant chemistries, and this helps account for their widespread use along with their obvious effectiveness. This still does not fully explain why so few other elements are used in flame retardant work. The answer to why so few other elements have been used is really that work on other elements for flame retardancy has really just begun. Since flame retardancy is an applied science, the majority of funding goes towards improving the existing chemistry and making it work for a current flame retardant problem. Only more recently has there been a demand for non-halogen flame retardancy due to perceptions of poor environmental impact with specific organohalogen compounds (*12,13*), and so this demand has begun to generate some funding for basic research in exploring other compounds which may provide flame retardancy.

This paper focuses on flame retardancy effects that have been found with transition metals which make up the a large portion of the periodic table. With that being said, only a few of these transition metals have been studied in detail, but in all these cases, it appears that it is the metal that does the work of the flame retardancy. The chemical structure of that metal is again important to the effectiveness of that flame retardant effect, but it is the metal that does the work of either providing vapor phase flame inhibition, or condensed phase char formation. The rest of this paper will focus on the transition metal oxides and metal complexes (including organometallics) that have shown flame retardancy.

## Metal Oxides/Salts

The term metal oxide/salt is quite broad on purpose since the examples in this section of the paper cover a wide range of chemistries. Where there are examples of metal oxides being used for flame retardancy, there are also examples of metal phosphates, borates, and halides which have enabled flame retardancy. Metal halides are a well understood class of flame retardants since they are often formed *in-situ* during burning with halogenated flame retardants. In fact, these metal halides work as synergists for the halogenated flame retardants in many polymeric systems, with antimony oxide ($Sb_2O_3$) being the most common added oxide for this purpose. Since the purpose of this paper is to focus on the uncommon versions of metal oxides and salts in flame retardancy, metal halides will not be discussed in this paper as there are already review papers on the subject (*1*).

Some recent results with metal oxides have looked at complex mixed oxides coming from fly ash or from spent refinery catalysts. Fly ash is a waste product which comes from power production plants using coal or oil, and while primarily composed of silicon and aluminum oxides, does have a variety of other metal oxides present depending upon the source of the fuel used to run the power plant. In one example, fly ash was used to flame retard polycarbonate with UL-94 V-0 results being obtained (*14*). However, 25wt% of the fly ash had to be used to obtain this result, making it a less than impressive flame retardant performance since polycarbonate can usually be flame retarded with charring salts in the 1wt% loading range, or halogenated aromatic compounds in the 2-4wt% range. Still, for a mixed oxide waste product, the result is interesting and results in the paper showed that the fly ash did change the decomposition products of the polycarbonate during pyrolysis of the polymer when compared to aluminum oxide and silicon oxide control samples. This suggests that the other metal oxides in the fly ash may be playing a role. Spent refinery catalyst, which is primarily composed of silicon and aluminum oxides in zeolite structures, also can contain varying amounts of transition metals, and in one paper where the spent catalyst was used to improve the performance of an intumescent formulation, the spent catalysts would contain low percentages of Fe, Ti, La, Ce, Pr, Nd, and Na oxides, along with trace amounts of Ni, V, Sb, Cu, Sn, and Bi oxides (*15*). These waste catalysts were ground up and combined with a classic ammonium polyphosphate/pentaerythritol system, melt compounded into a polyethylene-co-butyl acrylate system, and tested by UL-94 and cone calorimeter. What was found was that catalysts rich in rare earth oxides did not do as well in flammability reduction as those rich in transition metal oxides, but more importantly, the type of aluminosilicate support (zeolite structure) along with the transition metal oxides had the greatest effects in lowering heat release. This suggests that the structure of the metal oxide in the aluminosilicate matrix is very important, which makes sense if we consider the fact that these refinery catalysts only work when the metals are in specific confirmations. However, there are contradictory results about how transition metals affect flame retardant performance. Work published on various metals in zeolites in combination with an ammonium polyphosphate and pentaerythritol intumescent systems in polyethylene-co-butyl acrylate showed that the transition metals on zeolite supports had a negative effect on the limiting oxygen index (LOI) flammability measurement (*16*). It should be noted though that LOI, UL-94, and cone calorimeter all measure flammability in very different ways, and it may be that some of the contradiction in results is due to how the various metal improves or worsens a particular flammability performance (heat release, flame spread, char formation, dripping, etc.) under those particular tests. More study of transition metals on silicate supports is needed to further develop an understanding of how these materials may or may not be useful for flame retardant material development.

Another mixed oxide system studied, but not based on waste products, was the use of ZnO/MgO, ZnO/CaO, or ZnO/CaO/MgO glasses in combination with PVC filled with $CaCO_3$ and $Sb_2O_3$ (*17*). The glasses were made by mixing the various oxides in the appropriate ratio, calcining/fusing the oxides together at 1023 °C, followed by grinding the resulting product and then compounding it into PVC. While these mixed oxide glasses were not studied in detail for their chemical structure, it is clear from the obtained char yields and limiting oxygen index (LOI) data for the PVC systems containing these glasses that the zinc oxide is the active ingredient, but not only that, the zinc oxide is more effective when it is mixed with other oxides than used alone. The authors propose that the mixed oxides do both catalytic chemistry which promotes char formation as well as form ceramic protective layers, but the catalytic chemistry is easier to believe with the data presented in the paper and the supporting references for zinc catalyzing crosslinking between the C-Cl bonds on different PVC chains.

Combinations of transition metal oxides (V, Ta, Cu, Pd, Mo, Ti, Cr, Fe, Zn) and some silicon oxides (or silicone oils) were studied in poly(butylene terephthalate) (PBT) for flame retardant effects by cone calorimeter and UL-94 V testing (*18*). In this study it was found that combinations of zinc oxide and silicone oil (5wt% of each) gave the greatest reduction in heat release rate by cone calorimeter and the best burn times in UL-94 V, but it should be pointed out that none of the materials tested obtained any UL-94 V rating. So at first glance when considering the UL-94 V results, one would argue that the metal oxides had little effectiveness, but when looking at the reduction in peak heat release of this system it becomes clear that the zinc oxide is having an effect. The peak HRR of the PBT + 5wt% ZnO + 5wt% silicone oil was reduced to 850 $kW/m^2$ when compared to the base PBT which had a peak HRR of 2100 $kW/m^2$. Pyrolysis GC-MS of these materials showed significant changes in polymer decomposition products in the presence of the ZnO, which again points to condensed phase chemistry for a metal oxide, and again possible catalytic reactions. While ZnO showed an effect in PBT with silicone oils, it showed no effect on reducing heat release rate by cone calorimeter or in increasing LOI when it was compounded into thermoplastic polyurethane (TPU). However, it did show some intumescent effects of char formation, which suggests that the ZnO may be changing the TPU decomposition chemistry to allow for char formation with gas release that resulted in the observed carbon foam in the reported experiment (*19*).

Some more recently published work has recognized that metal oxides can work well with other flame retardants, and this includes work on nanoscale $TiO_2$ and $Al_2O_3$ in poly(methyl methacrylate) (PMMA) (*20*), and the use of $Al_2O_3$ and $Fe_2O_3$ in combination with red phosphorus in flame retarding recycled poly(ethylene terephthalate) (PET) (*21*). While $Al_2O_3$ is not a transition metal, its use in these two papers is informative when compared to the metal oxides. In the PMMA example, PMMA was combined with nanoscale $TiO_2$ or $Al_2O_3$ (3-12%) and an aluminum organophosphinate (3-12%) via melt compounding. These materials were analyzed by TGA and cone calorimeter, with the residual chars from cone calorimeter testing being studied further by SEM to investigate char structure. What was generally found in the study was that $Al_2O_3$ was generally better at reducing flammability than $TiO_2$, but the primary mechanism of flame retardancy appeared to be the fact that the $Al_2O_3$ formed chars with less cracks and openings in it when compared to chars which contained $TiO_2$. The only flammability category where the $TiO_2$ was better than $Al_2O_3$ was in smoke release. The one question unanswered (and unasked) in this paper is about the nanoparticle dispersion, which is well known to affect flammability performance. Therefore, was the flammability better for the $Al_2O_3$ system due to better dispersion of this particle, or was it because the particles helped form a more coherent char when reacting with the organophosphinate? The answer to this question has not been answered and the authors themselves point out that more work is needed to study how these oxides are working for flame retardancy. Still, it is clear from the cone calorimeter results in the paper that a metal oxide could reduce HRR more than just the phosphinate alone when the metal oxide was combined with the phosphinate. Some synergistic fire performance was noted and so this paper does support the idea that metal oxides have a role to play in flame retardancy. Supporting this is the work for the PET example mentioned above,[21] where amounts of red phosphorus flame retardant were replaced with small amounts of $Al_2O_3$ and $Fe_2O_3$ to obtain better flammability performance by LOI, although no real improvement in flammability was observed by cone calorimeter. As with the other paper, the chemical mechanism is not studied in detail and most of the work focuses on the $Al_2O_3$, but still, it is another example of metal oxides working with other flame retardants to enhance flammability performance, or aspects of flammability performance for a material.

The last three examples in this section are uses of metal salts, but this term can be misleading since the salts studied are not soluble in water. Instead they are complex inorganic compounds composed of a transition metal in a particular oxidation state with other "anions" countering the positive charge on the metal. One simple example was the use of iron carbonate, which was found to be as effective as $Mg(OH)_2$ and $Al(OH)_3$ in flame retarding low density polyethylene (*22*). However, the iron carbonate had to be used in high loadings since its primary mechanism of flame retardancy was release of $CO_2$ from the carbonate at fire temperatures. This is an example very different than many of the other results in this review where the metal compound would be used in very low loadings and would improve flame retardancy through potential catalytic effects of char formation and crosslinking.

Another example of a metal salt used for flame retardancy was the use of a wide range of iron phosphates, oxides, borates, and silicates in combination with a charring polymer (polyphenylene oxide, or PPO) and zinc borate which were formulated into polyamide-4,6 via melt compounding (*23*). By itself, even in loading levels up to 20wt%, the iron compounds do very little for flame retardancy by UL-94 and LOI testing, but in combination with PPO/zinc borate, the iron compounds become more effective although iron borate ($FeBO_3$) did not show much effectiveness. Interestingly, many of the formulations which used the iron containing compounds has final chars which were magnetic. X-ray diffraction analysis of the these chars found the presence of magnetic iron oxide ($Fe_3O_4$) which suggests some reduction/oxidation (redox) chemistry occurred during the combustion of these samples which would have led to this structure. Interestingly, if this form of iron oxide was used to provide flame retardancy, it was no better or worse than using non-magnetic iron oxide, and therefore the formation of $Fe_3O_4$ is an interesting byproduct formed during burning, but is not necessarily an important intermediate in providing flame retardancy to the system. While the paper presenting these results presents a great deal of data, like many others it only proposes mechanisms of flame retardancy for how the iron compounds improved the flammability of the polyamide system more than expected. The authors do however, provide an extensive series of references which mostly point to condensed phase (charring) mechanism of actions, as well as some potential vapor phase actions for these iron compounds to become active. The observed physical phenomena of burning seen in the paper points more to the condensed phase action for these metal salts, but the authors insightfully point out that conditions can easily occur within fires that would lead to *in-situ* formation of $Fe(CO)_5$, which is well known to be a very strong vapor phase flame retardant; as will be discussed in the next section of this paper.

The last example of a metal salt being combined with other flame retardants is the use of a nanoscale layered nickel phosphate in combination with a intumescent flame retardant formulation in polypropylene (*24*). For this example, an ammonium polyphosphate/pentaerythritol intumescent system was compounded into PP via two roll mill with other possible synergists, including zinc borate, a 4Å zeolite, an organically treated clay, and a layered nickel phosphate. The layered nickel phosphate was characterized by x-ray diffraction (XRD), transition electron microscopy, and thermogravimetric analysis (TGA). The diffraction and electron microscopy work turned out to be very helpful later in the presented research as it helped locate the nickel phosphate in final chars of the samples. Flammability was measured by LOI and UL-94, and TGA-FTIR data collected as well. For UL-94, the use of 2wt% nickel phosphate allowed for replacement of up to 7wt% (18wt% final loading down from 25wt%) of the intumescent FR system while maintaining a UL-94 V-0 result. For comparison, 2wt% of zeolite, zinc borate, or organically treated clay used in combination with 18wt% of the intumescent FR system failed the UL-94 test. Significant improvement in LOI was noted as well, increasing from 28 to 35 with the use of 2wt% nickel phosphate. However, higher loadings of the nickel phosphate between 3 and 5wt% caused LOI to decrease again down to 32. Pictures of the intumescent chars formed during burning are quite impressive, with huge intumescent chars being formed with the addition of the nickel phosphate. Analysis of these chars by XRD and scanning electron microscopy show that the nickel phosphate is helping form more coherent chars through enrichment of C-P bonds in the final char. This data, combined with the TGA-FTIR data suggests that the nickel phosphate helps accelerate the reaction between ammonium polyphosphate and pentaerythritol which would lead to faster intumescent foam formation. The mechanism is likely catalytic, with the nickel phosphate reacting with both the ammonium polyphosphate as well as helping form graphitic char through C-C bond formation during the burning of the sample. The presence of the graphitic char was confirmed also with XRD and electron microscopy, and since graphite has a higher thermal stability than other carbon chars, this may account for some of the higher LOI values found. For this example, it is clear that a small amount of a metal compound can greatly enhance the flame retardancy of a polymeric material, and further, it again suggests that metal compounds play catalytic roles in flame retardancy.

Related to the work above is work on layered double hydroxides (LDH), which are mixed metal oxides and fall into more than one category of flame retardant class. For the first category, they are distinctly ordered materials with layered structures containing metal (typically aluminum) hydroxides as one layer, followed by another layer of metal hydroxide (typically a transition metal) which can be ion exchanged with organic anions. This leads to the second category, which is that these organically treated double layer hydroxides can be mixed with polymer to form polymer nanocomposites. This in effect means that they will reduce polymer flammability by forming a nanostructure, and, can potentially reduce flammability further by any interactions between decomposing polymer and metal oxide. There have been several recent studies with LDH materials in epoxy (*25*), polyolefins (*26,27,28*), polyamide-6 (*29*), polystyrene (*30*), and poly(methyl methacrylate) (*31*). Because these LDH materials can form polymer nanocomposite structures, the mechanisms of flammability reduction are complex, but in many of the papers, not only does the level of nanoparticle dispersion affect flammability performance, but so does the chemical structure of the LDH itself. Specifically, the transition metal hydroxide appears to have effects on the amount of char formation and amount of heat release rate reduction that is observed by cone calorimeter, assuming that levels of nanoparticle dispersion are the same between LDH samples are the same. This is not always the case though, but the promising results with LDH materials in polymers shows that there is much more to be learned through study of this materials. This lack of knowledge though is not a drawback to this class of materials, but instead shows that this newly discovered class of metal oxide-based flame retardants has a great deal of potential due to the wealth of chemical structures available.

## Metal Complexes

Work of metal and metal complexes in flame chemistry has been studied as early as the 70s, and perhaps even earlier when one considers that tetraethyl lead used to be the common "anti-knock" additive in many grades of automotive fuel as far back as the 1940s. More recent work of metals to inhibit flames was begun in the 90s with efforts to find replacements to "Halon" type fire extinguishing agents (*32*), and many metal compounds were among the materials studied due to their recorded efficiency in catalyzing the recombination of low heat release radicals in the vapor phase. Metals such as Cr, Mn, Sn, U, Mg, Ba had been reported in the 70s as having this effect (*33*), and more recent work has studied the effect of Fe, Sn, and Mn complexes on flame inhibition (*34,35*). These studies used premixed methane flames to study the effect of metal compounds on flame inhibition, as well as to study their vapor phase flame chemistry. They found that Iron carbonyl ($Fe(CO)_5$) was the most effective metal-based flame inhibitor, followed by methylcyclopenadienyl manganese tricarbonyl (Figure 1) and tetramethyl tin (Figure 1) when compared to a Halon-type extinguishing agent, bromotrifluoromethane ($CBrF_3$). It was further found that the metals had complex vapor phase chemistry, with the metals going through various oxidation states and oxide/hydroxide intermediaries before reaching either final oxide or hydroxide states. While these results were only with methane flames and not burning polymer, their relevance to polymeric flame retardancy is clear; if the materials can be volatilized into the flame during polymer burning, they can inhibit vapor phase combustion of the burning polymer. Unfortunately, this particular manganese compound and tin compound reported here cannot be used as vapor phase flame retardants (nor can iron carbonyl) due to the high toxicity of these organometallic species, but still, they serve as useful scientific examples for understanding the potential of metals as flame retardant additives. Ferrocene has also shown effects on flame inhibition (*36*), and may be worth further studies since it does not have the toxicity problems of other organometallics.

The use of metal complexes, specifically pentanedione (acac) complexes of molybdenum, cobalt, chromium, and vanadium (Figure 1) was studied as dual flame retardant/smoke suppressant additive in PVC (*37*) and in intumescent coatings for polyvinylchloride (PVC) wiring (*38*). The use of these complexes did result in improvements in flame retardancy (either by limiting oxygen index (LOI) or cone calorimeter) as well as smoke release, but the results were complex and were system dependent. For example, in PVC plasticized with phthalate, the $Mo(acac)_2O_2$ complex yielded the best reduction in smoke release with highest increase in LOI, but for PVC plasticized with tricresyl phosphate, $Co(acac)_3$ gave the highest reduction in smoke release and the 2$^{nd}$ best increase in LOI. The addition of alumina trihydrate as a flame retardant along with these metal complexes also showed some system dependent effects, but again the $Mo(acac)_2O_2$ complex showed the greatest reduction in smoke release. When this same complex was combined with an intumescent coating formulation for PVC, smoke was again reduced, but had little effect on the heat release measured by cone calorimeter. The papers did not provide a clear mechanistic insight for how these complexes provided smoke and flame suppression, but they do serve as useful data points for future investigation. What is clear about the technology shown in these papers is that these complexes can be quite effective at smoke reduction in reducing heat release in very low levels. In the papers levels of 1 to 3 phr were found to have good effect which suggests that the effects these materials have is likely catalytic; not surprising since many transition metals are often used as catalysts for both breaking and forming C-C bonds.

Related to the above examples is work on copper-based flame retardants/ smoke suppressants for PVC where it was found that copper compounds, sometimes in combination with other metals such as molybdenum and tin, would greatly reduce smoke production while increasing char yield and reducing heat release as measured by cone calorimeter (*39,40*). For these systems, it appeared that copper compounds in both Cu(I) and Cu(II) oxidation states were equally favorable at smoke and heat release reduction, but in general the Cu(II) forms were more effective since they had better thermal stability and also provided some delays in time to ignition by cone calorimeter measurement. The proposed mechanism of flame retardancy for these copper compounds is that the metal crosslinks two polymer chains together by oxidatively adding across the carbon-chloride bonds in PVC, and then reductively eliminating itself while forming a C-C bond in its place. While such catalytic intermediates have not been isolated, the results found do suggest that copper's effects in PVC are catalytic, as this hypothesis can be further supported with known copper-based chemistry which forms carbon-carbon bonds between aryl halides (*41*).

For polymers other than PVC, a series of papers has studied the use of metal chelates in combination with Kaolin or magnesium hydroxide to provide flame retardancy. In the first example, a iron chelate (Figure 2, top) was synthesized with polyacrylonitrile as the main polymer component in this polymeric chelate, and then the chelate was absorbed onto Kaolin clay before being melt compounded into polypropylene (*42*). Cone calorimeter studies showed mixed

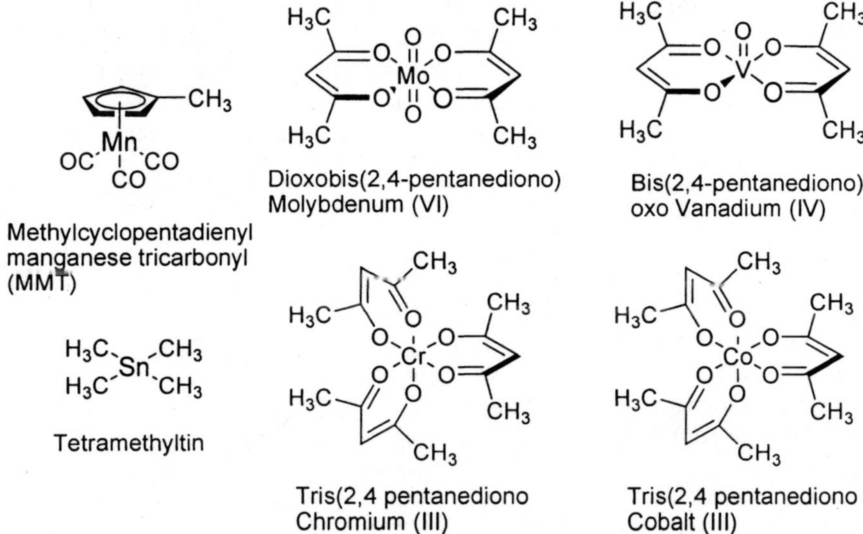


*Figure 1. Organometallics and Metal Complexes use for methane flame inhibition and smoke/flame reduction in PVC*

results for this system. Kaolin clay by itself was found to be the most effective at delaying time to ignition and reducing peak heat release rate, but when Kaolin clay was used in combination with the iron chelate treated clay in a nearly 1:1 ratio, the iron chelate enabled significant reductions in smoke release while keeping peak heat release rate low. However, this reduction in peak heat release rate by the iron chelate modified clay was still not as good as Kaolin alone. In a second example, similar polymer chelate chemistry to that used by the iron system was used to chelate cobalt and this chelate (Figure 2 middle) was used directly to flame retard polypropylene in combination with magnesium hydroxide (*43*). For this study, it was found that as there was a maximum level of cobalt chelate that could be added, as above this level the replacement of $Mg(OH)_2$ with the chelate would result in increases in heat release and smoke production. Still, about 10 phr of the cobalt chelate in combination with 40 phr of $Mg(OH)_2$ yielded a material with superior overall flammability performance as well as lowered smoke release and greatly lowered CO production. In a final example were iron, zinc, and cobalt chelates (Figure 2 bottom) that were used with kaolin clay. In this example. the iron chelate showed the best performance with dramatic delays in time to peak HRR, as well significant lowering of total HR and total smoke release (*44*). However, in none of these papers is a detailed mechanism described, although TGA data in each paper does suggest that the metal chelates do have an effect on the depolymerization kinetics of the polypropylene, but these results do not always correlate to the observed flammability behavior. For the example of Co, Zn, and Fe chelates used with Kaolin in polypropylene, the Co chelate delayed the peak mass loss of the polymer the most, and yet this did not translate into a delayed time to ignition or delayed time to peak HRR in the cone calorimeter. Instead the Co chelate was one of the poor performers in this system. Therefore there must be other chemistry ongoing in these systems which would explain the observed reductions in flammability and smoke release.

## Summary and Future Directions

From the examples presented in this paper, it is clear that transition metal materials, whether they be oxides, salts, organometallics or metal chelate complexes, can serve as flame retardants for polymers. Therefore we can draw the following conclusions about this very broad class of potential flame retardants:

- The use of transition metal compounds as flame retardants is not universal, and one must look at the thermal decomposition chemistry of the polymer to be flame retarded, as well as the form of the metal and how it can interact with that specific polymer decomposition chemistry.

- Depending upon the form of the transition metal compound, it can have either vapor phase flame retardancy (inhibition of radical formation/ recombination) or condensed phase flame retardancy (char formation). Both types of flame retardant mechanisms appear to be catalytic in nature.
- The transition metal compounds are most effective in combination with other flame retardants, and when used in small amounts.

These three points are however, all we can really say about these materials. While the vapor phase chemistry for some transition metal compounds has been worked out for controlled hydrocarbon flames, what the chemistry would be in the presence of a more complex polymer decomposition fuel source is not as clear. Further, how exactly each of these transition metals catalyzes char formation through crosslinking is in most cases supposed; it is not proven through isolation of catalyst cycle intermediates or through spectroscopic analysis. One can argue that this is for good reason, since measuring chemical pathways in the condensed phase of a polymer while it is on fire is a very difficult task, and to date only post-fire analysis has served to help identify what reactions may have occurred between polymer and transition metal. Still, there is a lot of support for the hypothesis that all of these transition metal compounds work in a catalytic manner to form graphitic types of chars which further resist pyrolysis and flame damage. A recent paper showed that multiwall carbon nanotubes could be grown from combusted polypropylene in the presence of activated nickel oxides, hydroxides, and carbonates on a zeolite catalyst or organically treated clay support (*45)*. In this paper, studies of the how the catalysts would work in converting ethylene, methane, and acetylene into nanotubes indicated that the metal had to be reduced *in situ* to its zero valent state to form these nanotube structures. Indeed, another paper where an organically treated clay was partly exchanged with $FeCl_3$ and $CuCl_2$ found that enhanced char formation was found when this clay was put into a polymer and the polymer exposed to flame. This paper also posited that the metals needed to be in a particular oxidation state to induce the observed additional carbon formation which was not pyrolyzed away by the flame (*46*). These two examples, as well as the other examples, make it clear that the metal has to be in the right form and coordination with other atoms (either chelates or oxide supports) to do catalytic char-formation chemistry, and therefore this presents an area for fundamental surface science and catalyst work. Research in determining what transition metals can rapidly convert hydrocarbons to graphite could yield very promising new flame retardant materials for modern material use. These materials may be even more promising from an environmental perspective since they likely would be useful in low loadings which could make recycling of the plastic easier.

There is much to be learned about the use of transition metals in flame retardancy of polymeric materials. The fundamental chemical pathways of flame

Iron Polymer Chelate on Kaolin

Crosslinked Cobalt Polyamide Chelate

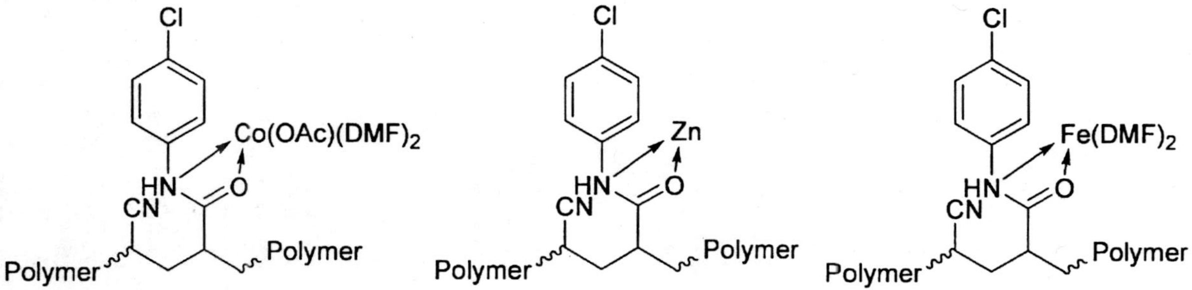

Cl

HN

CN

$Co(OAc)(DMF)_2$

O

Polymer

Polymer

Cl

HN

CN

Zn

O

Polymer

Polymer

Cl

HN

CN

$Fe(DMF)_2$

O

Polymer

Polymer

Metal Polyamide/Polyacrylonitrile Chelates

*Figure 2. Metal Chelates Used in Polypropylene Flame Retardancy*

retardancy need to be understood, as well as what chemical structure the metal atom needs to be in to induce the formation of graphitic char. Practical use of the material to pass flame retardant regulatory tests still needs to be done, as well as addressing all the other issues that comes with a new technology in the very regulatory-driven field of flame retardancy. Still, even with these unknowns there is a great deal of potential success in this part of the periodic table for new flame retardant chemistry to be developed, and it is hoped that flame retardant scientists, catalyst chemists, and inorganic/organometallic chemists could come together for such a project to find the new environmentally flame retardant additives that society is asking for.

## References


1. "Halogen-Containing Fire-Retardant Compounds" Georlette, P.; Simons, J.; Costa, L. in "Fire Retardancy of Polymeric Materials" ed. Grand, A. F.; Wilkie, C. A. Marcel Dekker, Inc. NY, NY. 2000. ISBN 0-8247-8879-6
2. Levchik, S. V.; Weil, E. D. *J. Fire Sci.* **2006**, *24*, 345-364.
3. Weil, E. D.; McSwigan, B *Plastics Compounding* **1994**, 31-39.
4. Costa, L.; Camino, G. *J. Thermal Analysis* **1988**, *34*, 423-429.
5. Costa, L.; Camino, G.; Bertelli, G.; Locatelli, R.; Luda, M. P. *Polym. Degrad. Stab.* **1991**, *34*, 55-73.
6. Inorganic Hydroxides and Hydroxycarbonates: Their Function and Use as Flame-Retardant Additives" Horn, W. E. in "Fire Retardancy of Polymeric Materials" ed. Grand, A. F.; Wilkie, C. A. Marcel Dekker, Inc. NY, NY. 2000. ISBN 0-8247-8879-6
7. "Silicon-Based Flame Retardants" Kashiwagi, T.; Gilman, J. W. in "Fire Retardancy of Polymeric Materials" ed. Grand, A. F.; Wilkie, C. A. Marcel Dekker, Inc. NY, NY. 2000. ISBN 0-8247-8879-6
8. Shen, K. K.; Kochesfahani, S.; Jouffret, F. *Polym. Adv. Technol.* **2008**, *19*, 469-474.
9. "Inorganic Tin Compounds as Flame, Smoke, and Carbon Monoxide Suppressants for Synthetic Polymers" Cusack, P. A.; Killmeyer, A. J. in "Fire and Polymers: Hazards Identification and Prevention" ACS Symposium Series #425, Ed. Nelson, G. L. American Chemical Society, Washington DC, 1990. Pp 189-210.
10. Morgan, A. B.; Tour, J. M. *Macromolecules* **1998**, *31*, 2857-2865.
11. Ellzey, K. A.; Ranganathan, T.; Zilberman, J.; Coughlin, E. B.; Farris, R. J.; Emrick, T. *Macromolecules* **2006**, *39*, 3553-3558.
12. "New Thinking on Flame Retardants" *Environmental Health Perspectives* **2008**, *116*, A210-213.
13. Tange, L.; Drohmann, D. *Polym. Degrad. Stab.* **2005**, *88*, 35-40,
14. Soyama, M.; Inoue, K.; Iji, M. *Polym. Adv. Technol.* **2007**, *18*, 386-391.

15. Estevao, L. R. M.; Le Bras, M.; Delobel, R.; Nascimento, R. S. V. *Polym. Degrad. Stab.* **2005**, *88*, 444-455.
16. Le Bras, M.; Bourbigot, S.; Delporte, C.; Siat, C.; Le Tallec, Y. *Fire Mater.* **1996**, *20*, 191.
17. Tian, C.; Wang, H.; Liu, X.; Ma, Z.; Guo, H.; Xu, J. *J. App. Polym. Sci.* **2003**, *89*, 3137-3142.
18. Ishikawa, T.; Ueno, T.; Watanabe, Y.; Mizuno, K.; Takeda, K. *J. App. Polym. Sci.* **2008**, *109*, 910-917.
19. Bourbigot, S.; Duquesne, S.; Fontaine, G.; Turf, T.; Bellayer, S. *18th Recent Advances in Flame Retardant Polymeric Materials Proceedings* **2007**, Stamford, CT, BCC Communications.
20. Laachaci, A.; Chochez, M.; Leroy, E.; Ferriol, M.; Lopez-Cuesta, J. M. *Polym. Degrad. Stab.* **2007**, *92*, 61-69.
21. Laoutid, F.; Ferry, L.; Lopez-Cuesta, J. M.; Crespy, A. *Polym. Degrad. Stab.* **2003**, *82*, 357-363.
22. "Evaluation of iron carbonate as a flame retardant for polyolefins" Gilman, J. W.; Brassell, L. D.; Davis, R. D.; Shields, J. R.; Harris, R. H.; Wentz, D.; Reitz, R.; Becker, A.; Chi, C. *FRPM 2005 Conference Proceedings*, Berlin, Germany, 7-9 September 2005.
23. Weil, E. D.; Patel, N. G. *Polym. Degrad. Stab.* **2003**, *82*, 291-296.
24. Nie, S.; Hu, Y.; Song, L.; He, S.; Yang, D. *Polym. Adv. Technol.* **2008**, *19*, 489-495.
25. Zammarano, M.; Franceschi, M.; Bellayer, S.; Gilman, J. W.; Meriani, S. *Polymer* **2005**, *46*, 9314-9328.
26. Ye, L.; Ding, P.; Zhang, M.; Qu, B. *J. App. Polym. Sci.* **2007**, *107*, 3694-3701.
27. Costa, R. R.; Wagenknecht, U.; Heinrich, G. *Polym. Degrad. Stab.* **2007**, *92*, 1813-1823
28. Zhang, G.; Ding, P.; Zhang, M.; Qu, B. *Polym. Degrad. Stab.* **2007**, *92*, 1715-1720
29. Du, L.; Qu, B.; Zhang, M. *Polym. Degrad. Stab.* **2007**, *92*, 497-502
30. Costache, M. C.; Heidecker, M. J.; Manias, E.; Camino, G.; Frache, A.; Beyer, G.; Gupta, R. K.; Wilkie, C. A. *Polymer* **2007**, *48*, 6532-6545.
31. Manzi-Nshuti, C.; Wang, D.; Hossenlopp, J. M.; Wilkie, C. A. *J. Mater. Chem.* **2008**, *18*, 3091-3102.
32. "Construction of an Exploratory List of Chemicals to Initiate the Search for Halon Alternatives" Pitts, W. M.; Nyden, M. R.; Gann, R. G.; Mallard, W. G.; Tsang, W. *NIST Technical Note 1279* August 1990, Washington D. C.
33. Bulewicz, E. M.; Padley, P. J. *Proc. Combust. Inst.* **1971**, *13*, 73.
34. Linteris, G. T.; Knyazev, V. D.; Babushok, V. I. *Combustion and Flame* **2002**, *129*, 221-238.
35. Linteris, G. T.; Katta, V. R.; Takahashi, F. *Combust. Flame* **2004**, *138*, 78-96.

36. Hirasawa, T.; Sung, C-J.; Yang, Z.; Joshi, A.; Wang, H. *Combustion and Flame* **2004**, *139*, 288-299.
37. Sharma, S. K. *Fire Technology* **2003**, *39*, 247-260.
38. Sharma, S. K. Saxena, N. K. *J. Fire Sci.* **2007**, *25*, 447-466.
39. Starnes, W. H. Jr.; Pike, R. D.; Cole, J. R.; Doyal, A. S.; Kimlin, E. J.; Lee, J. T.; Murray, P. J.; Quinlan, R. A.; Zhang, J. *Polym. Degrad. Stab.* **2003**, *82*, 15-24.
40. Pike, R. D.; Starnes, W. H. Jr.; Jeng, J. P.; Bryant, W. S.; Kourtesis, P.; Adams, C. W. *Macromolecules* **1997**, *30*, 6957-6965.
41. P.E. Fanta *Synthesis* **1974**: 9-21
42. Shehata, A. B.; Hassan, M. A.; Darwish, N. A. *J. App. Polym. Sci.* **2004**, *92*, 3119-3125.
43. Shehata, A. B.; *Polym. Degrad. Stab.* **2004**, *85*, 577-582
44. Hassan, M. A.; Shehata, A. B. *Polym. Degrad. Stab.* **2004**, *85*, 733-740.
45. Jiang, Z.; Song, R.; Bi, W.; Lu, J.; Tang, T. *Carbon* **2007**, *45*, 449-458.
46. Nawani, P.; Desai, P.; Lundwall, M.; Gelfer, M. Y.; Hsiao, B. S.; Rafailovich, M.; Frenkel, A.; Tsou, A. H.; Gilman, J. W.; Khalid, S. *Polymer* **2007**, *48*, 827-840.

# Chapter 20

# Salen Copper Complexes: A Novel Flame Retardant for Thermoplastic Polyurethane

**Gaëlle Fontaine*, Thomas Turf, and Serge Bourbigot**

**Equipe Procédés d'Elaboration des Revêtements Fonctionnels (PERF), LSPES – UMR/CNRS 8008, Ecole Nationale Supérieure de Chimie de Lille (ENSCL), Avenue Dimitri Mendeleïev – Bât. C7a, BP 90108, 59652 Villeneuve d'Ascq Cedex, France,**

Herein we present the synthesis and the use of salen copper complexes as flame retardants in TPU. Fire retardancy evaluated by mass loss calorimeter show that 10% loading in TPU decreases significantly the peak rate of heat released (pRHR). The diminution of pRHR depends of the structure of the salen copper complex. The first investigations of mechanism suggest condensed phase action.

## Introduction

It is well known that the flame retardancy of polymers can be achieved according to different mechanisms. Flame retardants can act either in the vapor phase or in the condensed phase (sometimes both) through a chemical and/or physical mechanism to interfere with the combustion process upon heating, pyrolysis, ignition or flame spread stages. Some of them afford radicals during thermolysis that interrupt and suppress the free radical combustion process.

Based on this idea, we have focused our attention on metal (Ti, Mn, Fe, Cr, Cu...) complexes of *N,N'*-(bis salicylidene)ethylenediamine compounds (Metallosalen). They have been used since a long time for various catalytic reactions, such as the epoxidation of alkenes or olefins (*1*). Metallosalen can also act as nuclease i.e. cleavage of RNA or DNA, copper salen or iron salen act as artificial nuclease and the DNA cleavage was explained by formation of hydroquinone system and free radicals (*2*). We assumed that such compounds might act as fire retardant by a radical mechanism. So, we have synthesized *N,N'*-bis(salicylidene)ethylenediamine and *N,N'*-bis(hydroxysalicylidene)ethylenediamine metal complexes. Herein we present results obtained with *N,N'*-bis(salicylidene)ethylenediamine copper (II) complex **CC1**, *N,N'*-bis(4-hydroxysalicylidene)ethylenediamine copper (II) complex **CC2** and *N,N'*-bis(5-hydroxysalicylidene)ethylenediamine copper (II) complex **CC3** (Scheme 1). Indeed, copper nanoparticles have provided a synergistic effect with ammonium polyphosphate in epoxy resin at amount as low as 0.1 wt.-% and might also enhance the thermal stability of polymers by itself (*3*). We have evaluated those compounds as a potential new class of halogen-free FR in thermoplastic polyurethane (TPU) which requires flame retardancy for numerous applications.

# Experimental

## Chemistry

Salen and hydroxysalens **2a-d** were obtained under air using a conventional procedure for salen synthesis based on the condensation of salicylaldehyde or its hydroxy-derivatives with ethylenediamine (*4,5*). Reagents were purchased from Aldrich or Lancaster and were used without further purification, except triethylamine which was distilled over $CaH_2$ under argon. Diethyl ether and ethanol were reagent grade commercial solvents and were used without further purification. The purity of all compounds was assessed by TLC, $^1$H- and $^{13}$C-NMR. TLC was carried out using silica gel 60F-254 (Merck, 0.25 mm thick) precoated UV-sensitive plates. Spots were visualized by inspection UV light at 254 nm. $^1$H- and $^{13}$C-NMR spectra were recorded on a Bruker AC 200 spectrometer. Chemical shifts ($\delta$) are referenced to internal solvent and given in ppm. Coupling constants (*J*) are given in Hz. The following abbreviations apply to spin multiplicity : s (singlet), d (doublet), t (triplet), q (quartet), m (multiplet) and bs (broad singlet).

*N,N'*-bis(salicylidene)ethylenediamine (**2a**). To a solution of ethylenediamine (6.7 mL; 6 g; 0.1 mol) in 600 mL of ethanol, salicylaldehyde (21.8 mL; 25 g; 0.2 mol) in 100 mL of ethanol was added dropwise with vigorous stirring. The product precipitated immediately and the mixture was refluxed for a half-hour, then kept at room temperature and filtered. The product was recrystallized from EtOH, filtered, washed with ethanol and ethyl ether, then dried under reduced pressure over $P_2O_5$ to give the desired Schiff base **2a** as yellow crystals, 20 g (75 %). mp = 127-128 °C ; IR (KBr, cm$^{-1}$) $\nu$ 3440 (OH), 1630 (C=N), 1570, 1490, 1280, 860, 750 ; MS (MALDI$^+$) m/z 290.9 $(M + Na)^+$;

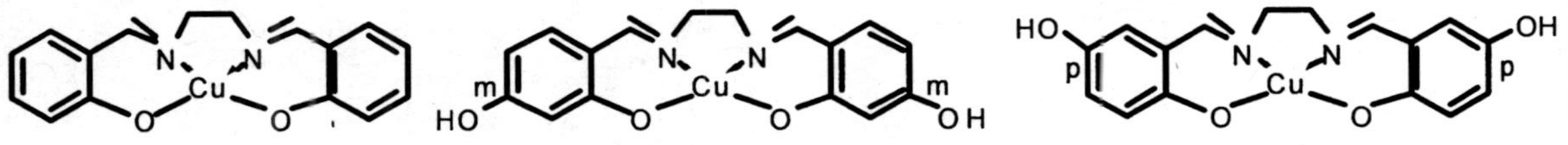

Scheme 1. CC1-CC3 Formula

$R_f$ ($CH_2Cl_2$/MeOH 90/10) 0.90 ; $^1$H NMR (DMSO-$d_6$) δ 3.91 (s, 4H), 6.88 (m, 4H), 7.35 (m, 4H), 8.57 (s, 2H) ; $^{13}$C NMR (DMSO-$d_6$) δ 58.8 ($CH_2$), 116.4 ($C_{Ar}$), 118.5 ($C_{Ar}$), 118.6 ($C_{Ar}$), 131.6 ($C_{Ar}$), 132.3 ($C_{Ar}$), 160.4 ($C_{Ar}$), 166.8 (CN).

*N,N'*-bis(4-hydroxysalicylidene)ethylenediamine (**2b**). According to the method described for **2a**, 28.2 g (0.2 mol) of 2,4-dihydroxybenzaldehyde (**1b**) and 6.7 mL (6 g; 0.1mol) of ethylenediamine yielded compound **2b** (28.1 g ; 94 %) orange solid ; mp > 260 °C ; IR (KBr, $cm^{-1}$) ν 3410 (OH), 2560, 1640 (C=N), 1580, 1470, 1230, 1160, 790 ; MS ($MALDI^+$) m/z 301.6 $(M + H)^+$, 323.5 $(M + Na)^+$, 339.5 $(M + K)^+$ ; $R_f$ ($CH_2Cl_2$/MeOH 80/20) 0.58 ; $^1$H NMR (DMSO-$d_6$) δ 3.77 (s, 4H), 6.15 (d, J = 2.4 Hz, 2H), 6.26 (dd, J = 2.4, 8.6 Hz, 2H), 7.16 (d, J = 8.6 Hz, 2H), 8.35 (s, 2H) ; $^{13}$C NMR (DMSO-$d_6$) δ 57.8 ($CH_2$), 102.5 ($C_{Ar}$), 106.9 ($C_{Ar}$), 111.2 ($C_{Ar}$), 133.4 ($C_{Ar}$), 161.7 ($C_{Ar}$), 164.3 ($C_{Ar}$), 165.7 (CN).

*N,N'*-bis(4-hydroxysalicylidene)ethylenediamine (**2c**). According to the method described for **2a**, 30.0 g (0.22 mol) of 2,5-dihydroxybenzaldehyde (**1c**) and 6.7 mL (6 g ; 0.10 mol) of ethylenediamine yielded compound **2c** (30.0 g ; 99 %) ochre solid ; mp > 260 °C ; IR (KBr, cm˙1) m 3300 (OH), 1640 (C=N), 1590, 1450,1410, 1300, 1260, 1210, 1150, 1030, 840, 780; MS (MALDI+) m/z 301.9 (M+H)+, 323.5 (M+Na)+;Rf($CH_2Cl_2$/MeOH 80/20) 0.84; $^1$H NMR (DMSO-d6) d 3.87 (s, 4H), 6.66–6.79 (m, 6H), 8.46 (s, 2H); $^{13}$C NMR (DMSO-d6) d 59.1 ($CH_2$), 116.5 (CAr), 116.8 (CAr), 118.5 (CAr), 119.8 (CAr), 149.2 (CAr), 152.8 (CAr), 166.5 (CN).

*N,N'*-bis(salicylidene)ethylenediamine copper (II) complex (**CC1**) *N,N'*-bis(salicylidene)ethylenediamine **2a** (70 g, 0.26 mol) was dissolved in 500 mL absolute ethanol then a solution of copper (II) acetate (52 g, 0.26 mol) in 250 mL of water was added dropwise and the mixture was refluxed under vigorous stirring for 2.5 h. The green precipitate was collected by filtration, washed thoroughly with ethanol then dried at reduced pressure over $P_2O_5$ to give *N,N'*-bis(salicylidene)ethylenediamine copper (II) complex **CC1** (72 g, 84%) as a fine green powder. NMR spectra of the complex was not recorded due to paramagnetism of Cu(II) (this well-known phenomenon leads to poor resolution of the spectra).

*N,N'*-bis(4-hydroxysalicylidene)ethylenediamine copper (II) complex (**CC2**). According to the method described for **CC1**, 50 g (0.17 mol) of *N,N'*-bis(4-hydroxysalicylidene)ethylenediamine **2b** and 33 g of copper acetate yielded compound **CC2** as a black powder (57 g ; 93 %). NMR spectra of the complex was not recorded due to paramagnetism of Cu(II) (this well-known phenomenon leads to poor resolution of the spectra).

*N,N'*-bis(5-hydroxysalicylidene)ethylenediamine copper (II) complex (**CC3**). According to the method described for **CC1**, 25.0 g (0.08 mol) of *N,N'*-bis(5-hydroxysalicylidene)ethylenediamine **2c** and 16.6 g of copper acetate yielded compound **CC3** as a brown powder (27.7 g ; 92 %). NMR spectra of the complex was not recorded due to paramagnetism of Cu(II) (this well-known phenomenon leads to poor resolution of the spectra).

### Materials and Methods

Thermoplastic Polyurethane (TPU) C85A was supplied by BASF. Formulations (Table I) were mixed using a Brabender laboratory mixer (type 350/EH, roller blades) equipped with a "nitrogen chamber" designed in our laboratory to avoid thermo-oxidation of the polymer during processing. TPU is dried for at least 12 hours at 100 °C before use, the temperature of the mixer was set at 180 °C and the shear was 50 rpm for 10 minutes. Sheets were then obtained using a Darragon press at 180 °C with a pressure of $2.10^6$ Pa for 3 minutes then $4.10^6$ Pa for 5 minutes.

FTT (Fire Testing Technology) Mass Loss Calorimeter was used to carry out measurements on samples following the procedure defined in ASTM E 906. The equipment is identical to that used in oxygen consumption cone calorimetry (ASTM E-1354-90), except that a thermopile in the chimney is used to obtain heat release rate (HRR) rather than employing the oxygen consumption principle. Our procedure involved exposing specimens measuring 100 mm x 100 mm x 3 mm in horizontal orientation. External heat flux of 35 $kW/m^2$ was used for running the experiments. This flux corresponds to common heat flux in mild fire scenario. The mass loss calorimeter was used to determine heat release rate (HRR). When measured at 35 $kW/m^2$, HRR is reproducible to within ±10%. The data reported in this paper are the average of three replicated experiments.

## Results and Discussion

### Chemistry

Salen copper complexes **CC1-CC3** (Scheme 2) were obtained in two steps by a conventional synthetic route from (di)hydroxybenzaldehyde **1a-c**. The reactions are fast and easy leading to **CC1-CC3** obtained in very good global yield, 87%, 92% and 86%, respectively.

### Fire retardant performance

In previous work it was shown that copper nanoparticles provide a significant fire retardant effect at concentration as low as 0.1wt.-% (*3*). Based on this, we have first evaluated the effect of compound **CC1** and **CC2** at 0.5 wt.-% in TPU which corresponds approximately to 0.1wt.-% of copper. **CC1** and **CC2** were chosen as candidates for the preliminary study to assess the effect of the presence of an additional hydroxyl group in the aromatic rings. Mass loss calorimeter experiments reveal that compound **CC1** (*Figure 1*) and **CC2** (*Figure 2*) does not reduce the peak rate of heat release (pRHR) at this concentration. **CC1** and **CC2** were also tested at 5, 10 and 25 wt.-%. It appears that **CC1** does not reduce RHR significantly at 5 or 10 wt.-% but at 25 wt.-% the pRHR is decreased by 25%.

**Table I. Composition of the Formulations**

| Sample | Copper Complex | % Copper Complex (-%wt) | % TPU (-%wt) |
|---|---|---|---|
| 1 | CC1 | 0.5 | 99.5 |
| 2 | CC1 | 5 | 95 |
| 3 | CC1 | 10 | 90 |
| 4 | CC1 | 25 | 75 |
| 5 | CC2 | 0.5 | 99.5 |
| 6 | CC2 | 5 | 95 |
| 7 | CC2 | 10 | 90 |
| 8 | CC2 | 25 | 75 |
| 9 | CC3 | 10 | 90 |

This first results show that we observe an effect of **CC2** at 5% loading but performance is not enough. At higher loadings, there is an effect due to the supplementary hydroxyl group in the aromatic rings, and the effect of the position of the hydroxyl group in the aromatic moiety should be investigated.

On the other hand, **CC3** is tested at 10% in TPU and RHR curve vs. time was compared with those of **CC1** and **CC2** at 10% loading in TPU (***Figure 3***). First of all, the introduction of an hydroxyl group permits a decrease in the pRHR. A reduction of 16% is observed when there is no additional hydroxyl group in the aromatic rings, in the case of **CC2** and **CC3** which contain hydroxyl group in position meta and para, respectively, the reduction of pRHR is 44%. The effect is also observed on the total of heat release (THR), the THR measured for neat TPU is 81.6 MJ/m², while it is only 67.6, 61.0 and 67.6 MJ/m² for TPU/**CC1-CC3** respectively. Furthermore, the position of the hydroxyl group seems to influence the time to ignition (TI), indeed, the TI for TPU is 111s whereas for **CC3** it increased to 116s but decreased to 59s and 79s for **CC1** and **CC2**.

Regarding the effect of the copper, we have evaluated salen **2c** by cone calorimeter at 10% loading in TPU (*Figure 4*). The comparison between the curves for **2c** and **CC3** (*Figure* 4) shown the influence of the copper chelate. Indeed, the salen alone had a fire retardant effect but when it is chelated by copper, this effect is higher. The pRHR is decreased by 29% in the case of compound **2c** whereas pRHR is decreased by 44% in the case of **CC3**. Moreover, the TI is 89s for **2c** while it is 116s for **CC3**. This result suggests that chelated copper might play a role improving flame retardancy of TPU.

To examine the mechanism of action of this copper complex salen, we have employed thermogravimetric analysis (TGA). We have run TGA experiments on

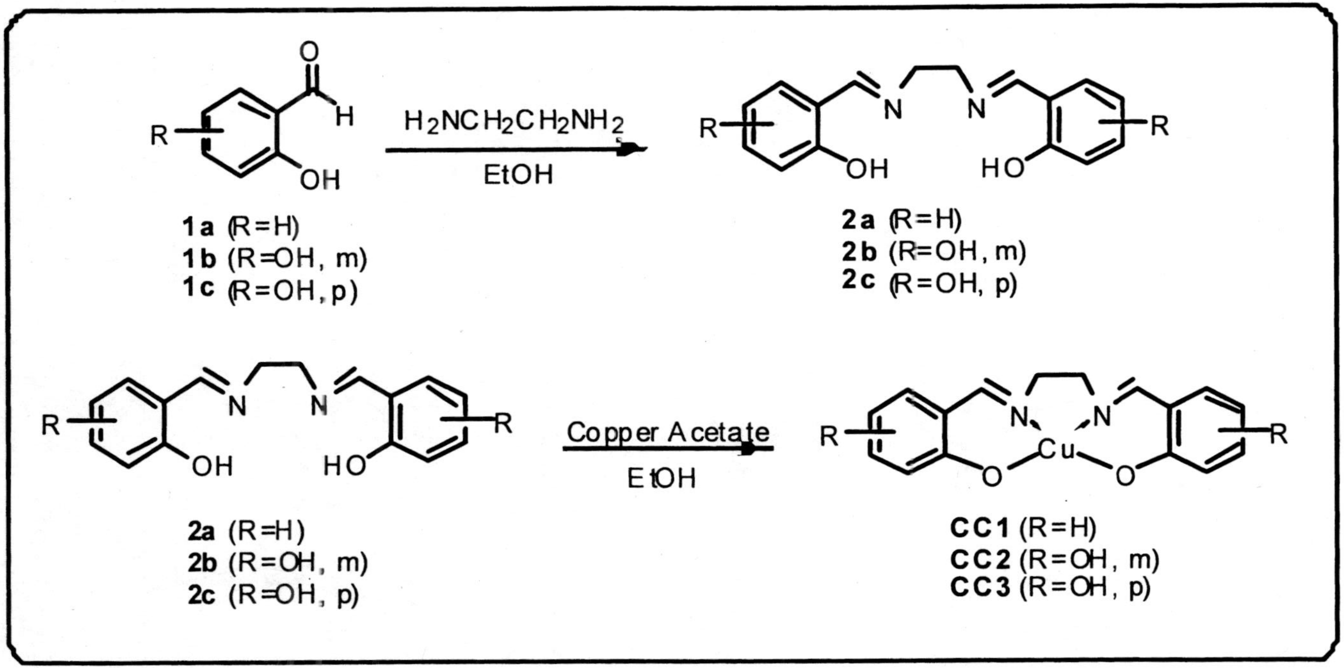

Scheme 2. Synthetic route for CC1-CC3

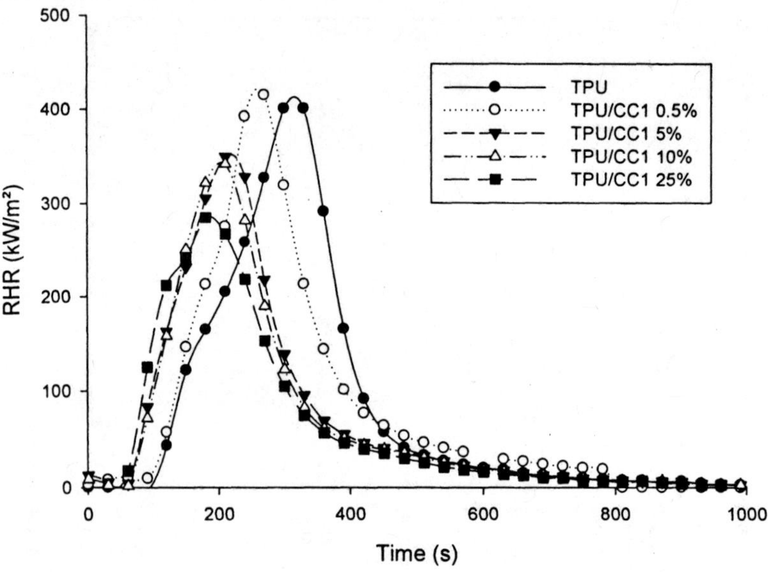


*Figure 1. RHR values (external heat flux = 35kW/m²) versus time of the formulations TPU/**CC1** in comparison with virgin TPU*

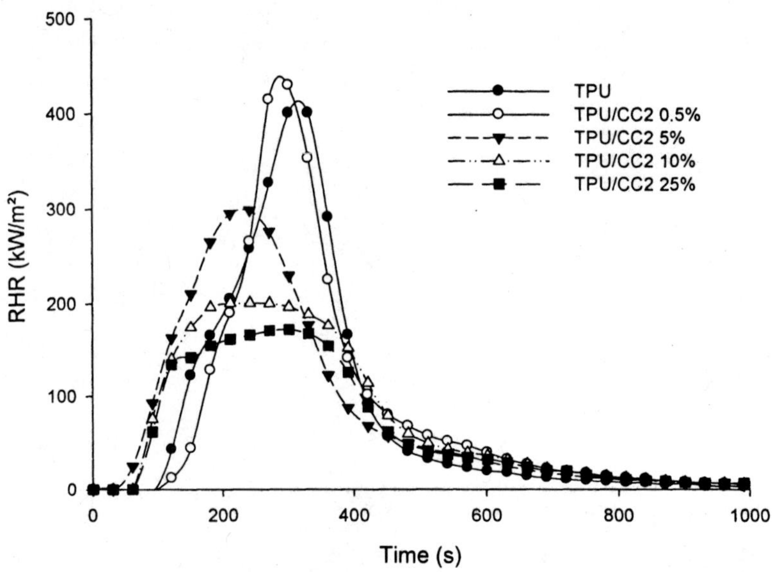


*Figure 2. RHR values (external heat flux = 35kW/m²) versus time of the formulations TPU/**CC2** in comparison with virgin TPU*

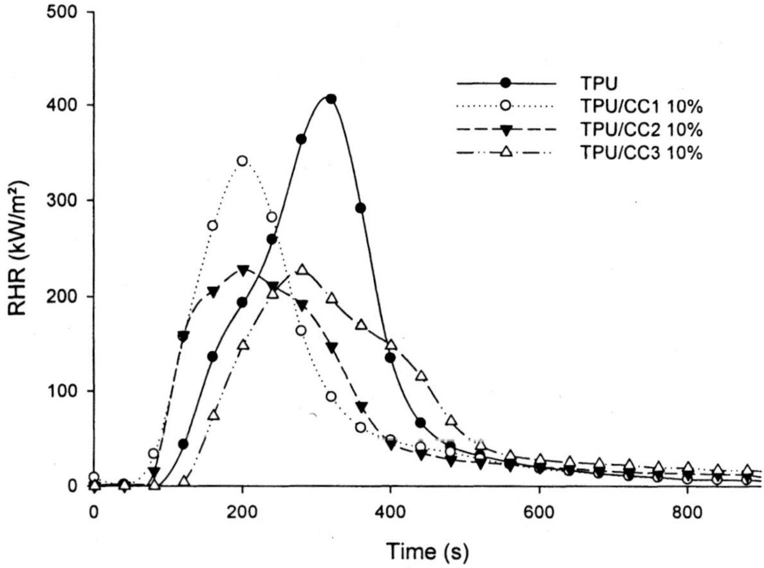


*Figure 3. RHR values (external heat flux = 35kW/m²) versus time of the formulations TPU/**CC1-CC3** (10% loading) in comparison with virgin TPU*

the pure products (*Figure 5*) to evaluate the stability of each copper complex in comparison with the TPU. We find that **CC2** and **CC3** have the same behavior and decompose in two apparent steps, the first is small (less than 10% of mass) and start at 80 °C while the second is larger (70% of mass) starting at 260 and 240 °C for **CC2** and **CC3** respectively. The **CC1** decomposes also in two apparent steps but it is more stable i.e. the first step starts at 290 °C and the second one at 370 °C. In comparison with the TPU, **CC2** and **CC3** decompose earlier but **CC1** decomposed at higher temperature.

We have also studied the formulations TPU/**CC1-CC3** (90/10). TGA curves (***Figure 6***) show that between 280 °C and 495 °C, TPU is less stable than the formulations with **CC1** and **CC2** while with **CC3** the formulation is stable until 520 °C.

The difference weight loss curves (***Figure 7***) present the same behavior for **CC1-CC3**, starting at 270 °C by positive interactions (stabilization area) and ending by small negative interactions (destabilization zone). The efficiency of **CC3** could be explained by the fact that it presents the biggest positive interaction and also by the fact that this zone ended at higher temperature (515°C) than **CC1** and **CC2** (490 °C).

The efficiency of these copper complexes in DNA cleavage was explained by free radical formation. We can envisage the reaction of **CC3** with oxygen leading to the formation of a quinone structure and water. This quinone structure is stabilized by resonance which could explain its stability and efficiency.

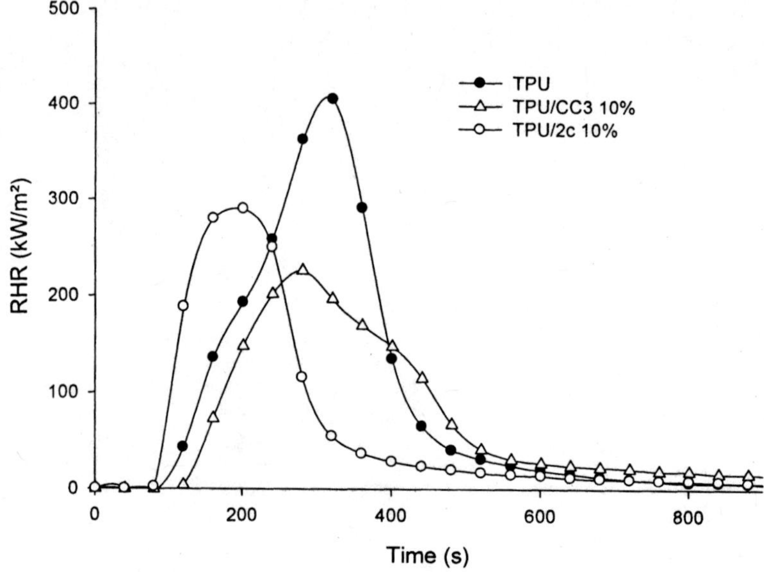


*Figure 4. RHR values (external heat flux = 35kW/m²) versus time of the formulations TPU/**CC3** and TPU/**2c** (10% loading) in comparison with virgin TPU*

The results obtained by TGA shows that the use of **CC1-CC3** in TPU at 10% loading permit to achieve positives interactions between 270 and 500 °C which correspond to the temperature of degradation of TPU. This result suggest than the FR properties seems to be due to a condensed phase mechanism instead of a gas phase as suggested before.

## Conclusion

In this study we have synthesized new flame retardants. An effect in TPU could be observed starting at 5% loading and a good effect is obtained at 10% loading which is a low loading compared to conventional FRs. The mass loss calorimeter experiments reveal that **CC2** and **CC3** are much more efficient than **CC1**. We might assign this effect to the presence of the second hydroxyl group which could enhance the fire retardancy. Moreover we have shown that the fire retardancy is not only due to the salen itself but it is due to the copper complexes. Further experiments are in progress to elucidate their mechanisms.

In the future, other metal complexes will be evaluated in TPU and also other polymers would be test. Moreover nanoparticles will be added for synergistic effect.

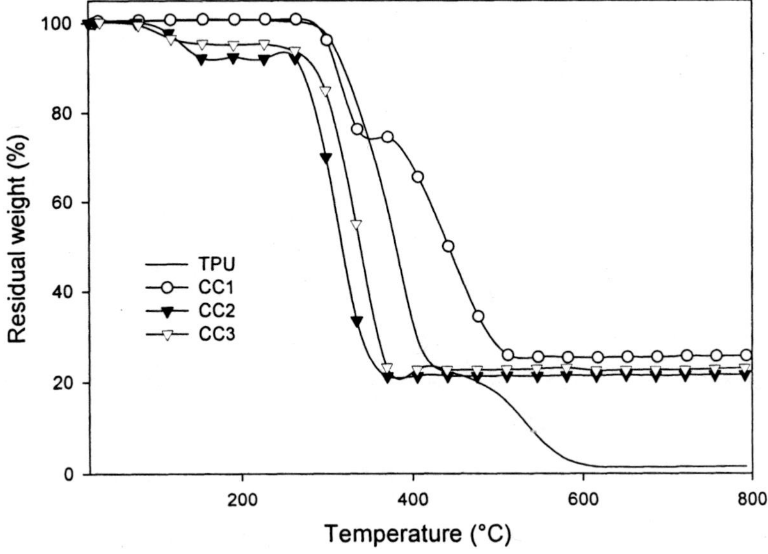


*Figure 5. TGA curves of pure products under air flow at 10°C/min*

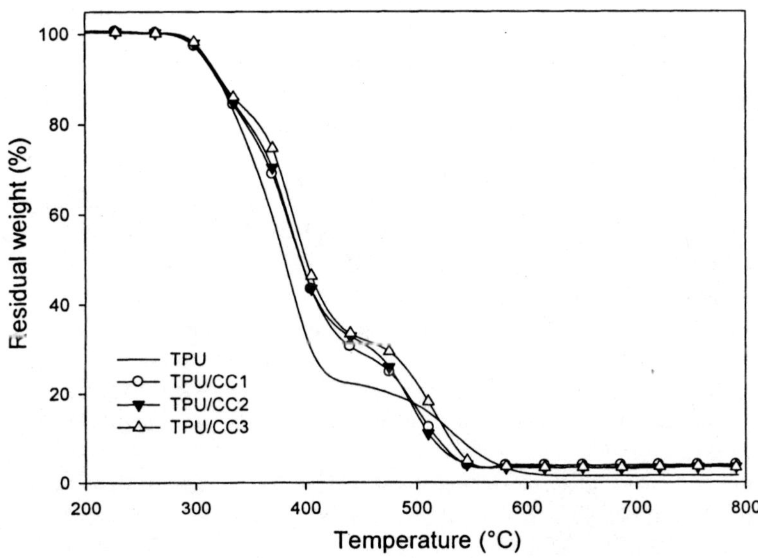


*Figure 6. TGA curves of the formulations TPU/**CC1-CC3** (90/10) in comparison with the virgin TPU.*

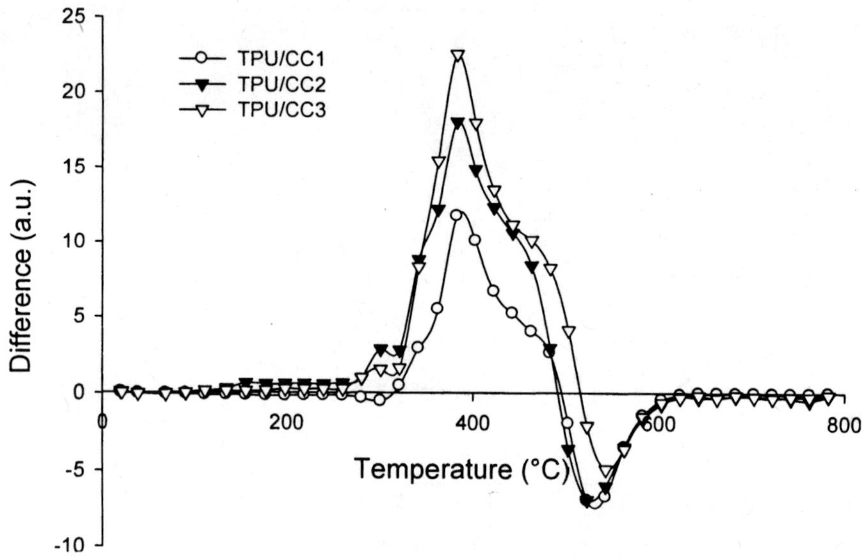


*Figure 7. Difference weight loss curve between experimental and theoretical TG curves of TPU/**CC1-CC3** (90/10)*

## References


1. (a) Canali, L. and Sherrington, D. C., *Chem. Soc. Rev.,* **1999**, *28*, 85 (b) Katsuki, T., Synlett, **2003**, 281 (c) Larrow J. F. and Jacobsen E. N., *Top. Organomet. Chem.*, **2004**, *6*, 123.(c) McGarrigle, E. M. and Gilheany, D. G. *Chem. Rev.,* **2005**, *105*(5), 1563-1602; (d) Martinez, A., Hemmert, C and Meunier, B. *J. Catal.,* **2005**, *234*(2), 250-255 (e) Cozzi, P. G. *Chem. Soc. Rev.,***2004**, *33*(7), 410-421.
2. Lamour E, Routier S, Bernier JL, Catteau JP, Bailly C, Vezin *H J. Am. Chem. Soc.* **1999**, *121*, 1862-1869.
3. Antonov, A.; Yablokova, M.; Costa, L.; Balabanovich, A.; Levchik, G.; Levchik, S. *Mol. Cryst. Liquid Cryst. Sci. Tech., Section A: Mol. Cryst. Liquid Cryst.* **2000**, *353*, 203-210.
4. Tanaka, T. *Bull. Chem. Soc. Jpn* **1960**, *33*, 259-61.
5. Joshi, V and Kaushik, NK *Indian J. Chem., Sect A* **1992**, *31*, 61-63.

# Toxicity

# Chapter 21

# Effects of Fire Retardants and Nanofillers on the Fire Toxicity

**Anna A. Stec[1], T. Richard Hull[1], José L. Torero[2], Ricky Carvel[2], Guillermo Rein[2], Serge Bourbigot[3], F. Samym[3], Giovanni Camino[4], Alberto Fina[4], Shonali Nazare[5], and Michael Delichatsios[6]**

**[1]Centre for Fire and Hazards Science, University of Central Lancashire, Preston PR1 2HE, United Kingdom**
**[2]BRE Centre for Fire Safety Engineering, The School of Engineering and Electronics, The University of Edinburgh, The King's Buildings, Mayfield Road, Edinburgh EH9 3JL, United Kingdom**
**[3]Procédés d'Elaboration des Revêtements Fonctionnels (PERF), LSPES–UMR/CNRS 8008, ENSCL, Avenue Dimitri Mendeleïev – Bât. C7a, B.P. 90108, 59652 Villeneuve d'Ascq Cedex, France**
**[4]Politecnico di Torino Sede di Alessandria-Centro di Cultura per l'Ingegneria delle Materie Plastiche, Viale Teresa Michel 5 15100, Alessandria, Italy**
**[5]Centre for Materials Research and Innovation, University of Bolton, Deane Campus, Bolton BL3 5AB, United Kingdom**
**[6]The Institute of Fire Safety Engineering Research and Technology (FireSERT), School of the Built Environment, University of Ulster, Newtownabbey, Co Antrim BT37 0QB, United Kingdom**

Four polymers, polyamide 6, polypropylene, ethylene vinyl acetate and polybutylene terephthalate have been prepared with fire retardants, nanofillers, (both separately and together). The toxic product yield of each material has then been measured under different fire conditions. The influence of polymer nanocomposites and fire retardants in the formulations, on the yields of toxic products from fire, have been studied using the ISO 19700 steady state tube furnace, and it was found that under early stages of burning more carbon monoxide may be formed in the presence of nanofillers and fire retardants, but, under the more toxic under-ventilated conditions, there was a relative reduction in the toxic product yields. In general, the differences between the samples containing fire retardants and nanofillers are very small compared to the differences obtained under different fire conditions or in the presence of certain heteroelements, such as nitrogen or halogens.

## Introduction

Published UK fire statistics show that most fire deaths are caused by inhalation of toxic gases *(1)*. While some fires may be represented by a single set of burning characteristics, or fire stages, most fires progress through several different stages *(2)*. Burning behaviour and particularly toxic product yields depend most strongly on a few of factors. Amongst them material composition, temperature and oxygen concentration are generally the most important.

Fire toxicity is both scenario and material dependent (*3*). The ideal small-scale method for assessing fire toxicity must allow the toxic product yields from each fire stage to be determined, allowing assessment of each material under different fire conditions. This appears to be the only way that the complexities of full scale burning behaviour can be replicated on a bench scale.

Comparing the toxicity of fire effluents containing gases, vapours, and aerosols, having fundamentally different effects on humans is very difficult. Although CO is most commonly given as the cause of death, incapacitation by irritant gases may have trapped the victim in the first place, different toxicants have different effects on different species, and multiple toxicants do not necessarily have additive effects. However, assessment of fire effluent toxicity is an essential component of fire hazard analysis. For example, performance based design approaches to fire safety require the estimated available safe escape time (ASET) before smoke, heat or toxic hazards prevent escape, to be greater than the required safe escape time (RSET), the time needed to evacuate . The hazard generally changes with fire scenario, and while the toxic product yields are typically higher for under-ventilated flaming, the much higher rate of burning can increase the rate of generation of toxic products in well-ventilated flaming making this the more hazardous fire scenario. Different materials have different burning behaviour and there are a number of factors which have to be taken into account in assessing the contribution of any material to fire development and hazard(*4*). Potentially polymer nanocomposites could deliver unique combinations of stronger, lighter, and more versatile polymer composites creating a new class of products fulfilling industry and consumer demands, since they are effective at ~1% filler loadings and generally enhance the mechanical properties of the polymer. Although numerous detailed studies on the preparation and characterisation of polymer nanocomposites have been undertaken, relatively little work has been done on their contribution to fire toxicity *(5,6)*.

**Fire Effluent toxicity**

The formation of CO, often considered to be the most toxicologically significant fire gas, is produced over a range of conditions from smouldering to fully developed flaming. CO results from incomplete combustion, which can arise from:

- Insufficient heat in the gas phase (e.g. during smouldering).
- "Quenching" of chemical reactions in a flame (e.g. when halogens are present in the flame, or excessive ventilation cools the flame.
- The presence of stable molecules, such as aromatics, which survive longer in the flame zone, giving relatively higher CO yields in well-ventilated conditions, but lower than yields than non-aromatic polymers in under-ventilated conditions *(7)*.
- Insufficient oxygen (e.g. in under-ventilated fires, large radiant heat fluxes continue to produce fuel which is partially oxidised, but there is not enough oxygen to complete the conversion to $CO_2$.

The high yields of the asphyxiant gas CO from under-ventilated fires are held responsible for most of the deaths through inhalation of smoke and toxic gases, but this under-ventilated burning is the most difficult to create on a bench-scale, because the heat release and hence the rate of burning is generally dependent on the oxygen concentration. Figure 1 illustrates, in general terms, the change in toxic product yields during the growth of a fire, through its various stages, from non-flaming through to well-ventilated flaming and restricted ventilation. Although the toxic product yields are often highest for non-flaming combustion, the rates of burning and the rate of fire growth are much slower, so under-ventilated flaming conditions are generally considered the most toxic fire stage. Other toxic species include hydrogen cyanide, HCN (another asphyxiant gas), and incapacitating irritants, which may cause blinding pain to the eyes and flooding of the lungs (oedema) and respiratory tracts, inhibiting breathing and preventing escape. The wide variety of these irritants has led to groupings such as acid gases, organo-irritants and particulates, in order to estimate incapacitation *(8)*.

Data from large scale fires *(9, 10)* shows much higher concentrations of the two asphyxiant gases (CO and HCN) under conditions of reduced ventilation than under well-ventilated conditions. It is therefore essential to the assessment of toxic hazard from fire that these different fire stages can be adequately replicated, on the small scale and preferably the individual fire stages treated separately. This allows a single data set to be applied to multiple fire scenarios. Assessment of fire hazard requires both data describing the rate of burning of the material, and data describing the toxic product yield of the material. This

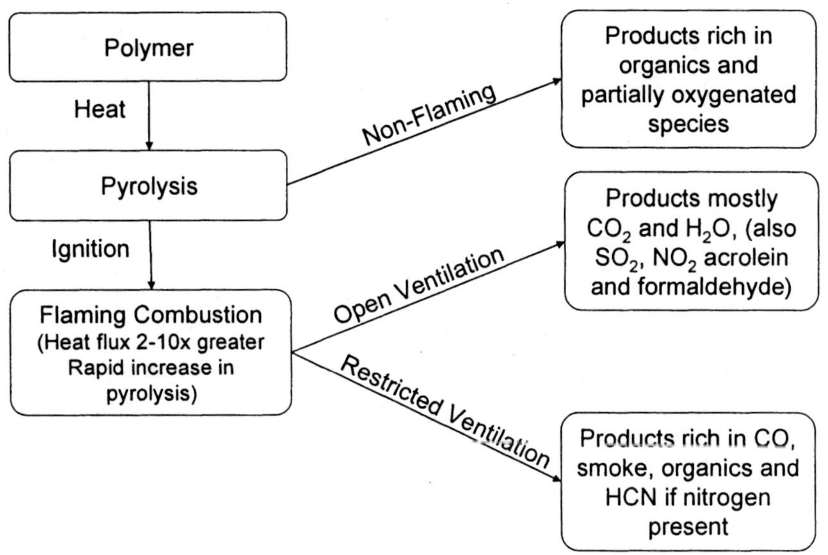


*Figure 1. Effect of fire stage on toxic gas production*

is best achieved using the steady state tube furnace *(7)*, in which the air supply and rate of burning are fixed, during the "steady state" burning period as the sample is driven into a furnace, and subjected to an increasing applied heat flux. Although the temperature of the furnace is constant – the increasing heat flux comes about through the gradual exposure of the sample to the temperature profile in the furnace. Using this technique, a clear relationship has been demonstrated between the yield of toxic products (for example in grams of toxicant per gram of polymer) and the fire condition, for a given material composition *(11)*. A more detailed account of other methods for assessment of fire toxicity has recently been published. (15)

## Calculation of toxic hazard

Fire effluent toxicity can be assessed by quantification of the yield of each toxic product and the smoke from the fire, which can then be related to toxicity using expressions for fractional effective dose (FED) as shown in Equation 1 which may be normalised by expressing the toxicity as an $LC_{50}$ (the mass which when burnt under the stated conditions will kill 50% of the population in a unit volume of air, usually expressed in g $m^{-3}$ *(12,13,14)*.

$$FED = \left\{ \frac{[CO]}{LC_{50,\,CO}} + \frac{[HCN]}{LC_{50,\,HCN}} + \frac{[HCl]}{LC_{50,\,HCl}} + .... \right\} \times V_{CO_2} + A + \frac{21-[O_2]}{21-5.4}$$

$$V_{CO_2} = 1 + \frac{\exp(0.14[CO_2]) - 1}{2} \qquad (1)$$

A is an acidosis factor equal to $[CO_2] \times 0.05$.

The calculation of FED based on the Pursèr model *(12)*, presented in Equation 1 uses $V_{CO_2}$ as multiplication factor for $CO_2$ driven hyperventilation, to account for the increased respiration rate resulting from inhalation of $CO_2$ on the harmful effect of the other toxic species, therefore increasing the FED contribution from all the toxic species. It also incorporates an acidosis factor A, to account for the toxicity of $CO_2$ in its own right.

**Fire Scenarios (Fire stages)**

A key concept in characterising and predicting the gas phase flaming combustion conditions, and the yields of products such as carbon monoxide (CO), carbon dioxide ($CO_2$), hydrogen cyanide (HCN) and hydrocarbons, is the global equivalence ratio. This relates the equivalence ratio ($\phi$), presented in Equation 2, to the fire effluent composition taken over both the fire plume ($\phi_p$) and the hot (sometimes reactive) upper layer to give a global equivalence ratio ($\phi_g$).

$$\phi = \frac{\text{Actual fuel / Air ratio}}{\text{Stoichiometric fuel / Air ratio}} \qquad (2)$$

As previously outlined fires can be divided into a number of stages: from smouldering combustion and early well-ventilated flaming, through to fully developed under-ventilated flaming *(2)*. Table I shows the typical conditions of each fire stage.

Table II shows how the fire conditions, particularly the upper layer temperature effect the toxic product yields.

This work has been carried out as part of the Predfire Nano Project Framework 6 (with support from European Union), aiming to develop a tool to understand and predict fire behaviour of fire retardant polymer nanocomposites. Although most of the Predfire Nano research effort has been focused on prediction of fire growth rate, most fire deaths are attributed to inhalation of toxic gas and smoke. For a particular fire scenario, fire smoke toxicity depends

**Table I. ISO Classification of fire stages (Adapted) *(2)***

| *Fire Stage* | *Max Temp /°C* | | *Equivalence ratio* $\phi$ | *CO/ CO₂ ratio* |
|---|---|---|---|---|
| | *Fuel* | *Smoke* | | |
| **Non-flaming** | | | | |
| 1a. Self sustained smouldering | 450-800 | 25 - 85 | - | 0.1-1 |
| **Well-ventilated flaming** | | | | |
| 2. Well-ventilated flaming | 350-650 | 50 - 500 | <1 | < 0.05 |
| **Under-ventilated Flaming** | | | | |
| 3a. Small under-ventilated | 300-600 | 50-500 | > 1 | 0.2-0.4 |
| 3b. Large Underventilated (Post-flashover) | 350-650 | >600 | > 1 | 0.1-0.4 |

**Table II. Applicability of the global equivalence ratio *(15)***

| *Upper layer Temperature* | *Upper layer composition* | *Comments* |
|---|---|---|
| < 425 °C | exclusively from plume. | Upper layer non-reactive. Relative production rates of combustion products very similar for full range of $\phi_g$ . |
| > 625 °C | Fuel lean[explain] | Similar to fully ventilated fires. CO levels low unless soot present to catalyse $CO_2 \rightarrow CO + O\cdot$ . |
| > 625 °C | exclusively from plume. | Prediction of combustion products possible over the full range of $\phi_g$ . When $\phi_g > 1.5$ CO yields are relatively constant. |
| 425-625 °C | exclusively from plume. | Relative generation rates for combustion products dependent on upper layer temperature and upper layer residence time. Global equivalence ratio concept needs to be modified to correct for temperature dependence. |
| > 625 °C | direct entry of air to upper layer | This situation is commonly reported in real enclosure fires. The remaining fuel is oxidised to CO in preference to $CO_2$ in the upper layer. CO likely to be greater than predictions based on $\phi_g$ . |

both on the fire conditions (temperature and ventilation) and on the material. If either is changed, this can result in an increase in toxicity sometimes by a factor of ten or more, encroaching on the safety margins employed in typical fire hazard calculations. Fire smoke toxicity requires sophisticated measurement techniques so no data is available from real unwanted fires, while the data available from large scale tests is generally fairly limited, and certainly to obtain such data with current methods would prove too expensive for product development and routine materials testing.

This paper describes the measurement of yields of toxic components of fire effluents from polyamide 6 (PA 6), polypropylene (PP), polybutylene terephthalate (PBT) and ethylene-vinyl acetate copolymer (EVA) and fire retarded and nanocomposite modifications of these materials, using a steady state tube furnace for fire toxicity over a range of fire conditions *(16,17)*.

## Fire Retardancy

Strategies range from blocking the fire physically, to promoting alternative chemical reactions to stop the material from burning.

### *Physical action*

There are several ways in which the combustion process can be retarded by physical action:

- *By cooling* - Endothermic reactions cool the material.
- *By forming a protective layer* – Obstructing the flow of heat and oxygen to the polymer, and fuel to the vapour phase.
- *By dilution.* Release of water vapour or carbon dioxide may dilute the free radicals responsible for the reactions in the flame, so it is extinguished.

The most widely used fire retardant, aluminium hydroxide ($Al(OH)_3$), and a related compound magnesium hydroxide ($Mg(OH)_2$) both of which are reported in this work, break down endothermically (absorbing a significant quantity of heat, 1.3 kJ $g^{-1}$), forming water vapour, diluting the radicals in the flame, while the residue (of alumina ($Al_2O_3$) or magnesia (MgO)) builds up to form a protective layer on the surface of the burning polymer. Unfortunately relatively large amounts may be needed to be effective (up to 70%) and the freshly formed alumina can lead to afterglow *(18)*.

$$2\,Al(OH)_3\,(s) \xrightarrow{180\text{-}200^\circ C} Al_2O_3(s) + 3\,H_2O(g)$$

$$\Delta H = +1.3\ kJ\ g^{-1}$$

*Chemical action*

- ***Reaction in the gas phase*** - the radical reactions of the flame can be interrupted by a suitable flame retardant. If the radical concentration falls below a critical value, the flame becomes extinguished. The processes releasing heat are thus reduced, and the polymer formulation cools down. However, interfering with the flame reactions can result in the production of toxic and irritant partially burnt products, including carbon monoxide, generally increasing the toxic potency of the fire gases while reducing fire growth. Most gas phase *flame* retardants are based on halogens (chlorine or bromine) or phosphorus compounds.

- ***Reaction in the solid phase*** - *Char Formation*- condensed phase *fire* retardants are those which cause a layer of carbonaceous char to form on the polymer surface. This can occur, for example, by the fire retardant removing the side chains and thus generating double bonds in the polymer. Ultimately, these may form a carbonaceous layer following the formation of aromatic rings. Char formation usually reduces the formation of smoke and other products of incomplete combustion. Char formation is often initiated by an acidic species, such as phosphoric acid, which is produced by the condensed phase decomposition of phosphorus-based fire retardants.
  *Intumescence*- The incorporation of blowing agents, causes swelling behind the surface layer, providing much better insulation under the protective barrier. The same technology is used for coatings, for protecting wooden buildings or steel structures. Ammonium polyphosphate is often a key component in such systems, decomposing into ammonia (the "blowing agent) and phosphoric acid.

## Polymer Nanocomposites

In order to produce properly dispersed polymer nanocomposites, it is generally necessary to add a compatibilising agent, such as a surfactant, to the polar nanofiller surface, in order to insert it between the polymer chains. In polar polymers, such as nylon, dispersion is much easier than in hydrophobic, crystalline polymers such as isotactic polypropylene (PP). In these cases it is generally necessary to attach a grafting agent, such as maleic anhydride onto the polymer to ensure adequate dispersion. While the mechanical properties of the polymer depend on adequate dispersion at ambient temperatures, the fire behaviour is a function of the dispersion of the nanofiller in the molten bubbling polymer. In many cases, the surfactant will have decomposed, leaving the polar nanofiller. There has been intense speculation as to whether this results in incompatibilisation, and migration of the filler to the surface, or whether the preferential loss of the first few hundred nanometres of polymer results in accumulation of filler at the surface. In some cases no fire retardant effect is observed without adequate dispersion. In others it is evident, suggesting that dispersion occurs in the molten, decomposing, bubbling polymer *(6)*.

# Experimental

## Apparatus

The steady state tube furnace, shown in Figure 2, where, during the steady state burning period, the rate of burning is controlled by the sample feed rate, allows the ventilation to be controlled without affecting the rate of burning *(16,19)*. The apparatus is able to investigate the production of smoke (aerosol or particulate matter) and combustion gases for different fire stages, but can also be used to study different characteristics such as ignitability and heat release by oxygen depletion. (*20*).

Conditions with different equivalence ratios are created by varying the mass of sample, and keeping the air flow rate across the sample constant for the range of polymers studied. The equivalence ratio for a particular polymer may then be varied by adjustment of the primary (furnace tube) air flow rate, making up to 50 litre $min^{-1}$ with secondary air in the effluent dilution chamber, or by varying the mass feed rate. In this way, toxic gas yields can be directly compared for a range of fire conditions and a range of polymers *(17,19)*.

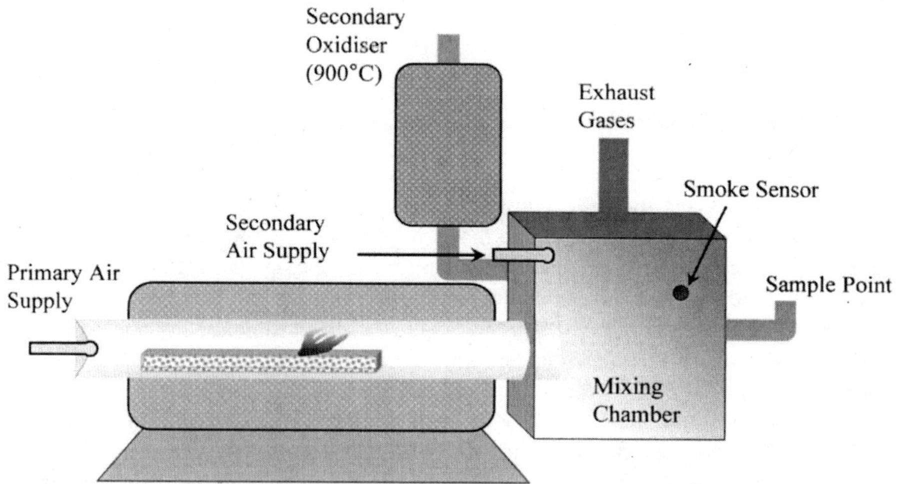


*Figure 2. The steady state tube furnace ("Purser" furnace)*

During an experiment, the sample is pushed into the furnace tube, at a constant and controlled rate, by a suitable drive mechanism, typically over a period of about 20 minutes. A fixed stream of primary air passes through the tube furnace, and over the sample, into a mixing and measuring chamber where is diluted with a fixed and controlled secondary air supply. Primary air passing over the specimen supports the combustion process. Secondary air increases the volume of effluent for analysis and is used to adjust the concentration of effluents from different ventilation conditions, within the same analytical range. A light-beam photocell system is used to determine smoke density across the mixing and measurement chamber. The requirement in each test run is to obtain a steady state period of at least 5 minutes during which the concentrations of effluent gases and smoke particulates remain constant and can be measured. For this study, on-line FTIR *(21)* measures a range of the combustion products with additional analysis to confirm the concentrations found by FTIR using a combination of electrochemical sensors and non dispersive infrared (NDIR) analysers for CO and $CO_2$. Oxygen depletion is measured using a paramagnetic analyser.

The relationship between type of the fire, the ventilation conditions, and the toxic combustion products yields over a range of variables affecting conditions (temperature, fuel chemistry) are reported.

## Materials

Different materials were investigated, using the formulations presented in Table III. The rigid materials used for the preparation of nanocomposites were commercial polypropylene (PP) added with maleic anhydride grafted PP as a compatibiliser (PP, Moplen HP500N-Basell), blended wth 5% PPgMA Polybond 3200 by Crompton) polyamide 6 (Technyl S27 BL from Rhodia), unreinforced polybutylene terephthalate (PBT, Celanex 2000-2 by Ticona) and ethylene-vinyl acetate (a polymer blend of EVA and LDPE supplied by Kabelwerk Eupen). The fire retardant (FR) for PP was Exolit AP 760 (ammonium polyphosphate), for PA6 was Exolit OP1311 (an organic aluminum phosphinate combined with melamine polyphosphate in the ratio 2 to 1 (wt:wt)), for PBT was Exolit OP1240, all supplied by Clariant. For EVA, two different fire retardants were used, aluminium hydroxide $Al(OH)_3$ (ATH) supplied by Kabelwerk Eupen and magnesium hydroxide $Mg(OH)_2$ (MH) Micronex 5C supplied by Minelco. The nanoclay (NC) was Cloisite 20A for PP and Cloisite 30B for PA6 and PBT supplied by Southern Clay Products. In addition sepiolite amine supplied by Tolsa as nanoclay was used for PBT. The preparation and characterisation of these materials have been reported elsewhere (*22,23,24*).

**Table III. Formulation of PP, PA 6, PBT and EVA polymer materials, with and without FR and NC (in wt %)**

| *Material* | *Polymer %* | *Fire Retardant %* | *Nanoclay %* |
|---|---|---|---|
| Polypropylene (PP) | 100* | | |
| PP+FR | 70* | 30 | |
| PP+NC | 95* | | 5 |
| PP+FR+NC | 65* | 30 | 5 |
| | | | |
| Polyamide 6 (PA 6) | 100 | | |
| PA 6+FR | 82 | 18 | |
| PA 6+NC | 95 | | 5 |
| PA 6+FR+NC | 77 | 18 | 5 |
| | | | |
| PBT | 100 | | |
| PBT+FR | 82 | 18 | |
| PBT+NC (Cloisite) | 95 | | 5 |
| PBT+NC (Sepiolite) | 95 | | 5 |
| PBT+FR+NC (Cloisite) | 77 | 18 | 5 |
| PBT+FR+NC (Sepiolite) | 77 | 18 | 5 |
| | | | |
| EVA | 100 | | |
| EVA+FR (ATH) | 32 | 68 | |
| EVA+FR (MH) | 32 | 68 | |
| EVA+NC | 95 | | 5 |
| EVA+FR (ATH) +NC | 32 | 63 | 5 |
| EVA+FR (MH) +NC | 32 | 63 | 5 |

* containing 5% PPgMA

## Results and Discussion

### Effluent from burning PA 6 based materials

Carbon monoxide yields were measured, together with hydrogen cyanide (HCN), nitric acid (NO) and nitrogen dioxide ($NO_2$) yields for PA 6 materials, for a series of characteristic fire types from well-ventilated to large vitiated (Table III). The yields are all expressed on a mass loss basis (i.e. mass of product per unit mass of sample.)

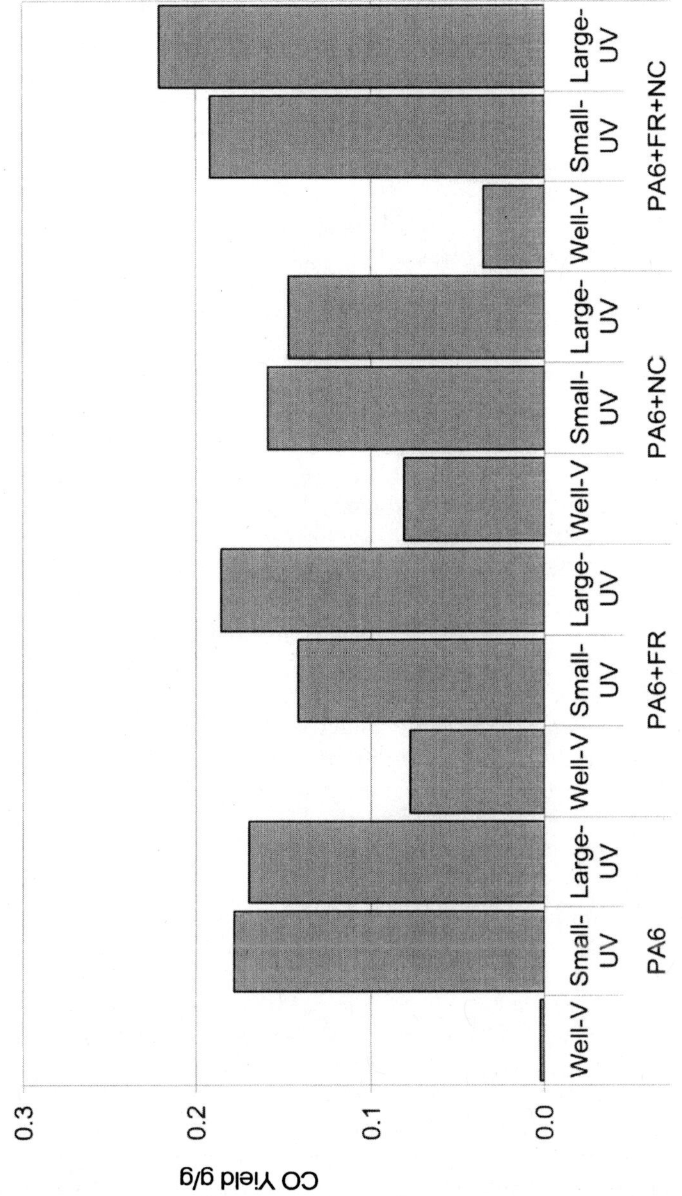


*Figure 3. Carbon monoxide yields for PA 6 based materials (V - ventilated, UV – underventilated)*

Figure 3 shows the CO yields from the PA 6 based materials under different fire conditions. This shows consistently lower CO yields for well-ventilated burning compared with small or large under-ventilated conditions. Under well-ventilated conditions it shows increased CO yields for materials including nanoclay or a fire retardant, but surprisingly the combined effect of both FR and NC result in lower CO yield. Under the relatively more toxic under-ventilated conditions, overall the yields of CO are much higher, but there is little difference between small and large under-ventilated conditions, or with incorporation of either fire retardant or nanoclay (*25*).

A similar trend to carbon monoxide is observed for hydrogen cyanide. HCN yields increase with reduced ventilation, but is less sensitive to the furnace temperature. The $NO_2$, NO and HCN yields are presented in Figure 4.

The higher yield of HCN for PA6+FR, for well-ventilated flaming corresponds to a higher yield of CO for the same conditions. It is interesting to note that the HCN yield increases with severity of the fire condition, whereas the CO yield levels off or even decreases

## Effluent from burning PP based materials

The yields of CO from burning PP under different fire conditions are presented in Figure 5. As with PA 6, carbon monoxide yields are much lower under well-ventilated conditions ($\phi$ < 1) compared to the under-ventilated conditions (*26*).

In contrast to the CO yields for PA 6, there is a progressive increase in CO yield in going from small to large under-ventilated flaming. Although the addition of either FR or NC has no significant effect on the CO yield, it appears that the presence of both NC and FR results in noticeably higher CO yields for under-ventilated flaming.

## Effluent from burning EVA polymer based materials

Ethylene-vinyl acetate copolymer (EVA) is a widely used material in the cable industry, which undergoes thermo-oxidative decomposition with the formation of a protective layer, and exhibits higher stability to thermal decomposition compared to the base polymer low density polyethylene (LDPE) (*7*).

The CO yields from burning EVA based materials are presented in Figure 6. The large percentage of aluminium hydroxide (ATH), which leaves a residue of about 40% alumina, (aluminium oxide) and the presentation of the data on a mass loss basis gives the false impression that the ATH had little effect on the CO yield. Since almost half the mass loss will be water from ATH, there seems

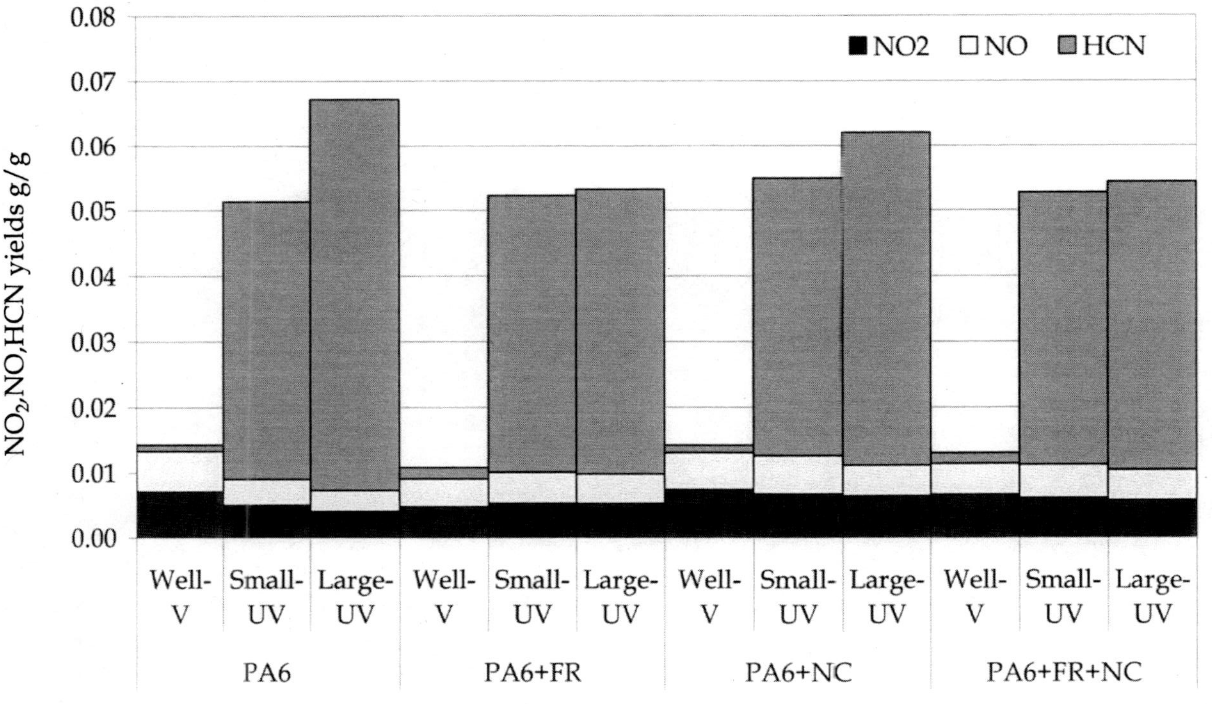


*Figure 4. HCN, $NO_2$, and NO yields for PA 6 based materials*

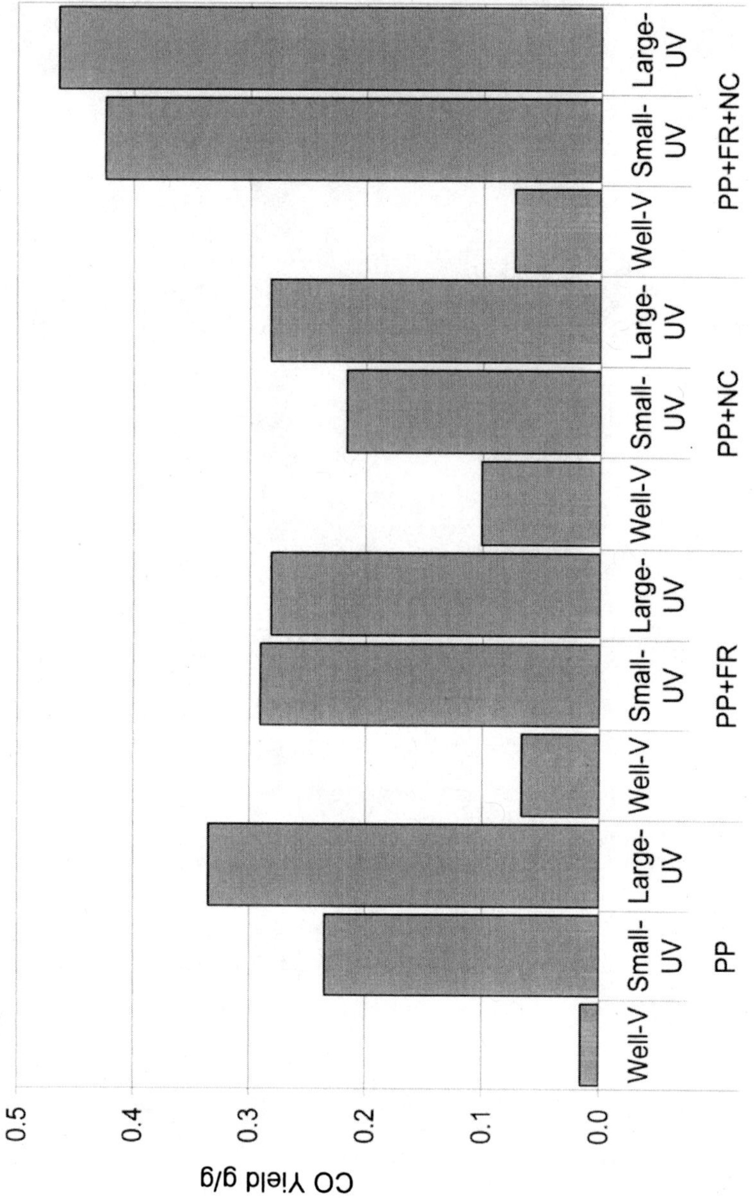


*Figure 5. Carbon monoxide yields for PP based materials*

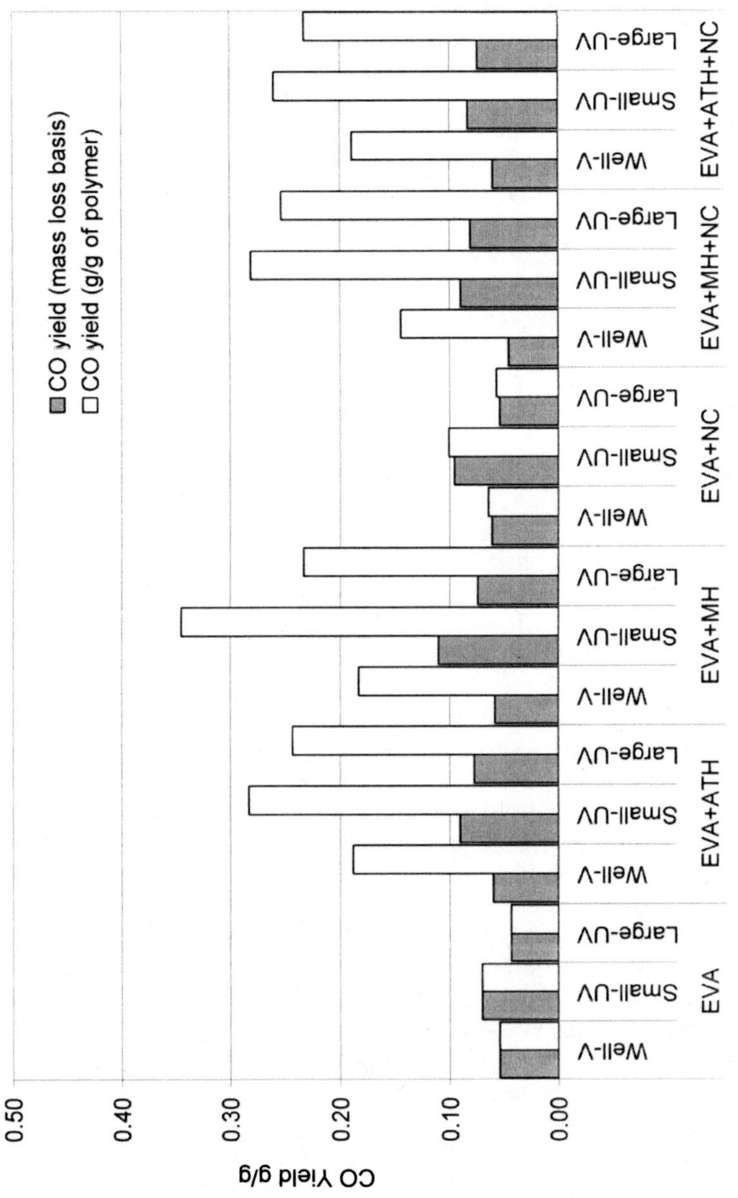


*Figure 6. Carbon monoxide yields for EVA based materials*

to be little effect on the CO yield. Therefore the data for the materials containing significant amounts of filler have been replotted on a yield on a g/g of polymer basis. This shows that ATH and magnesium hydroxide enhance the CO yield compared to the pure polymer, or polymer plus NC which have little effect. This is only important for materials which contain large amounts (~70%) of filler which decompose to produce water or other inert gas.

There is surprisingly little variation of CO yield with change in fire conditions for these proprietary EVA formulations although in earlier work lower yields of CO were found for well-ventilated conditions (*7*). However, this effect would best be described as low yields in under-ventilated conditions, since the "accepted" value of CO yield for under-ventilated fires is 0.2 g/g. The samples containing ATH show the apparently catalytic effect of alumina in increasing CO yields at low ventilation, as reported in earlier work on EVA*(18)*.

### Effluent from burning PBT based materials

The quantified combustion products for PBT under different fire conditions are presented in Figure 7.

These show a slight increase in CO yield from well-ventilated to under-ventilated for PBT and PBT with sepiolite, and PBT+FR combinations. However, the presence of the FR dramatically increases the CO yield when used alone, but has little effect when used in the presence of nanoclay. The nanoclay alone has little effect on the CO yield.

### Impact of fire retardants and nanofillers on toxicity

To illustrate the relative importance of the toxic product yields on the overall effluent toxicity, the contribution of the individual toxicants for each material towards the overall toxicity is considered. In each case, the toxicity is expressed as fractional effective dose (FED) for an arbitrary dilution of 1g of fuel loss per 50 litres of fire effluent. This is a rather concentrated fire effluent, and such exposure would normally only occur very close to a fire. However, the data is useful in assessing the relative importance of the different toxicants, and the relative effects of incorporation of fire retardants and nanofillers.

Figure 8 shows the low toxicity of well-ventilated burning, where the major toxicant is $NO_2$, and there are only minor differences between the FR and NC materials compared to the virgin polymer.

During small or large under-ventilated burning HCN is the major toxicant, showing a consistent, but modest increase from small to large under-ventilated flaming, and a reduction in toxicity on incorporation of either FR or NC or both.

Figure 9 shows toxicants from burning PP materials. The relative toxicity is much lower than that for the nylon samples in under-ventilated conditions

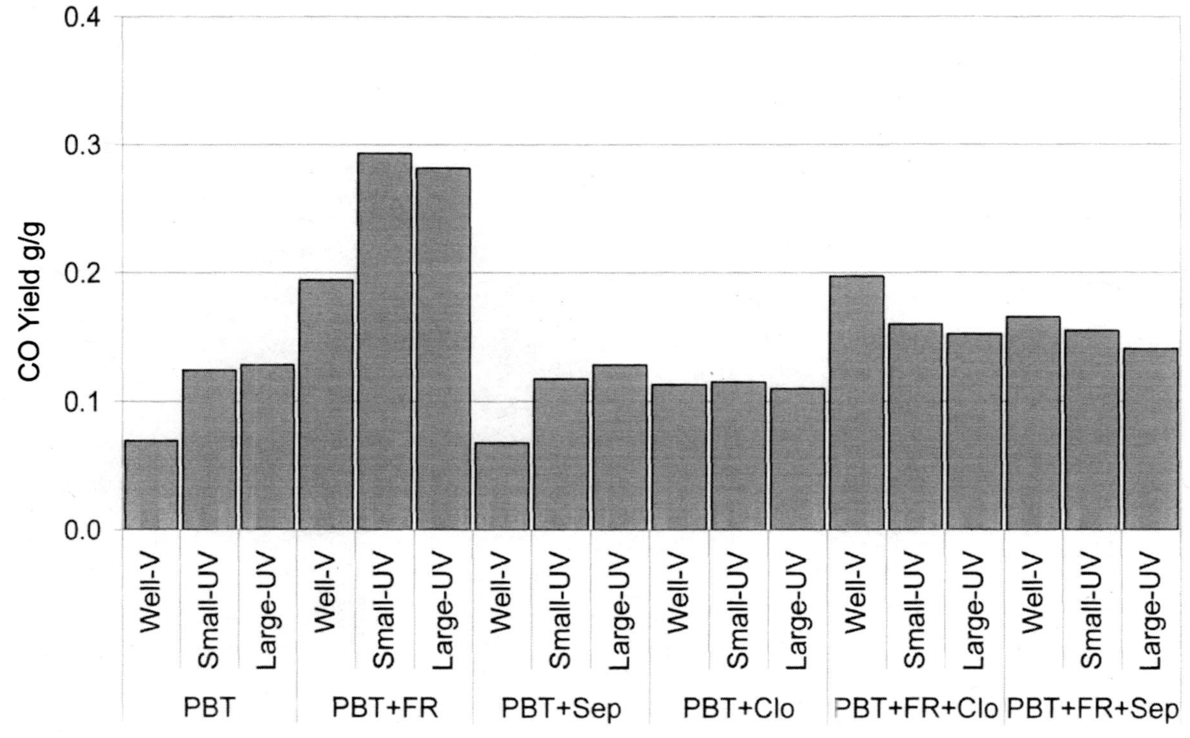


*Figure 7. Carbon monoxide yields for PBT based materials*

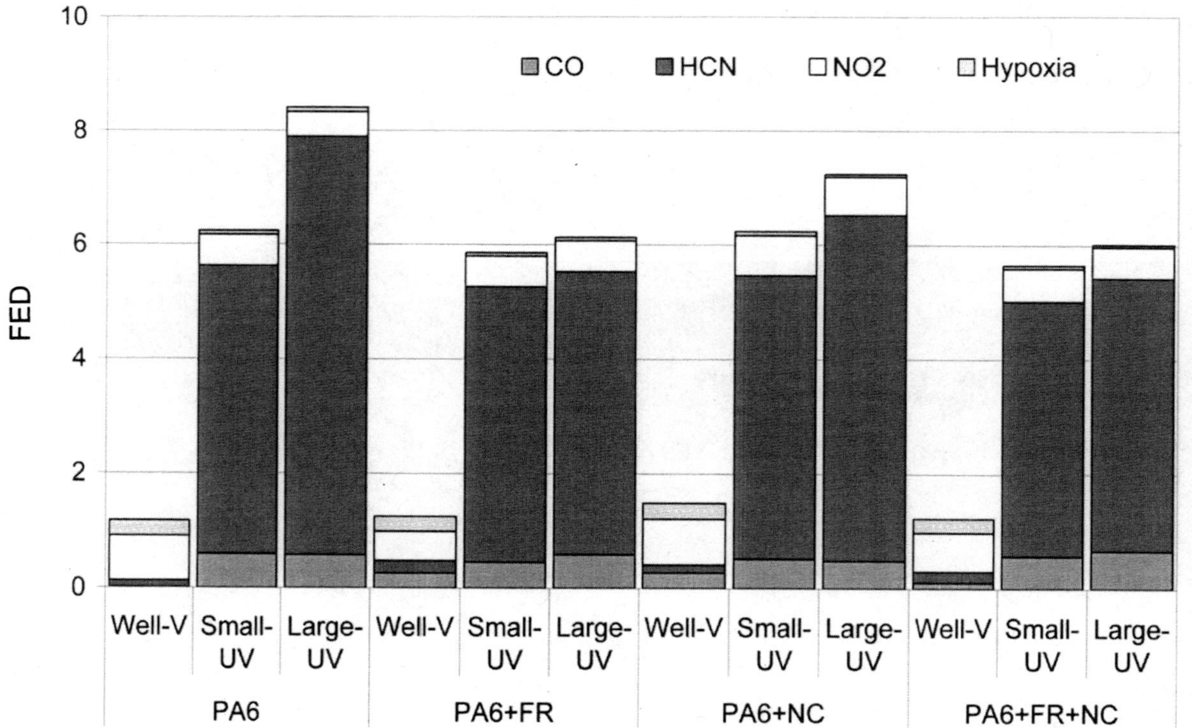


*Figure 8. Contribution of individual gases to toxicity from burning PA6 materials under different fire conditions*

which produced HCN. Unlike HCN under well-ventilated conditions, it is clear that both NC and FR increase the toxicity of the effluent, by the increasing the CO yield. Again there is a progressive increase in toxicity from small under-ventilated to large under-ventilated which is shown consistently across the samples. There is also a decrease in CO for all the materials under the most dangerous under-ventilated conditions on incorporation of either NC or FR, and only a slight increase when both are present.

The predicted fire toxicity for EVA based materials is shown in Figure 10, again shows the surprisingly high CO yields in well-ventilated conditions. However, in comparison with other polymers all of the FED values are rather low, even in this concentrated fire effluent.

Figure 11 shows the consistently higher toxicity from burning PBT under different fire conditions, and the addition increase in toxicity resulting from the use of fire retardant. It is interesting to see this effect disappear on incorporation of either sepiolite or cloisite.

## Conclusions

The presence of fire retardants and the incorporation of nanoclay reduces flammability, but this work has shown that there is no systematic increase in fire toxicity through the use of these additives, particularly under the most lethal under-ventilated fire conditions.

Predicting the toxicity of fire effluents, where individual toxicants have fundamentally different impact on human physiology is very difficult. However, quantification of the individual components, giving rise to toxic potency, and assessment of overall fire effluent toxicity is clearly an essential part of fire hazard analysis. Comprehensive post mortem studies have shown the majority of fatalities (50 to 80%) in fires had excessive levels of carboxyhaemoglobin, consistent with exposure to high levels of CO (*27*). However, in most cases HCN is not included in the post-mortem analysis. CO and HCN are likely to have an additive effect on depriving the body of oxygen, rather than synergistic, (greater than additive) or antagonistic (less than additive) effect, as the main mechanism of either gas is to reduce the ability the body to deliver oxygen to the cells.

In some cases CO, which is often assumed to be the most, or even the only toxicologically significant fire gas, is of less importance than HCN, when nitrogen is present, and CO may even be less important than organo-irritants. For the PA 6 samples the yield of the main toxic gases (CO and HCN) increases by more than a factor of 10 as the fire grows from well-ventilated to large-vitiated. PP is a hydrocarbon polymer, without any heteroelements to increase the fire toxicity. Although it has some structural similarity to EVA, it shows higher CO yields in under-ventilated conditions.

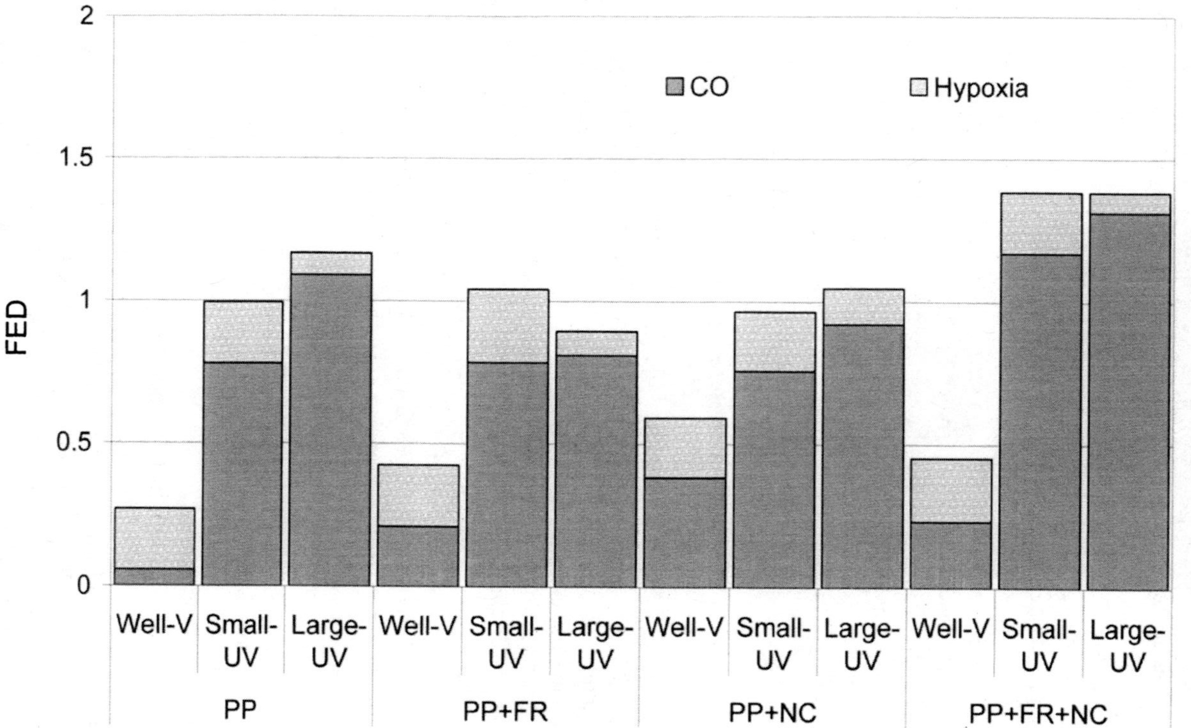


*Figure 9. Contribution of individual gases to toxicity from burning EVA materials under different fire conditions*

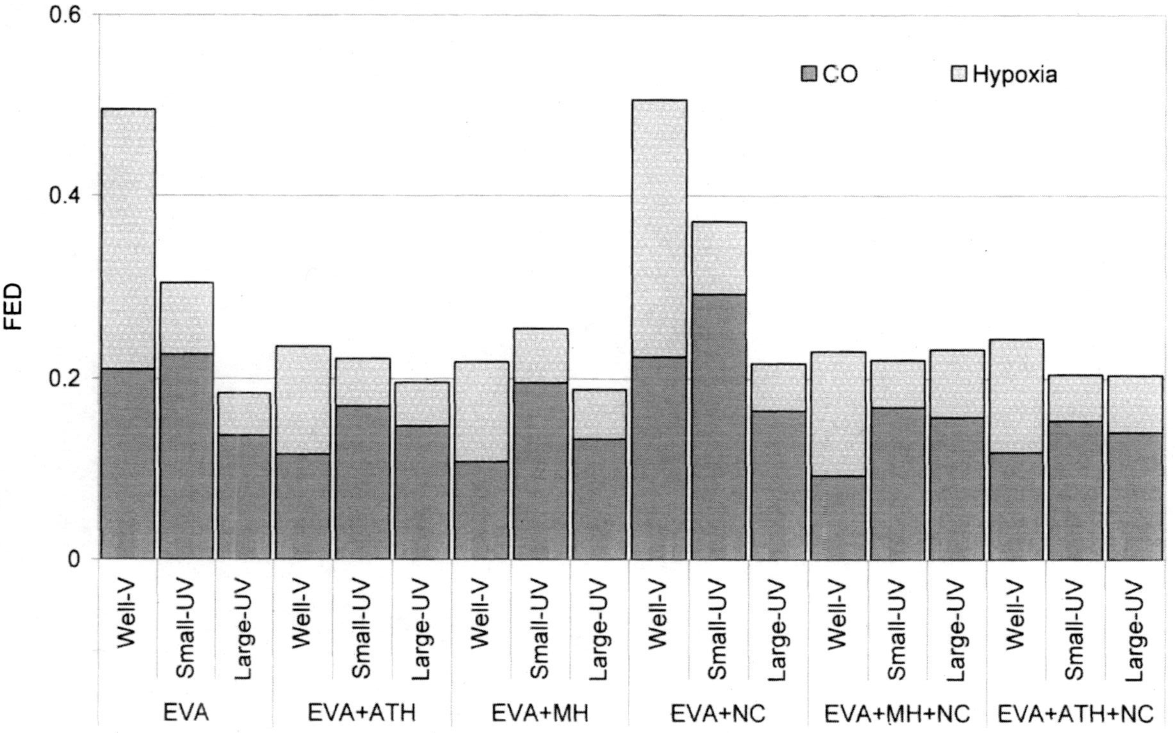


*Figure 10. Contribution of individual gases to toxicity from burning PBT materials under different fire conditions*

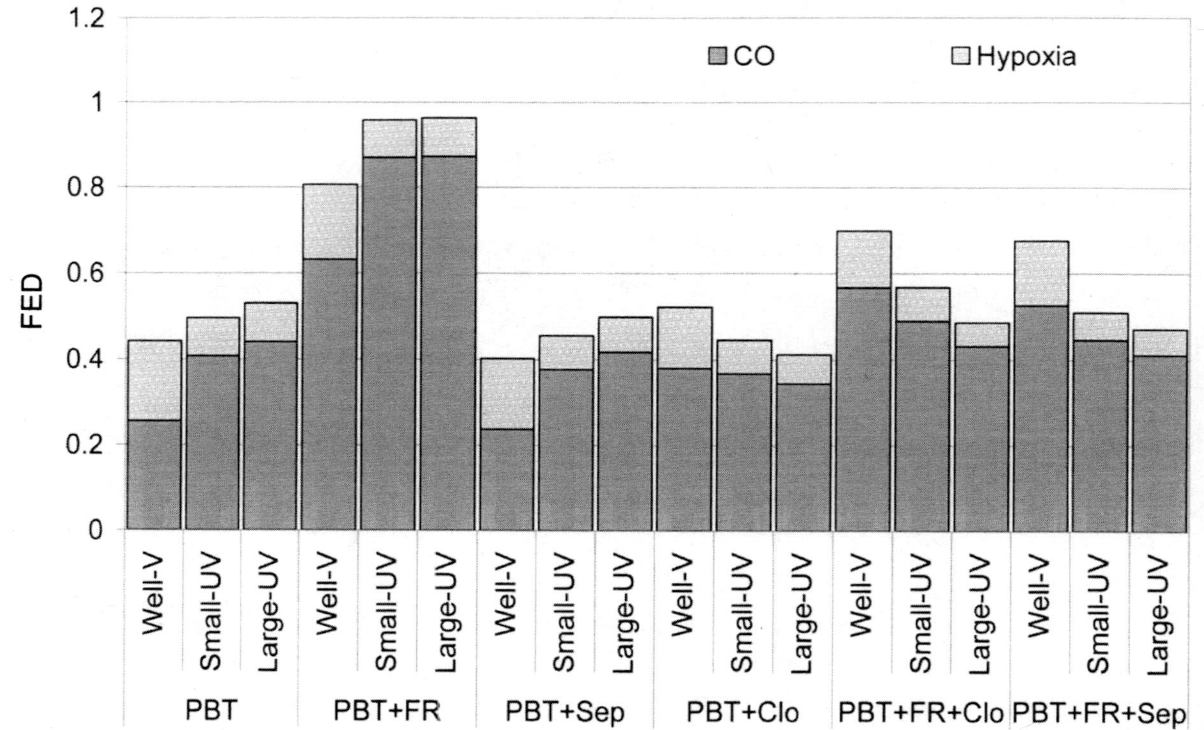


*Figure 11. Contribution of individual gases to toxicity from burning PBT materials under different fire conditions*

# References


1. United Kingdom Fire Statistics, Home Office , London, **2006**.
2. ISO 19706:2007 Guidelines for assessing the fire threat to people, **2007**.
3. Stec, A.A.; Hull, T.R.; Lebek, K.; Purser, J.A.; Purser D.A.; **Fire Mater.**, **2008**, *32*, 49-60.
4. Hull, T.R.; Stec, A.A.; Lebek, K.; and Price, D.; *Polym. Degrad. Stab.*, **2007**, *92*, 2239-2246.
5. Schartel, B.; Bartholmai, M.; Knoll, U.; *Polym. Adv.Tech.,* **2006**, *17*, 9-10, 772.
6. Hull, T.R.; Price, D.; Liu, Y.; Wills, C.L.; and Brady, J.; *Polym. Degrad. Stab.*, **2003**, Vol. 82, p. 365.
7. Hull, T.R.; Carman, J.M.; and Purser, D.A.; *Polym. Int.,* **2000**, *49*, 1259.
8. ISO 13571:2007 Life-threatening components of fire - Guidelines for the estimation of time available for escape using fire data, **2007**.
9. Blomqvist, P.; and Lonnermark, A.; *Fire Mater.*, **2001**, *25*, 71.
10. Andersson, B.; Markert, F.; and Holmstedt, G.; *Fire Safety J.,* **2005**, *40*, 439.
11. Stec, A.A.; Hull, T.R.; Lebek, K.; Purser, J.A.; Purser, D.A.; *Fire Mater.*, **2008**, *32*, 49.
12. ISO 13344:2004 Estimation of lethal toxic potency of fire effluents, **2004**.
13. ISO-TR 9122-5:1993 Toxicity testing of Fire Effluents-Part 5, **1993**.
14. Purser, D.A.; SFPE Handbook of Fire Protection Engineering, National Fire Protection Association, Third Edition p. 2-83, **2002**.
15. Pitts, W.M.; *Prog. Energy Combustion Sci.*, **1995**, *21*, 97.
16. ISO 19703:2005 Generation and analysis of toxic gases in fire - Calculation of species yields, equivalence ratios and combustion efficiency in experimental fires, **2005**.
   17 BS 7990:2003 Tube Furnace method for the determination of toxic product yields in fire effluents, **2003**.
18. Hull, T.R.; Quinn, R.E.; Areri, I.G.; and Purser, D.A.; *Polym. Degrad. Stab.* **2002**, *77*, 235.
19. ISO TS 19700:2006 Controlled equivalence ratio method for the determination of hazardous components of fire effluents, **2006**.
20. Stec, A.A.; Lebek, K.; and Hull, T.R.; *Polym. Degrad. Stab.,* in press
21. ISO 19702:2004 Analysis of fire gases using Fourier infra-red technique-FTIR, **2004**.
22. Samyn, F.; Bourbigot, S.; Jama, C.; Bellayer, S.; Nazare, S.; Hull, T.R.; Fina, A.; Castrovinci, A.; Camino, G.; *Eur. Polym. J.*, **2008**, *44*, 1631.
23. Samyn, F.; Bourbigot, S.; Jama, C.; Bellayer, S.; Nazare, S.; Hull, T.R.; Castrovinci, A.; Fina, A.; and Camino G.; *Eur. Polym. J.*, **2008**, *44*, 1642.
24. Nazare, S.; Hull, T.R.; Biswas, B.; Samyn, F.; Bourbigot, S.; Jama, C.; Castrovinci, A.; Fina, A.; and Camino, G.; *Fire Retardancy of Polymers: New Strategies and Mechanisms*, Royal Society of Chemistry, in press.

25. Stec, A.A.; Hull, T.R.; New Strategies and Mechanisms, Royal Society of Chemistry, in press.
26. Stec, A.A.; Hull, T.R.; Proceedings of the Fifth International Seminar on Fire and Explosion Hazards, Edinburgh, in press.
27. Punderson, J.O.; *Fire Mater.*, **1981**, *5*, 41.

# Thermal Degradation Studies

## Chapter 22

# Modelling Thermal Degradation of Flame-Retarded Epoxy Resin Formulations under Different Heating Conditions

**Everson Kandare[1], Baljinder K. Kandola[1], Richard A. Horrocks[1], and John E. J. Staggs[2]**

**[1]Centre for Materials Research and Innovation, University of Bolton, Deane Campus, Bolton BL3 5AB, United Kingdom**
**[2]School of Process, Environmental and Materials Engineering, University of Leeds, Leeds LS2 9JT, United Kingdom**

The thermal degradation kinetics of epoxy resin, triglycidyl-p-aminophenol (TGAP), and its flame-retarded formulations are investigated by thermogravimetric analysis (TGA) under oxidative conditions and are modelled using a simple global multi-step mechanistic scheme. Experimental TGA data acquired under non-isothermal conditions is used to optimise the kinetic parameters for selected boundary linear heating rates. These optimised kinetic parameters which adequately describe the thermal degradation process of epoxy resin formulations over the defined range of experimental conditions are then applied in the prediction of mass loss profiles of investigated materials under different heating conditions.

 

# Introduction

In recent years, interest in flame-retardants' (FRs) use in thermoplastics and thermosetting resin formulations has been focused on nitrogen and phosphorous containing additives due to their low toxicity and low smoke volume productions (*1-4*). The addition of FRs to polymer matrices, considerably affect the thermal stability of the resultant polymer-additive composite material. The thermal degradation behaviour of the resin formulations containing FR additives can be assessed via thermogravimetric analysis from which kinetic parameters describing the decomposition process such as apparent activation energy ($E_a$), Arrhenius pre-exponential factor ($A$), apparent order of reaction ($n$) and hence the rate constant ($k$) can be derived. There is a vast amount of literature investigating the thermal (inert atmosphere) and thermo-oxidative degradation kinetics of flame-retarded polymeric materials but the interpretation of the extracted kinetic parameters in relation to degradation mechanisms still remains a challenge. However, these parameters may reveal the flame-retarding mechanism during solid state degradation reactions of polymeric materials from the view point of their relative apparent magnitudes.

Thermogravimetric data collection is achieved via isothermal and non-isothermal methods with the earlier method less preferred as it does have some evident disadvantages that include; (i) samples undergoing side reactions during the process of raising the temperature to a desired value and (ii) the reaction being restricted to a single temperature (i.e. no temperature history). There are two approaches via which the kinetic parameters can be extracted from non-isothermal thermogravimetric data which are model free and model fitting methods. While the latter method has been extensively criticised for non-isothermal applications due to resultant indistinguishable fits between different models, (*5, 6*) its importance can not be overestimated as it may provide information on energy barriers and offer mechanistic clues. On the other hand, model free approaches allow one to examine the dependence of apparent activation energies on the conversion fraction, information which may give a clue as to the possible numbers of physical stages involved during thermal degradation. The inevitable need for cost effective ways to evaluate the thermal stability of resin formulations under different heating conditions over wide temperature ranges has prompted researchers to develop simplified kinetic schemes to predict thermal degradation of polymeric materials combining both model free and model fitting methods (*7-9*).

Despite the absence of an obvious experiential relationship between the chemical processes occurring during thermal degradation of polymeric materials and the calculated corresponding kinetic values, the latter can be useful in an attempt to predict the rate of mass loss under different heating conditions when the collection of experimental data is time consuming; i.e. screening for effective resin/flame retardant formulations for applications under different service temperatures or environments. The burning behaviour including the temperature profiles and mass loss of polymeric materials under experimental conditions that simulate real fire conditions such as cone calorimetry can be predicted via extrapolation and application of kinetic parameters to faster heating rates experienced in such cases.

In this paper we aim to determine global intrinsic kinetic model parameters that can reproduce accurately the profiles of mass-temperature dependence during the thermal degradation of epoxy formulations containing flame retardants, melamine phosphate (MP) and melamine pyro-phosphate (MPP). For a given formulation, a multi-step kinetic scheme is evaluated at determined boundary heating rates and the obtained average kinetic constants then used to predict the mass loss profiles of the same resin formulation under different heating rates (within the pre-determined heating rate boundaries) with the assumption that the degradation mechanism is invariant of heating conditions. Also, the variation of kinetic parameters with heating rates is evaluated in order to assess the applicability of the predictive tool at slower and faster heating rates.

## Experimental

### Materials and Sample Preparation

The samples used in this study were kindly provided by Dr Bhaskar Biswas and were prepared in partial fulfilment of his doctoral thesis (*3*). These samples show interesting fire behaviours under cone calorimetry thus we decided to explore their thermal degradation kinetics in detail. Epoxy resin samples with and without flame retardants were formulated via a hot-melt method (*3, 10*). The sample compositions and identities are given in Table 1.

**Table I. Mass percentages of various components in resin formulations**

| *Sample composition* | *Sample code* | *Mass (%)* | |
|---|---|---|---|
| | | *Resin* | *FR* |
| Epoxy | EP | 100 | - |
| Epoxy + 8% MP | EP-MP | 92 | 8 |
| Epoxy + 8% MPP | EP-MPP | 92 | 8 |

## Thermogravimetric Analysis

Thermogravimetric analysis (TGA) of uncured resin systems was performed on an SDT 2960 simultaneous DTA–TGA instrument from room temperature - 900 °C using 10 ± 1 mg samples heated at constant heating rates of 5, 10, 12.5, 15, 17.5, 20 and 50 °C/min in air flowing at 100 ± 5 mL/min. The experiments were performed in triplicates and show good reproducibility.

## Mathematical Modelling

The thermal degradation of polymeric materials such as epoxies inescapably follow multifaceted reaction mechanisms and revealing the kinetics of every single step while it is ultimately possible, entails an extensive understanding of the thermo-chemical processes involved and demands a rigorous knowledge of computational dynamics. The kinetic fitting of thermo-analytical curves may yield vast amounts of information about the overall degradation process but can not differentiate between partially overlapping chemical reactions. However, in cases where only the thermo-physical property degradation history of epoxy resin formulations is important; for example the effect of mass loss in examining mechanical strength retention under prescribed high temperature service conditions, the use of simplified multi-step kinetic degradation schemes in global kinetics modelling is acceptable.

The rate of thermal decomposition of a reactant can be approximated via a kinetic scheme; $\frac{d\underline{m}}{dt} = \underline{f}(\underline{m})$, where $\underline{m}$ is a vector of mass fractions, and $\underline{f}$ is a vector-valued function representing the degradation mechanism. First order decomposition rate constants are usually assumed to have Arrhenius dependency on temperature as shown by Eq. 1 below;

$$k_i = A_i \exp(-E_{ai}/RT),\ i = 1..\infty, \qquad [1]$$

where $A$ is the pre-exponential factor, $E_a$ is the activation energy, $R$ is the gas constant and $T$, is the sample temperature. The sample temperature is known as a function of time, $t$: $T = T_0 + \beta t$:, where $T_0$ is the initial temperature and $\beta$ is the heating rate.

Sets of experimental data in form of $m$, $T$ (or $m$, $t$) are collected and kinetic parameters are generated from anticipated thermal degradation chemistries based on proposed mechanisms found in literature (*3,11,12*). The set of calculated or predicted values of mass fractions is compared with experimental values, and improved values of the kinetic parameters are searched for until a minimum value of the sum of least squares for $N$ data sets is obtained:

$SS=\sum_{i=1}^{N}(\alpha_{meas}-\alpha_{calc})^2$ where $\alpha_{meas}$ and $\alpha_{calc}$ are experimental and predicted mass fractions respectively at each of the $N$ data points. The stopping point for the search is henceforth defined by the objective function above in accordance to a predefined experimental error range anticipated. For more details about the optimisation procedure the reader is referred to our previous work published elsewhere (*13*). Optimised kinetic parameters were obtained for the epoxy resin alone and also for resin formulations containing 8% wt. MP (EP-MP) and MPP (EP-MPP) using thermogravimetric data collected at 10 and 20 °C/min. The optimised kinetic constants obtained using these two heating rates were averaged and used to predict mass loss profiles at intermediate heating rates of 12.5, 15 and 17.5 °C/min in Maple 6.0® using an ODE solver for stiff equations (lsode) assuming that the kinetic parameters are invariant of the heating rate. We note here, differences between kinetic constants obtained at heating rates between 10 and 20 °C/min are subtle, Table 2. On another hand, kinetic parameters, $E_a$ and $A$, were plotted against the heating rates (10-20 °C/min) for the epoxy resin and their respective gradients extrapolated to obtain corresponding values at heating rates of 5 and 50 °C/min. These estimated kinetic parameters were used to predict the mass loss profiles which were then validated against experimental data.

# Results

## Chemical Kinetics and Predictive Modelling

### *Epoxy Resin*

The decomposition behaviour of amine cured epoxide resins has been reported to occur via at least three overlapping stages (*3,11-16*). The first step at temperatures around 200 – 250 °C may be attributed to homolytic scission of chemical bonds within the polymer chains. Even though these reactions do not result in loss of mass, they however, adversely affect the physical properties including the mechanical behaviour (*17*). The first major weight loss occurring in the second stage may be caused by the elimination of water molecules from the oxypropylene group, $-CH_2-CH(OH)-$, which may lead to the formation of organic species with double bonds. The dehydration of the epoxy networks occur simultaneously with the volatilisation of some small molecular species leading to the formation of a primary residual char. The final stage involve the thermo-oxidative degradation of the primary char which may proceed via reactions such as isomerisation, chain transfer and radical transfer to yield negligible residual char (*9*).

A multi-step scheme to describe the degradation mechanism of an amine-cured epoxy resin under a constant heating rate is shown in Scheme 1 below:

resin ($m_1$) → $r_1$ dehydrated resin ($m_2$) + ($1$-$r_1$) water [2]

$r_1$ dehydrated resin ($m_2$) → primary char ($m_3$) + volatiles [3]

primary char ($m_3$) → volatiles [4]

*Scheme 1. A simplified multi-step degradation scheme for an amine-cured epoxy resin.*

The degradation steps represented in Scheme 1 above can be presented as a coupled system of ordinary differential equations with 4 pairs of unknown kinetic parameters ($E_a$ and $A$) governing the rate at which the volatilisation process proceeds at defined heating rates:

$$\frac{dT}{dt} = \beta$$

$$\frac{dm_1}{dt} = -k_1 m_1$$

$$\frac{dm_2}{dt} = r_1 k_1 m_1 - k_2 m_2 - k_3 m_2$$

$$\frac{dm_3}{dt} = k_3 m_2 - k_4 m_3$$

where $r_1$ is the dehydrated fraction of the epoxy resin recorded starting at temperatures around 100 °C, $k_i$ ($i$ = $1..4$) are the kinetic constants associated with the degradation steps involved and $m_j$ the variations of mass for species $j$ = $1..3$ with time, $t$. Each of the reaction rates $k_i$ is assumed to have Arrhenius type dependence as given in Eq. 1 with initial conditions defined as follows; $T(0)$ = $T_0$, $m_1$ = 1, $m_2$ = 0, $m_3$ = 0 and $\beta$ = defined heating rates in °C/min. Since each group of molecular species evolved during thermal degradation has an independent variation with time, the overall mass loss is estimated from a linear combination of individual mass loss profiles (*13*).

Apparent kinetic constants for the degradation of epoxy resin calculated for all heating rates were optimised using the algorithm previously developed in our laboratories (*13*) and are presented in Table 2. While the obtained kinetic parameters are unique in relation to describing the degradation profile of a particular material, we however, found out that the values obtained in this work are similar to those obtained in our previous work (*13*) even though we were using a different epoxy resin. We are therefore led to conclude that while the degradation of epoxies follow a series of complex chemical reactions, the global physical processes (i.e. mass loss) occurring in each case can be defined by essentially a handful of grossly simplified mass loss kinetics models.

**Table II. Optimised kinetic parameters for epoxy resin, EP at different heating rates**

| $k_i$ | *Kinetic parameters* | *Heating rates (°C/min)* | | | | | | |
|---|---|---|---|---|---|---|---|---|
| | | *5.0*[a] | *10.0* | *12.5* | *15.0* | *17.5* | *20.0* | *50.0*[a] |
| $k_1$ | E/R($\times 10^3$) [K] | 15.0 | 15.0 | 15.0 | 15.0 | 15.0 | 15.0 | 15.0 |
| | ln A | 17.1 | 17.3 | 17.5 | 17.5 | 17.8 | 17.7 | 17.9 |
| | CF[b] | 876 | 868 | 857 | 859 | 842 | 849 | 839 |
| $k_2$ | E/R($\times 10^3$) [K] | 10.9 | 10.9 | 10.9 | 10.9 | 10.9 | 10.9 | 10.9 |
| | ln A | 10.8 | 10.8 | 11.0 | 11.0 | 11.1 | 11.1 | 11.2 |
| | CF[b] | 1006 | 1002 | 986 | 986 | 981 | 980 | 985 |
| $k_3$ | E/R($\times 10^3$) [K] | 16.0 | 16.0 | 16.0 | 16.0 | 16.0 | 16.0 | 16.0 |
| | ln A | 19.0 | 18.9 | 18.9 | 18.9 | 18.9 | 18.9 | 19.0 |
| | CF[b] | 844 | 846 | 847 | 847 | 847 | 848 | 848 |
| $k_4$ | E/R($\times 10^3$) [K] | 14.2 | 14.2 | 14.2 | 14.2 | 14.2 | 14.2 | 14.2 |
| | ln A | 11.2 | 11.3 | 11.2 | 11.4 | 11.6 | 11.3 | 11.2 |
| | CF[b] | 1268 | 1259 | 1263 | 1249 | 1227 | 1255 | 1245 |

[a] Kinetic values at 5 and 50 °C/min were obtained via extrapolation.
[b] CF = Compensation factor calculated as the ratio of *E/R* to ln *A*.

Optimised kinetic parameters obtained for the boundary heating conditions of 10 and 20 °C/min were averaged and used to predict mass loss profiles of the neat epoxy resin at intermediate heating rates of 12.5, 15 and 17.5 °C/min in Maple 6.0® using an ODE solver for stiff equations (lsode). The experimentally obtained and predicted mass fractions as a function of time for the epoxy resin are shown in Fig. 1(a) while Fig 1(b) shows plots of experimental versus theoretically predicted mass fractions. There is a good agreement between experimental and predicted data sets suggesting that the derived mathematical predictive model employed herein is reliable. The calculated objective functions are 0.07, 0.09, and 0.21 for heating rates of 12.5, 15 and 17.5 °C/min respectively. These values are lower than 0.3 which would be the maximum permissible if an error range of ±3% is considered.

While kinetic constants are argued to have no physical meaning, it is evident from this study that they still may be useful in the rapid assessment of the thermal behaviour of polymeric materials for a given range of heating conditions over which the degradation mechanism is "invariant". This mathematical predictive tool thus can provide a very cost effective method of generating mass loss profiles under different heating conditions which may prove very much pertinent in the extraction of effective kinetic parameters using multiple heating rates kinetics (MHRKs), applying the Flynn-Wall-Ozawa (*18*) method specifically derived for heterogeneous chemical reactions under linear heating rates (*19-21*).

The exceptional agreement between experimental and predicted data sets dictates that complete sets of optimised kinetic parameters determined for boundary heating rate conditions (10 and 20 °C/min) can be employed in order to predict the degradation profiles for intermediate heating rates (12.5, 15 and 17.5 °C/min). While the authors report a success in predicting mass loss profiles within a defined range of heating rates, this does not suffice to say that the predictive models are applicable for any other heating rate outside the boundary conditions.

To assess the applicability of the degradation scheme for heating rates outside the defined range, the mass loss profile of the neat resin was predicted for a slow heating rate of 5 °C/min and fast heating rate of 50 °C/min using the extrapolated kinetic parameters obtained from linear heating rates between 10 and 20 °C/min inclusively. Figs. 2 (a) and (b) show the variation of experimental and predicted mass fractions for the neat epoxy resin at heating rates of 5 and 50 °C/min respectively under an air atmosphere. There are subtle to considerable mass variations at the same temperature points over the entire conversion region. The predicted mass fractions of the neat resin are either under or overestimated relative to the experimentally obtained data. Overall, the attempt to use the predictive model to simulate the thermo-physical history (mass loss–time profiles) of the neat resin at very slow or fast heating rates can be described as somewhat successful. The proposed mechanism may not be applicable at very different heating due possibly to changes in the degradation chemistry. Further work needs to be done to improve the prescribed degradation mechanism in order to capture all events.

### *Flame-retarded Epoxy Resin Formulations*

The presence of phosphorus and nitrogen containing additives in epoxy resin formulations offer enhanced thermal stability and flame retardancy at low fractional loadings and are becoming increasingly popular (*1-4, 22*). Thermogravimetric mass loss profiles of the epoxy resin and its flame-retarded

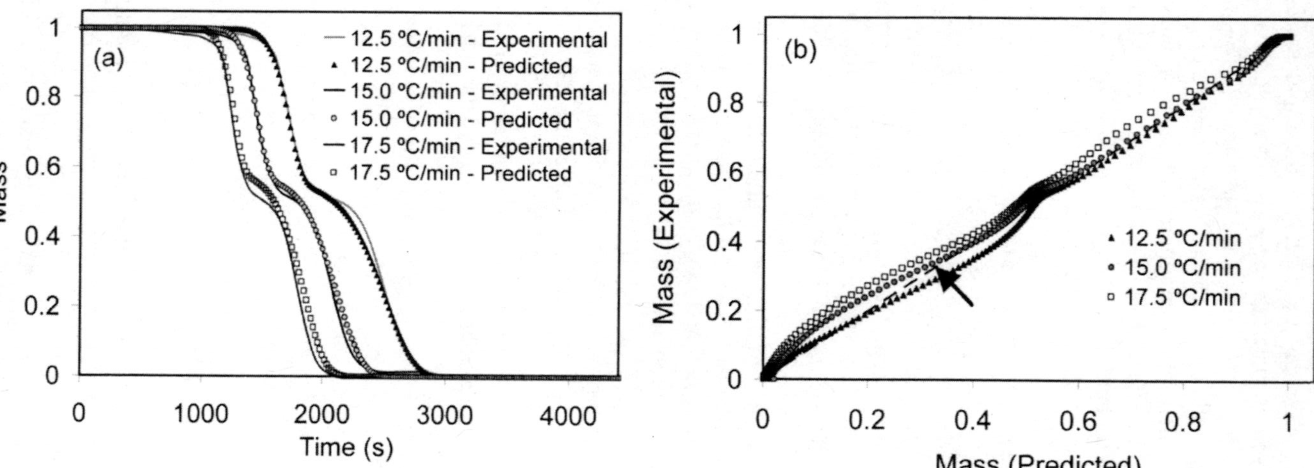


*Figure 1. (a) TGA experimental (lines) and predicted (data points) mass loss-time profiles, (b) experimentally obtained mass loss data plotted against predicted mass loss data points for EP at 12.5, 15.0 and 17.5 °C/min in air; the dashed line indicated by an arrow in 1 (b) represents the line $x = y$.*

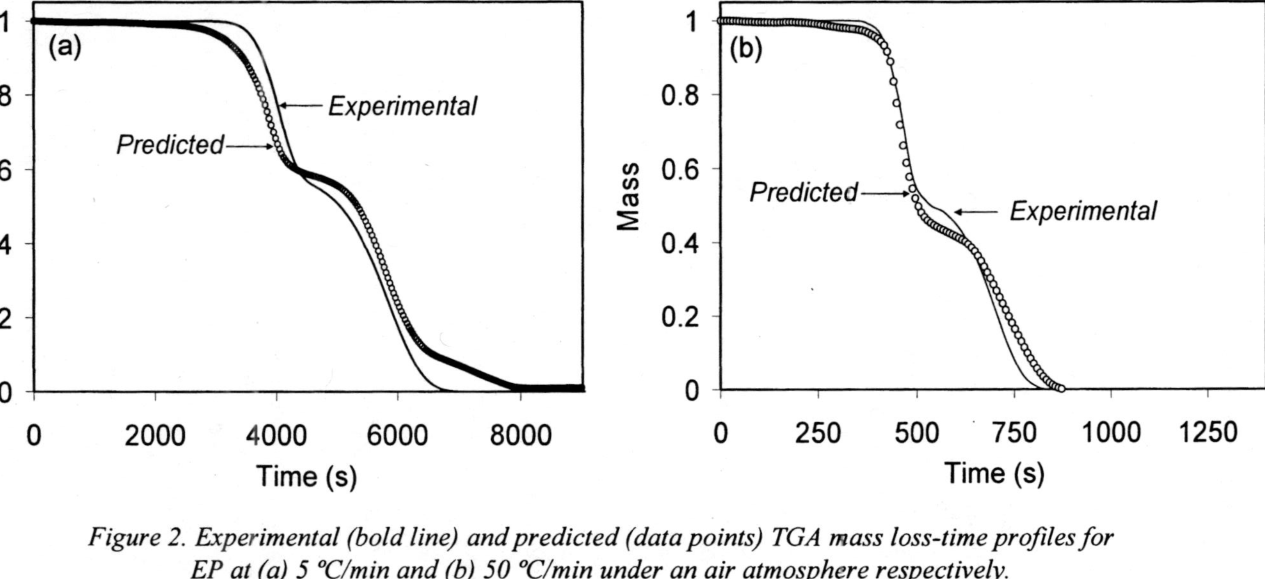


*Figure 2. Experimental (bold line) and predicted (data points) TGA mass loss-time profiles for EP at (a) 5 ºC/min and (b) 50 ºC/min under an air atmosphere respectively.*

formulations, EP-MP and EP-MPP, as a function of temperature at a heating rate of 20 °C/min in an air atmosphere are reported and extensively discussed in Biswas' PhD thesis (*3*). The inclusion of flame-retardant additives is not expected to significantly change the first two stages of thermal degradation of the epoxy resin but to promote char formation at higher degradation temperatures. Thus the schematic describing the degradation mechanism of epoxy resin shown above (Scheme 1) requires a subtle adjustment to factor in the subsequent thermo-oxidative degradation of the primary char into a more thermally stable carbonaceous char at higher temperatures. We note here that the degradation schemes presented represent overall physical processes or stages emanating from multifaceted chemical reactions which are otherwise too many and very complex to include in our analytical solution. Thus, while the participation of $O_2$ in the thermo-oxidative degradation of primary char into secondary and final residual char is envisaged, its inclusion in the degradation scheme renders the resultant coupled set of ODEs complex and impossible to solve within our means. A simplified multi-step kinetics scheme to describe the degradation mechanism of an amine-cured epoxy resin containing flame retardants, MP and MPP, under a constant heating rate in air is shown in Scheme 2 below:

resin matrix ($m_1$) → $r_1$ dehydrated resin matrix ($m_2$) + ($1$-$r_1$) water [5]

$r_1$ dehydrated resin matrix ($m_2$) → primary char ($m_3$) + volatiles [6]

primary char ($m_3$) → final char ($m_4$) + volatiles [7]

*Scheme 2. A simplified multi-step degradation scheme for an amine-cured epoxy resin containing phosphorous - based flame retardants.*

The degradation steps represented in Scheme 2 above are presented as a coupled system of ordinary differential equations with 5 pairs of unknown kinetic parameters ($E_a$ and $A$) governing the rate at which the volatilisation process proceeds at defined heating rates:

$$\frac{dT}{dt} = \beta$$

$$\frac{dm_1}{dt} = -k_1 m_1$$

$$\frac{dm_2}{dt} = r_1 k_1 m_1 - k_2 m_2 - k_3 m_2$$

$$\frac{dm_3}{dt} = k_3 m_2 - k_4 m_3 - k_5 m_3$$

$$\frac{dm_4}{dt} = k_5 m_3$$

where $r_1$ is the dehydrated fraction of the epoxy resin matrix recorded at temperatures above 100 °C, $k_i$ ($i = 1..5$) are the kinetic constants associated with the degradation steps involved and $m_j$ the variations of mass for species $j = 1..4$ with time, $t$. Each of the reaction rates $k_i$ is assumed to have Arrhenius type dependence as given in Eq. 1 with initial conditions defined as follows; $T(0) = T_0$, $m_1 = 1$, $m_2 = 0$, $m_3 = 0$, $m_4 = 0$ and $\beta$ = defined linear heating rates. Since each of the molecular species has an independent variation with time, the overall mass loss rate is estimated from a linear summation of individual rates (*13*).

Optimised apparent kinetic constants for the thermo-oxidative degradation of flame-retarded epoxy resin formulations were obtained for heating rates between 10 and 20 °C/min and are presented in Table 3. There is no significant difference between corresponding kinetic parameters from the two systems, EP-MP and EP-MPP, suggesting that the degradation mechanism is similar in both cases. However, head to head comparison of the same physical processes for the neat resin and flame-retarded formulations via the use of compensation factors defined herein as the ratio of *E/R* to ln *A* reveals the following:

i. the physical desorption of water molecules (dehydration process), ($k_1$), has a higher compensation factor value for the neat resin (Table 2) than for the flame-retarded resins (Table 3) implying that the interaction between the polymeric surface and adsorbed water molecules is stronger with the former than the latter. The presence of flame retardant additives may alter the curing kinetics hence the surface chemistry of the polymer matrix reducing its hydrophilic nature, (*23*)

ii. compensation factors for the second stage, ($k_2$ and $k_3$), suggests that the rate at which the dehydrated resin fraction is converted into a primary char is similar for all samples. However, we have to note here that despite the similarities in the volatilisation rates, the degradation of the neat resin is delayed relative to that of the flame-retarded composites possibly due to the reason suggested in (i) above, and,

iii. the rate at which the primary char volatilises, ($k_4$), is lower for flame-retarded composites as shown by slightly higher compensation factor values when compared to those obtained for the neat resin. This is consistent with the higher residual char yields for flame-retarded epoxy composites observed at temperatures greater than 450 °C. The similarity between compensation factor values for final char formation or condensed phase cross linking ($k_5$) for EP-MP and EP-MPP suggests no significant difference between the flame retardancy mechanism of melamine phosphate and melamine pyrophosphate. This parameter is not there in the neat epoxy resin degradation mechanism hence a comparative discussion is not possible.

**Table III. Optimised kinetic parameters for epoxy resin containing flame retardants, EP-MP and EP-MPP, at different heating rates**

| $k_i$ | *Kinetic parameters* | *Heating rates (°C/min)* | | | | |
|---|---|---|---|---|---|---|
| | | *10.0* | *12.5* | *15.0* | *17.5* | *20.0* |
| $k_1^a$ | $E/R(\times 10^3)$ [K] | 14.5 | 15.1 | 15.2 | 15.2 | 15.2 |
| | ln $A$ | 20.3 | 21.4 | 21.8 | 21.7 | 21.7 |
| | CF[c] | 715 | 706 | 697 | 699 | 699 |
| $k_2^a$ | $E/R(\times 10^3)$ [K] | 10.9 | 10.8 | 10.9 | 10.9 | 10.8 |
| | ln $A$ | 10.3 | 10.4 | 10.5 | 10.6 | 10.6 |
| | CF[c] | 1057 | 1045 | 1035 | 1027 | 1021 |
| $k_3^a$ | $E/R(\times 10^3)$ [K] | 16.1 | 16.0 | 16.1 | 16.0 | 15.9 |
| | ln $A$ | 18.6 | 18.6 | 18.7 | 18.6 | 18.4 |
| | CF[c] | 863 | 862 | 861 | 860 | 863 |
| $k_4^a$ | $E/R(\times 10^3)$ [K] | 14.2 | 14.2 | 14.4 | 14.1 | 14.1 |
| | ln $A$ | 10.7 | 10.7 | 10.8 | 10.6 | 10.3 |
| | CF[c] | 1319 | 1326 | 1329 | 1339 | 1376 |
| $k_5^a$ | $E/R(\times 10^3)$ [K] | 15.0 | 14.8 | 14.8 | 14.8 | 14.4 |
| | ln $A$ | 8.4 | 8.2 | 8.2 | 7.8 | 7.7 |
| | CF[c] | 1779 | 1800 | 1806 | 1903 | 1885 |
| $k_1^b$ | $E/R(\times 10^3)$ [K] | 14.9 | 15.1 | 15.1 | 15.4 | 15.0 |
| | ln $A$ | 20.3 | 21.1 | 21.7 | 22.0 | 20.9 |
| | CF[c] | 732 | 716 | 695 | 702 | 717 |
| $k_2^b$ | $E/R(\times 10^3)$ [K] | 10.8 | 10.8 | 10.8 | 10.9 | 10.8 |
| | ln $A$ | 10.3 | 10.4 | 10.4 | 10.5 | 10.7 |
| | CF[c] | 1053 | 1044 | 1045 | 1030 | 1012 |
| $k_3^b$ | $E/R(\times 10^3)$ [K] | 16.0 | 16.0 | 16.0 | 16.0 | 15.9 |
| | ln $A$ | 18.6 | 18.6 | 18.5 | 18.6 | 18.7 |
| | CF[c] | 859 | 859 | 865 | 859 | 848 |
| $k_4^b$ | $E/R(\times 10^3)$ [K] | 14.2 | 14.1 | 14.1 | 14.1 | 14.1 |
| | ln $A$ | 10.8 | 10.7 | 10.4 | 10.6 | 10.6 |
| | CF[c] | 1318 | 1316 | 1355 | 1339 | 1324 |
| $k_5^b$ | $E/R(\times 10^3)$ [K] | 14.9 | 14.8 | 14.9 | 14.9 | 14.7 |
| | ln $A$ | 8.5 | 8.6 | 8.5 | 9.0 | 8.6 |
| | CF[c] | 1750 | 1720 | 1761 | 1652 | 1707 |

[a] Kinetic values for epoxy resin containing melamine phosphate, EP-MP.
[b] Kinetic values for epoxy resin containing melamine pyro-phosphate, EP-MPP.
[c] CF = Compensation factor calculated as the ratio of $E/R$ to ln $A$.

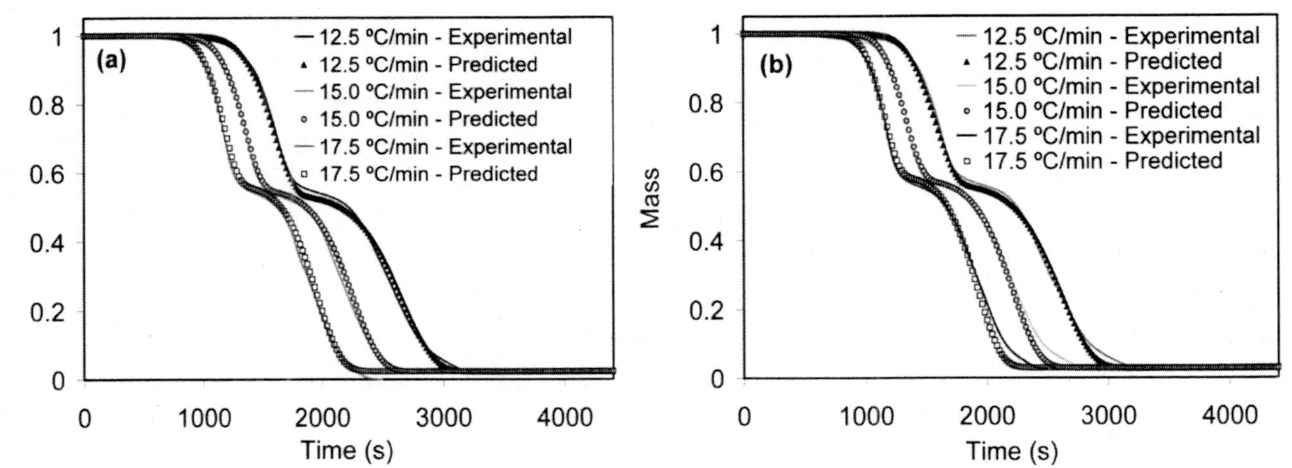


*Figure 3. Experimental (lines) and predicted (data points) TGA mass loss-time profiles for (a) EP-MP and (b) EP-MPP at 12.5, 15.0 and 17.5 °C/min under an air atmosphere.*

Optimised kinetic parameters obtained for the boundary heating conditions of 10 and 20 °C/min were averaged and used to predict the mass loss profiles for EP-MP and EP-MPP at intermediate linear heating rates of 12.5, 15 and 17.5 °C/min. The predicted and experimentally obtained mass fractions as a function of time for EP-MP and EP-MPP measured at constant linear heating rates of 12.5, 15 and 17.5 °C/min in air are shown in Figs. 3(a) and (b). There is an excellent agreement between experimental and predicted data. The calculated objective functions from EP-MP are 0.06, 0.06, and 0.12 for heating rates of 12.5, 15 and 17.5 °C/min respectively. Similar values of 0.06, 0.07, and 0.10 were calculated for heating rates of 12.5, 15 and 17.5 °C/min from EP-MPP.

To corroborate these conclusions Figs. 4(a) and (b) show plots of highly agreeing experimental versus theoretically predicted mass fractions while their mass differences plotted as a function of time in Figs. 5(a) and (b) are within acceptable data scatter thresholds of ±3% for large proportions of the reaction time. Minimal deviations form the equivalence line $x = y$ when experimental data points are plotted against predicted data points in Figs. 4(a) and (b) and the fact that there are very small temperature ranges within which the mass difference falls outside the experimental error range of ±3%, (Figs. 5(a) and (b)), suggests that the degradation kinetics of flame-retarded epoxy resins are reliably modelled albeit the use of grossly simplified kinetic schemes.

While we have shown from a mathematical point-of-view that the use of a global kinetics model fitting method and a multi-step degradation scheme can describe the course of a solid state reaction (polymer degradation) there are some limitations on how broadly one can apply this semi-global quantitative mathematical predictive tool. Obviously, one has to define a heating rate window for which optimised kinetic parameters obtained from independently fitting experimental data for each of the boundary heating rate are the same within experimental error. Averaged and/or extrapolated kinetic parameters from the boundary heating rate values can then be used to predict for intermediate value heating rates assuming that the degradation mechanism does not change. Despite the highlighted limitations, the predictive modelling tool developed herein, is vital in various areas of technological research and development including especially the prediction of service life of polymer composites at elevated temperature environments.

## Conclusions

The degradation kinetics of flame-retarded epoxy resin formulations under oxidative conditions were reliably modelled using a simplified kinetics model. The optimised kinetic parameters calculated for boundary heating rates of 10 and 20 °C/min were used to predict the mass loss profiles of the same materials under

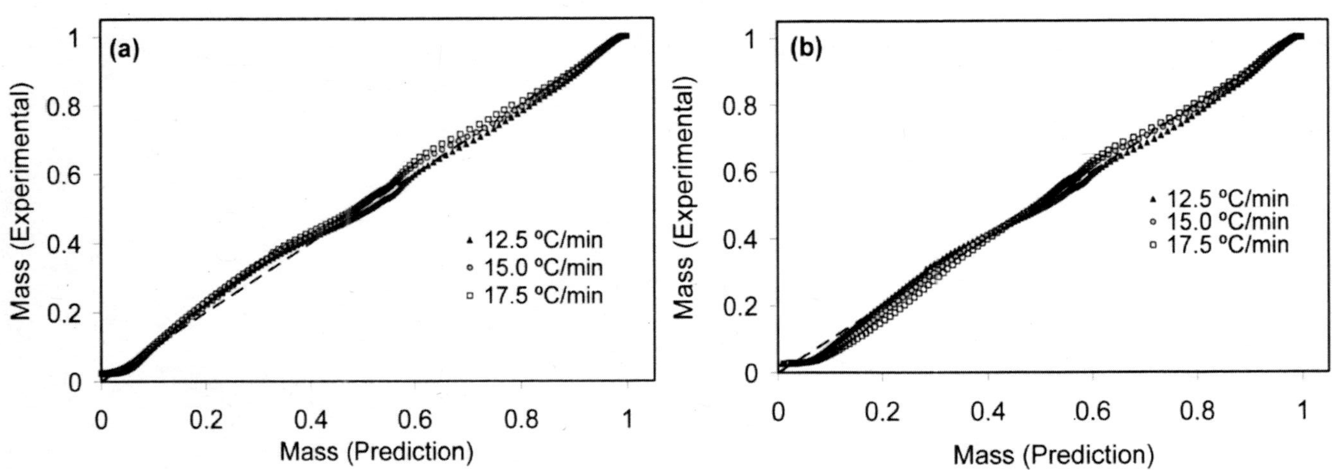


*Figure 4. Experimentally obtained TGA mass loss data points plotted against predicted mass loss data points for (a) EP-MP and (b) EP-MPP at 12.5, 15.0 and 17.5 °C/min under an air atmosphere; the dashed line represents the equivalence line $x = y$.*

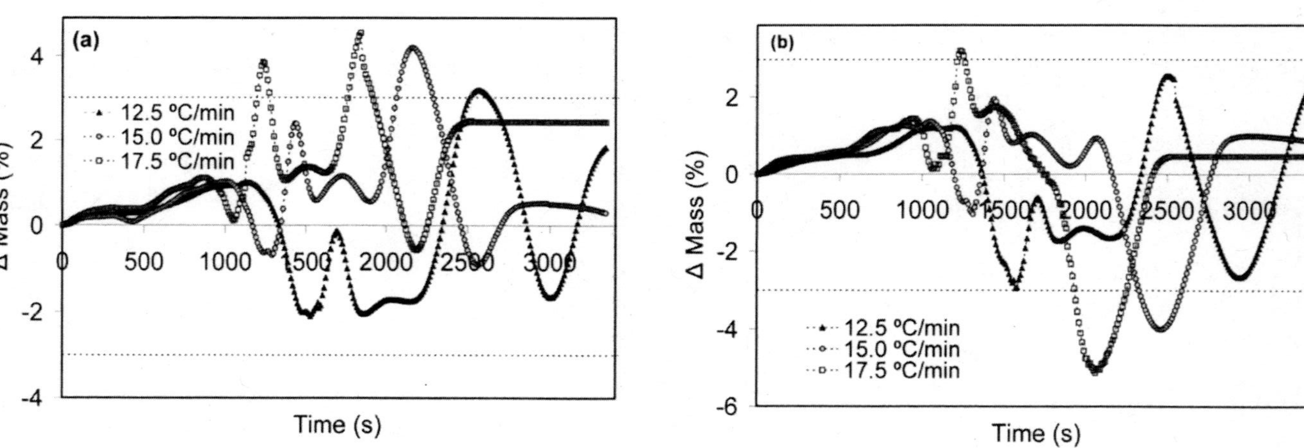


*Figure 5. Mass difference curves (experimental mass– predicted mass at same temperature or reaction time) for (a) EP-MP and (b) EP-MPP at 12.5, 15.0 and 17.5 °C/min under an air atmosphere; dashed lines represent the acceptable deviations in TGA data, i.e. error ranges of ± 3% are often reported.*

different linear heating rates; 12.5, 15, 17.5 °C/min while extrapolated kinetic parameters were fairly successful in predicting the mass loss behaviour at heating rates of 5 and 50 °C/min. The complete set of optimised kinetic parameters is sufficient to adequately simulate the full degradation history of all materials investigated in this study. While the relative thermal stability of flame-retarded formulations with respect to that of the neat resin was not the subject of this study, the authors observed that the presence of the flame retardant additives promote char formation at elevated temperatures. Despite the complexity of the degradation mechanism of epoxy-amine systems, the semi-global multi-step kinetics model sufficiently captured and collated the many envisaged chemical reactions during the degradation process into few physically solvable differential equations.

## Acknowledgements

The authors would like to acknowledge the financial support from the Engineering and Physical Sciences Research Council (EPSRC) and Dr. Bhaskar Biswas for providing samples for this study which were prepared for his PhD thesis.

## References


1. Lu, S. Y.; Hamerton, I. *Prog. Polym. Sci.* **2002**, *27*, 1661.
2. Liu, Y-L.; Chou, C-I. *Polym. Degrad. Stab.* **2005**, *90*, 515.
3. Biswas, B. Ph.D. Thesis, University of Bolton, Bolton, UK, **2007**.
4. Kim, J.; Yoo, S.; Bae, J-Y.; Yun, H-C.; Hwang, H.; Kong, B-S. *Polym. Degrad. Stab.* **2003**, *81*, 207.
5. Maciejewski, M. *J. Thermal. Anal.* **1988**, *33*, 1269.
6. Brown, M. E.; Maciejewski, M.; Vyazovkin, S.; Nomen, R.; Sempere, J.; Burnham, A.; Opfermann, J.; Strey, R.; Anderson, H. L.; Kemmler, A.; Keuleers, R.; Janssens, J.; Desseyn, H. P.; Li, C-R.; Tang, T. B.; Roduit, B.; Malek, J.; Mitsuhashi, T. *Thermochim. Acta.* **2000**, *355*, 125.
7. Budrugeac, P.; Segal, E.; Pe'rez-Maqueda, L-A.; Criado, J. M. *Polym. Degrad. Stab.* **2004**, *84*, 311.
8. Budrugeac, P. *Polym. Degrad. Stab.* **2005**, *89*, 265.
9. Vyazovkin, S.; Wight, C. A. *Inter. Rev. Phys. Chem.* **1998**, *17*, 407.
10. Biswas, B.; Kandola, B.; Horrocks A. R.; Price, D. *Polym. Degrad. Stab.* **2007**, *92*, 765.
11. Kandola, B. K.; Horrocks, A.R.; Myler, P.; Blair, D. In *Fire and Polymers;* Nelson, G.L.; Wilkie, C.A., Eds.; ACS Symp. Ser., **2001**, Vol 799, pp 344 - 360.

12. Rose, N.; Le Bras, M.; Delobel, R.; Costes, B.; Henry, Y. *Polym. Degrad. Stab.* **1993**, *42*, 307.
13. Kandare, E.; Kandola, B. K.; Staggs, J. E. J. *Polym. Degrad. Stab.* **2007**, *92*, 1778.
14. Macan, J.; Brnardić, I.; Orlić, S.; Ivanković, H.; Ivanković, M. *Polym. Degrad. Stab.* **2006**, *91*, 122.
15. Patterson-Jones, J. C.; Smith, D. A. *J. Appl. Polym. Sci.* **1968**, *12*, 1601.
16. Rose, N.; Le Bras, M.; Bourbigot, S.; Delobel, R. *Polym. Degrad. Stab.* **1994**, *45*, 387.
17. Zakordonskiy, V. P.; Hnatyshin, S. Y.; Soltys, M. M. *Polym. J. Chem.* **1998**, *72*, 2610.
18. Ozawa, T. *Bull. Chem. Soc. Jpn.* **1965**, *38*, 1881.
19. Pielichowski, K.; Kulesza, K.; Pearce, E. M. *J. Appl. Polym. Sci.* **2003**, *88*, 2319.
20. Wan, C.; Tian, G.; Cui, N.; Zhang, Y.; Zhang, Y. *J. Appl. Polym. Sci.* **2004**, *92*, 1521.
21. Kandare, E.; Deng, H.; Wang, D.; Hossenlopp, J. M. *Polym. Adv. Tech.* **2006**, *17*, 312.
22. Kandare, E.; Chigwada, G.; Wang, D.; Wilkie, C. A.; Hossenlopp, J. M. *Polym. Degrad. Stab.* **2006**, *91*, 1209.
23. Olmos, D.; Lo´pez-Moron´, R.; Gonza´lez-Benito, J. *Compos. Sci. Tech.* **2006**, *66*, 2758.

## Chapter 23

# Effect of Ambient Oxygen Concentration on Thermal Decomposition of Polyurethanes Based on MDI and PMDI

**Kenneth L. Erickson[1] and John Oelfke[2]**

[1]Sandia National Laboratories*, Albuquerque, NM 87185
[2]ORION International Technologies, Inc., Albuquerque, NM 87106

The presence of $O_2$ accelerated and increased char formation during the decomposition of polyurethanes based on MDI and PMDI. This appeared to occur by exothermic reaction between $O_2$ and polymer. The extent of char formation and exothermic heat release depended strongly on the ambient $O_2$ concentration. The char appeared to be chemically similar to the char that formed during decomposition in $N_2$. The increased char formation appeared to result from increased H abstraction due to reaction of $O_2$ with polymer. Ultimate reaction of the char to gaseous products depended strongly on the $O_2$ concentration and temperature.

NOTE: Sandia is a multi-program laboratory operated by Sandia Corporation, a Lockheed Martin Company, for the United States Department of Energy's National Nuclear Security Administration under Contract DE-AC04-94AL85000.

## Introduction

Thermal decomposition of organic polymers is important in many scientific and engineering applications. Thermal decomposition of polymers has been studied from a scientific perspective to gain insight into molecular structure, and from an engineering perspective to determine how specific materials behave at elevated temperatures. A particular area of interest is the behavior of polymer materials in fire environments.

Organic polymer materials are used frequently in structures and transportation systems. In an oxidizing environment, polymer materials can provide the fuel that propagates a fire. In a non-oxidizing environment, polymer materials can be damaged catastrophically as a result of an incident heat flux. Modeling the response of such structures and systems in fire environments has important applications in safety and vulnerability analyses. In either oxidizing or non-oxidizing environments, the thermal decomposition chemistry of the organic polymer materials is often an important factor. Specific applications include predicting fluxes of volatile species to a flame region, predicting the extent of thermally induced mechanical damage in structural composite materials, predicting pressure growth in closed containers, modeling liquefaction and flow of decomposing materials (particularly foams), determining the toxicity of evolved gases and vapors, and characterizing char formation.

To provide input to numerical models for hazard and vulnerability analyses involving polymer materials in inert environments subjected to external heat fluxes, thermal decomposition of several organic polymers in $N_2$ atmospheres was investigated previously using TGA-FTIR, pyrolysis-GC-FTIR, DSC, and infrared microprobe analysis of solid residues (*1-3*). Results were used to determine decomposition mechanisms and to develop rate expressions for use in numerical simulations. Materials studied included poly(methyl methacrylate) (PMMA), poly(diallyl phthalate) (DAP), poly(vinyl chloride) (PVC), polycarbonate (PC), poly(phenylene sulphide) (PPS), and two rigid polyurethanes. One polyurethane (referred to below as MDI RPU) was based on methylene-4,4'diphenyl diisocyanate (MDI) and polyhydroxy polyethers. The other polyurethane (referred to below as PMDI RPU) was based on polymeric diisocyanate (PMDI) and polyhydroxy polyethers.

TGA results and FTIR analysis of gaseous decomposition products from MDI RPU and PMDI RPU in $N_2$ atmospheres indicated two decomposition steps: (1) reaction at the urethane moieties to form isocyanates and alcohols or reaction to form anilines and $CO_2$, either directly or by secondary reaction, and (2) fragmentation and secondary reaction of the polyether moieties. These results were consistent with results from previous studies (*4, 5*).

Thermal decomposition of the same polymers in air atmospheres was subsequently investigated (*6*). To varying degrees, the presence of $O_2$ appeared to alter the decomposition mechanisms in all of the materials studied. In constant-heating-rate TGA experiments, the initial stage of decomposition of each polymer in air generally proceeded similarly to the initial stage of decomposition in $N_2$. Small deviations between decomposition in $N_2$ and decomposition in air were noted with PMMA and DAP. However, with the exceptions of PMMA and PVC, decomposition of each polymer in air involved an intermediate stage in which $O_2$ appeared to react with the decomposing condensed phase to form a more thermally stable product or char. That product then decomposed slowly until the temperature increased sufficiently to substantially increase the rate of reaction between the condensed phase and $O_2$. Samples were ultimately consumed by reaction with $O_2$ to form $H_2O$, CO, or $CO_2$. This behavior was most pronounced with polymers that formed a substantial amount of carbonaceous char during decomposition in $N_2$ atmospheres. In the case of both polyurethanes, complete consumption of the condensed-phase in air did not occur until temperatures of 700° C or higher.

Both MDI and PMDI based polyurethanes are frequently of interest in hazard and vulnerability analyses. Those analyses can involve a variety of scenarios in which $O_2$ may or may not be readily available to interact with the thermally decomposing polymers. Formation of an intermediate char could provide resistance to heat transfer to un-reacted polymer, could reduce the rate of pressurization in sealed containers, and could reduce liquefaction and flow.

To gain insight into the mechanism for formation of the intermediate char, the interaction of $O_2$ with MDI and PMDI based polyurethanes during thermal decomposition was investigated using several methods. Experiments were done with samples in atmospheres of $N_2$, air, or intermediate mixtures of $O_2$ in $N_2$. Chemical analyses of gas-phase and condensed-phase decomposition products were done using multiple techniques.

## Experimental

The experimental techniques used to examine thermal decomposition of the polymers have been discussed in detail previously (*1-3, 6*). The experimental data that will be discussed were obtained as described below.

Thermal gravimetric analysis (TGA) and simultaneous evolved gas analysis by Fourier transform infrared spectroscopy (FTIR) were used to obtain rate data, to examine gas-phase decomposition products as a function of time, and to obtain condensed-phase samples for postmortem analyses. The furnace purge gas exhaust from the TGA (TA Instruments Model 2950) was connected by a heated stainless steel transfer line to the TGA interface module of the FTIR spectrometer (Nicolet Magna 750). The purge gas was UHP $N_2$, air, or intermediate mixtures of UHP $O_2$ in UHP $N_2$ flowing at 50 to 60 ml/min. The transfer line temperature was set at 300° C. The TGA-FTIR interface module in the auxiliary experiment compartment of the FTIR spectrometer also was set at

300° C. The spectrometer provided concurrent chemical analysis of evolved gases. Multiple spectra were collected and averaged over consecutive 30-second intervals to provide time-averaged spectra as a function of time.

Both differential scanning calorimetry (DSC) and simultaneous DSC-TGA (SDT) were used to examine enthalpy changes during decomposition. A TA Instruments model 2920 DSC was used for temperature ranges from ambient to 600° C. The purge gas was UHP $N_2$ or air flowing at 60 ml/min. A TA Instruments Q600 SDT was used for temperature ranges from 200° C to 1200° C. The purge gas was UHP $N_2$, air, or intermediate mixtures of UHP $O_2$ in UHP $N_2$ flowing at 100 ml/min.

Partially decomposed samples of MDI RPU, as well as PMDI RPU, were prepared in the TGA by lowering the furnace and quenching the samples at preselected values of $m/m_0$ from 90% to 10%. Samples were prepared using $N_2$ and air as purge gases. During postmortem analyses, condensed-phase FTIR spectra were obtained using a Thermo Scientific 6700 FTIR spectrometer equipped with a Thermo Scientific "Smart Orbit" ATR accessory with a diamond objective. Elemental analyses for total C, H, and N were obtained using a Perkin Elmer 2400 Elemental Analyzer. Certified acetanilide was used as a standard.

Results from experiments with MDI RPU and PMDI RPU were similar. Results for MDI RPU are discussed first in detail. Results for PMDI RPU are then summarized, and small differences relative to MDI results are noted.

## Results: MDI Based Polyurethane

Figure 1 shows residual mass ratio $m/m_0$ (ratio of instantaneous sample mass to initial sample mass) versus temperature from replicate TGA experiments that were done with MDI RPU samples heated at 20° C/min. Results are shown from experiments using $N_2$, 1% $O_2$ (99% $N_2$), 2% $O_2$ (98% $N_2$), 5% $O_2$ (95% $N_2$), or air as purge gas. Figure 2 shows rate of mass loss corresponding to one of the TGA experiments in Figure 1 done with $N_2$ purge gas, one of the experiments with 2% $O_2$ (98% $N_2$) purge gas, and one of the experiments with air purge gas.

Relative to experiments using $N_2$ purge gas, the presence of $O_2$ had little effect on initial mass loss. In the temperature range between 250° C and 320° C, the curves representing residual mass versus temperature (Figure 1) essentially overlaid each other for each of the purge gases used. This is further illustrated in Figure 2, where the first peak in each of the rate-of-mass-loss curves occurred between about 320° C and 330° C.

With $N_2$ as the purge gas, the rate of mass loss steadily increased until $m/m_0$=54% (T=330° C). The rate of mass loss then gradually decreased until $m/m_0$=45% (T= 337° C). Between $m/m_0$=45% and $m/m_0$=40% (T=346° C), the

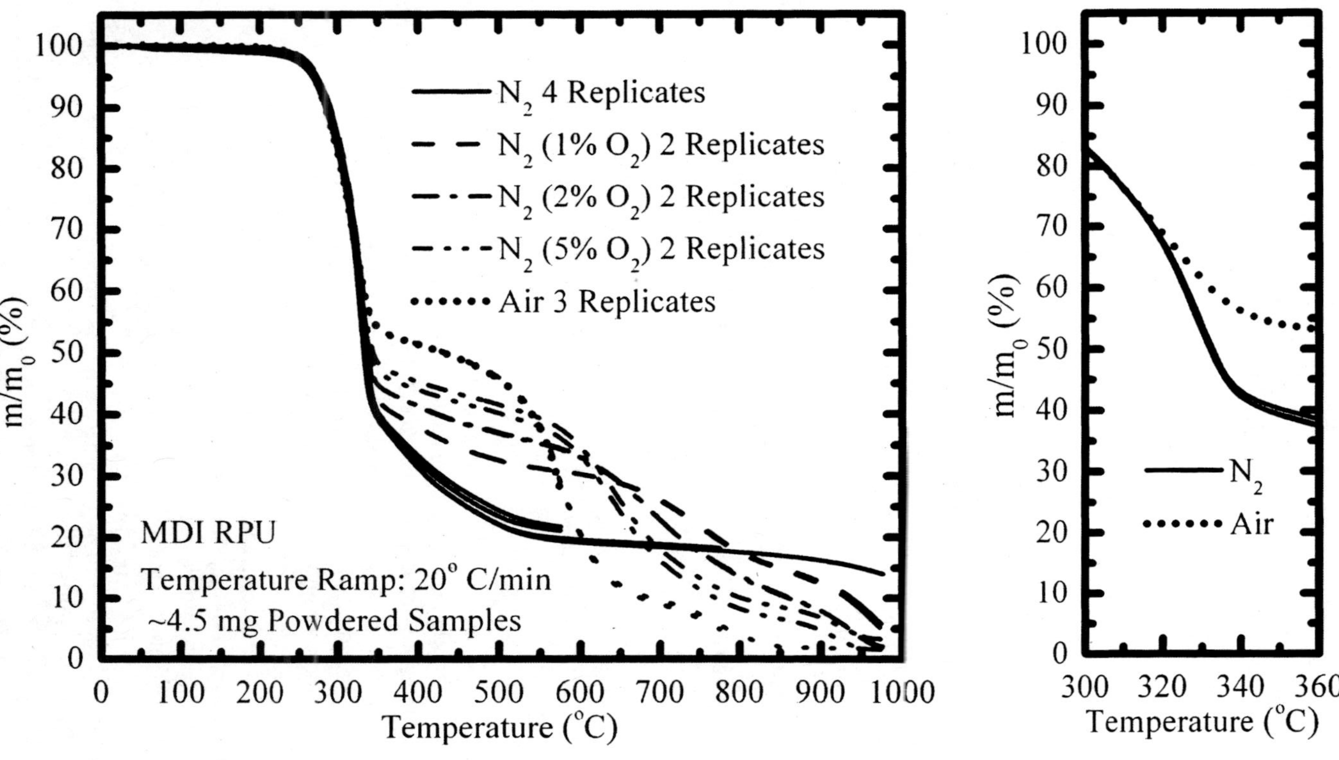


*Figure 1. TGA results from MDI RPU and several purge gases.*

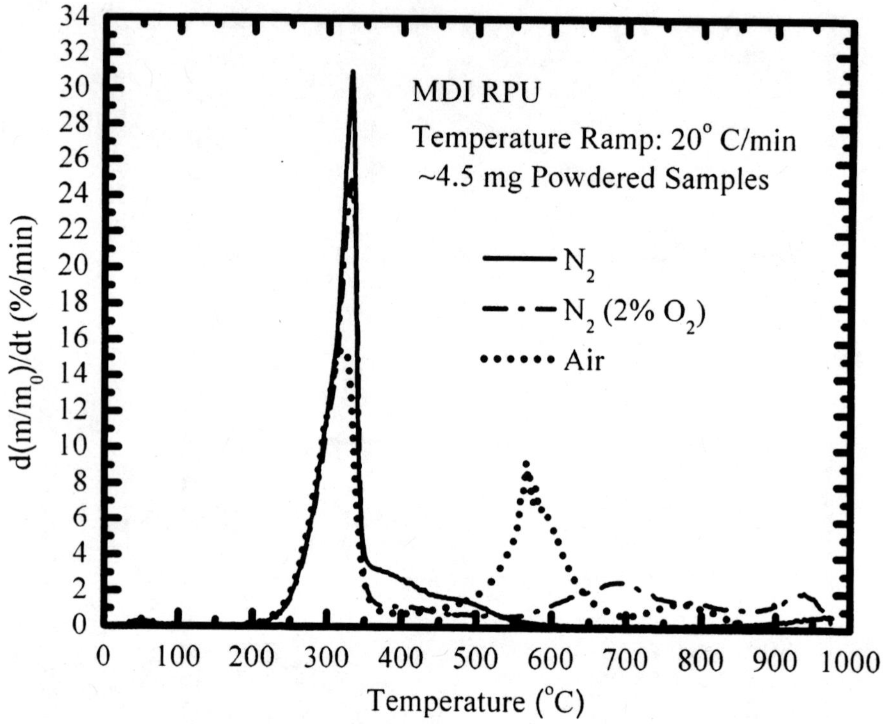


*Figure 2. Rate of mass loss for selected curves shown in Figure 1.*

rate of mass loss decreased abruptly. Below $m/m_0$=40%, the rate of mass loss continued to steadily decrease and became small at $m/m_0$=20% (T=547° C).

As the temperature increased above 320 °C ($m/m_0$=70%), decomposition in the presence of $O_2$ deviated substantially from decomposition in $N_2$. The decomposing condensed phase appeared to form a more thermally stable product, or char, having a much larger residual mass fraction than the char that formed in experiments using $N_2$. The mass fraction of the char that formed increased with the $O_2$ concentration in the purge gas. A concentration of only 1% $O_2$ approximately doubled the amount of char that formed relative to the char that formed with $N_2$ purge gas.

In air, the rate of mass loss began to gradually deviate from that in $N_2$ between $m/m_0$=70% (T=318 °C) and $m/m_0$=60% (T=332 °C). Between $m/m_0$=60% and $m/m_0$=54% (T=351 °C), the rate of mass loss decreased abruptly. Between $m/m_0$=54% and $m/m_0$=50% (T=435 °C), the rate of mass loss was slow. Between $m/m_0$=50% and $m/m_0$=40% (T=542 °C), the rate of mass loss gradually increased with temperature. For $m/m_0$ less than 40%, the rate of mass loss increased significantly. These results are consistent with previous work reported by Pielichowski, et al. (*7*) who examined thermal degradation of polyurethanes based on MDI and several polyhydroxy polyethers. During their TGA experiments using air purge gas and samples that were heated at 10° C/min to 500 °C, about 50% to 60% of the original sample mass remained at 500 °C.

Ultimate reaction of the char to gaseous products was also highly dependent on $O_2$ concentration in the purge gas, as well as temperature. As the $O_2$ concentration in the purge gas increased, the temperature range over which the char decomposed decreased and shifted to lower temperatures. This is illustrated in Figure 2. The second peak and third peaks in the curves for 2% $O_2$ and for air correspond to reaction of char to gaseous products. The second and third peaks in the curve for 2% $O_2$ are broad and occur at T=695 °C ($m/m_0$=23%) and at T=938 °C ($m/m_0$=5%), respectively. In the curve for air, however, the second and third peaks are sharper, larger, and occur at T=567° C ($m/m_0$=33%) and T=769 °C ($m/m_0$=6%), respectively.

Figure 3 shows heat flow versus temperature from DSC and SDT experiments using $N_2$, 2% $O_2$ (98% $N_2$), or air for purge gas. Experiments were done with open pans. DSC results obtained with $N_2$ showed primarily endothermic behavior during decomposition. A slight exothermic peak occurred at about 340 °C. The SDT results obtained using 2% $O_2$ or air for purge gas showed three exothermic peaks during decomposition. With 2% $O_2$ purge gas, the two most prominent peaks were at 337 °C and 674 °C, which were similar to the temperatures corresponding to the peaks in the rate of mass loss curves in Figure 2. The exothermic energy associated with decomposition of the char was greater than the energy associated with the preceding decomposition of the polymer. With air for purge gas, the two most prominent peaks were at about 337 °C and 572 °C, which also were similar to the temperatures corresponding to the peaks in the rate of mass loss curves in Figure 2. In this case, the exothermic energy associated with decomposition of the char was much greater than the energy associated with the preceding decomposition of the polymer.

The DSC and SDT results indicated that although the presence of $O_2$ had little effect on the initial mass loss in the TGA experiments, the presence of $O_2$ resulted in exothermic behavior throughout decomposition. The onset of exothermic behavior coincided with initial mass loss, and the exothermic heat flow increased during decomposition of the char. Furthermore, the exothermic heat flow that occurred throughout decomposition increased with the $O_2$ concentration in the purge gas.

Gas-phase FTIR spectra obtained during a TGA experiment using $N_2$ purge are shown in Figure 4 and correspond to $m/m_0$=80% (T=305 °C), $m/m_0$=60% (T=326 °C), and $m/m_0$=40% (T=346 °C). Spectra obtained using air purge are shown in Figure 5 and correspond to $m/m_0$=80% (T=304 °C), $m/m_0$=60% (T=332 °C), and $m/m_0$=40% (T=542 °C).

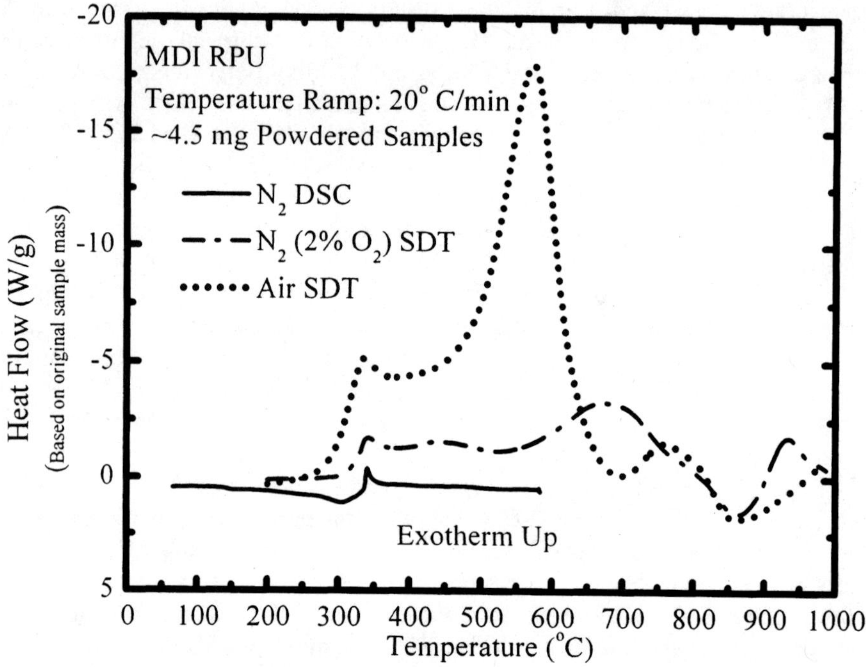


*Figure 3. DSC and SDT results from MDI RPU and three purge gases.*

In Figure 4 and Figure 5, the gas-phase spectra corresponding to $m/m_0$=80% and $m/m_0$=60% are similar. The large peak between about 1200 $cm^{-1}$ and 1000 $cm^{-1}$, C-O-C stretching (*8*), indicates evolution of fragments from polyether moieties in the polymer. The large peak between about 2400 $cm^{-1}$ and 2200 $cm^{-1}$ is due to overlapping absorptions by $CO_2$ and isocyanates that evolve from decomposition involving the urethane moieties (*3*). This signal is relatively more intense in Figure 5. Increased $CO_2$ would be consistent with reaction of oxygen with polymer during initial decomposition, which was indicated by the initial exotherm in the DSC results in Figure 3.

In Figure 5, the small peak at 1750 $cm^{-1}$, C=O stretching, in the spectra for $m/m_0$=80% and $m/m_0$=60% indicated a carbonyl moiety (*8*). This is consistent with aldehydes from oxidative decomposition of polyethers (*9, 10*). Traces of water were evident and were most prominent for $m/m_0$ between 80% and 60%.

In Figure 4 and Figure 5, the gas-phase spectra corresponding to $m/m_0$=40% are very different. In Figure 4, the small peak at 3402 $cm^{-1}$ and the three sharp peaks at 1625 $cm^{-1}$, 1514 $cm^{-1}$, 1267 $cm^{-1}$ indicate anilines (*8*). The two small peaks at 968 $cm^{-1}$ and 933 $cm^{-1}$ indicate traces of $NH_3$ (*11*). In Figure 5, the spectrum for $m/m_0$=40% is dominated by the $CO_2$ peaks between 2400 $cm^{-1}$ and 2250 $cm^{-1}$ and the CO peaks between 2250 $cm^{-1}$ and 2000 $cm^{-1}$ (*11*). The CO signal first appeared when $m/m_0$ = 57% (T=338 °C).

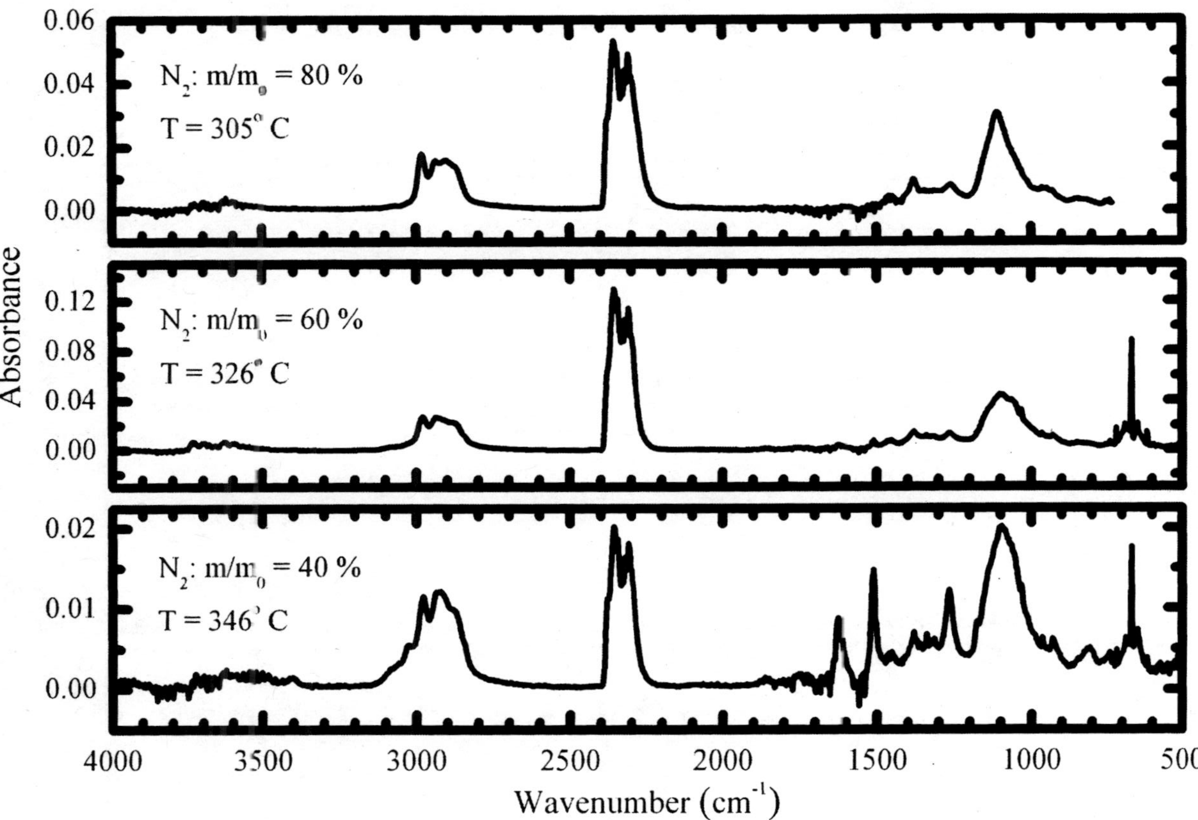


*Figure 4. FTIR spectra from TGA experiment using $N_2$ for purge gas.*

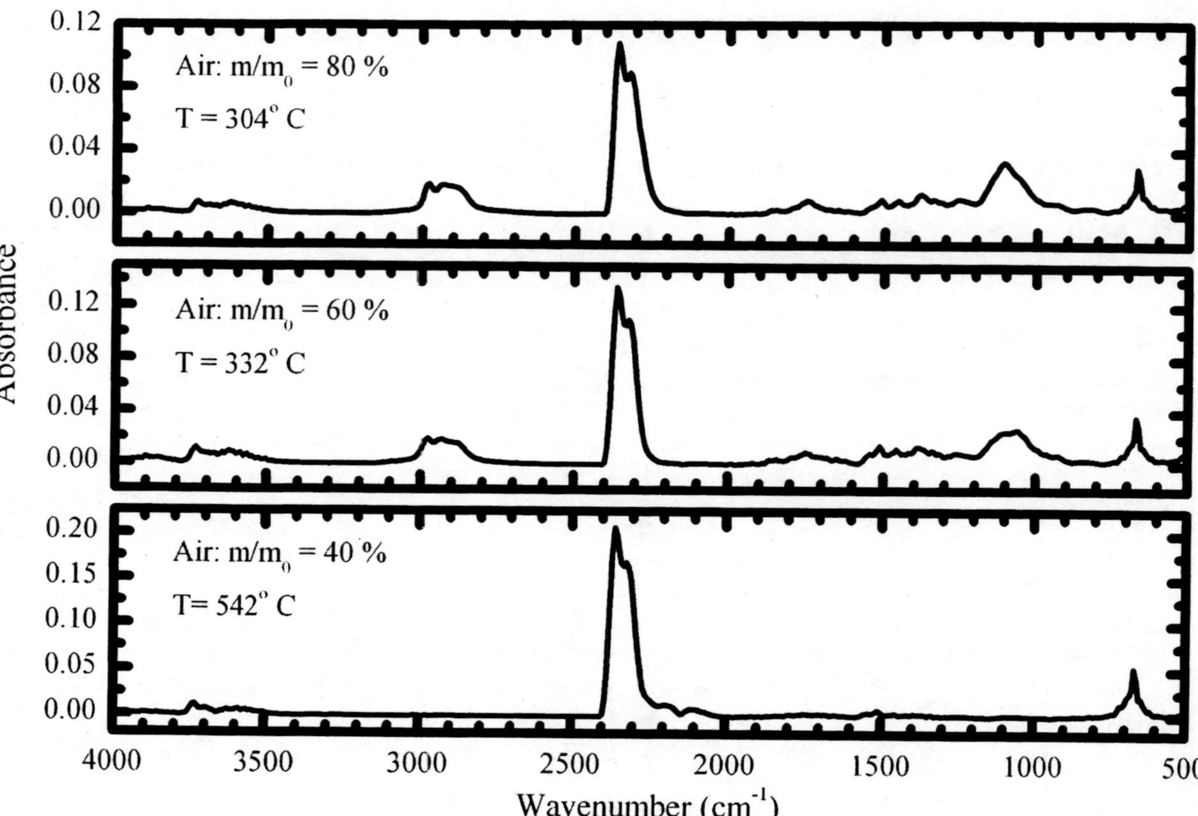


*Figure 5. FTIR spectra from TGA experiment using air for purge gas.*

To complement the gas-phase FTIR spectra, elemental analyses for C, H, and N, and condensed-phase FTIR spectra were obtained from partially decomposed samples of MDI RPU that were prepared in the TGA using $N_2$ and air as purge gases. Results from the elemental analyses are shown in Figure 6. The FTIR results are shown in Figure 7 ($N_2$ purge) and Figure 8 (air purge).

The concentrations of H, versus $m/m_0$, in samples partially decomposed in $N_2$ and in air are shown in the upper portion of Figure 6. The corresponding concentrations of N are shown in the lower portion. Error bars corresponding to the standard deviation between replicate samples (usually 3) were on the order of the size of the symbols in Figure 6, and were omitted.

Concentrations of H in partially decomposed samples prepared using $N_2$ initially varied little with the extent of decomposition $m/m_0$. At values of $m/m_0$ of 50% and less, the concentration of H decreased as $m/m_0$ decreased. Between $m/m_0$=40% and $m/m_0$=20%, the concentration of H decreased substantially from about 5.9 mass percent to 3.5 mass percent. This decrease in the concentration of H corresponded to the substantial decrease in the rate of mass loss that occurred between $m/m_0$=0.4 and $m/m_0$=0.2 in TGA experiments (Figure 1 and Figure 2).

Concentrations of H in partially decomposed samples prepared using air varied little with the extent of decomposition for $m/m_0$=80% and greater. At $m/m_0$=70%, the concentration of H began to decrease. Between $m/m_0$=60% and $m/m_0$=50%, the concentration decreased abruptly from 5.5 mass percent to 3.1 mass percent. This decrease in the concentration of H corresponded to the abrupt decrease in the rate of mass loss that occurred between $m/m_0$=60% and $m/m_0$=54% in the TGA experiments (Figure 1 and Figure 2). Also in Figure 6, the concentration of H for $m/m_0$=20% from decomposition in $N_2$ is similar to the concentration of H for $m/m_0$=50% from decomposition in air. Analogously, the condensed-phase FTIR spectrum for $m/m_0$=20% in Figure 7 (decomposition in $N_2$) is similar to the spectrum for $m/m_0$=50% in Figure 8 (decomposition in air).

Concentrations of N in partially decomposed samples prepared using $N_2$ steadily increased as $m/m_0$ decreased from 100% to 30%. The concentration of N then decreased at $m/m_0$=20%. Concentrations of N in samples prepared in air were very close to the values for samples prepared in $N_2$ for values of $m/m_0$ from 100% to 60%. The concentration steadily decreased for values of $m/m_0$ less than 60%, though not as abruptly as the concentration of H decreased.

In Figure 7 ($N_2$ purge) and Figure 8 (air purge), FTIR spectra from the initial polymer and from partially decomposed samples are shown. In Figure 7 and Figure 8, the major spectral features (*8*) in the initial material include: (1) the broad O-H and N-H stretching region between 3700 $cm^{-1}$ and 3100 $cm^{-1}$, due to un-reacted OH groups in the polyhydroxy polyether moieties and to the NH groups in the urethane moieties; (2) the broad C-H stretching region between about 3100 $cm^{-1}$ and 2700 $cm^{-1}$, due to aromatic and aliphatic H; (3) the C=O stretching peak at 1726 $cm^{-1}$, due to the C=O group in the urethane moieties; (4) the aromatic ring vibrations at 1595 $cm^{-1}$ and 1520 $cm^{-1}$; (5) the absorption at

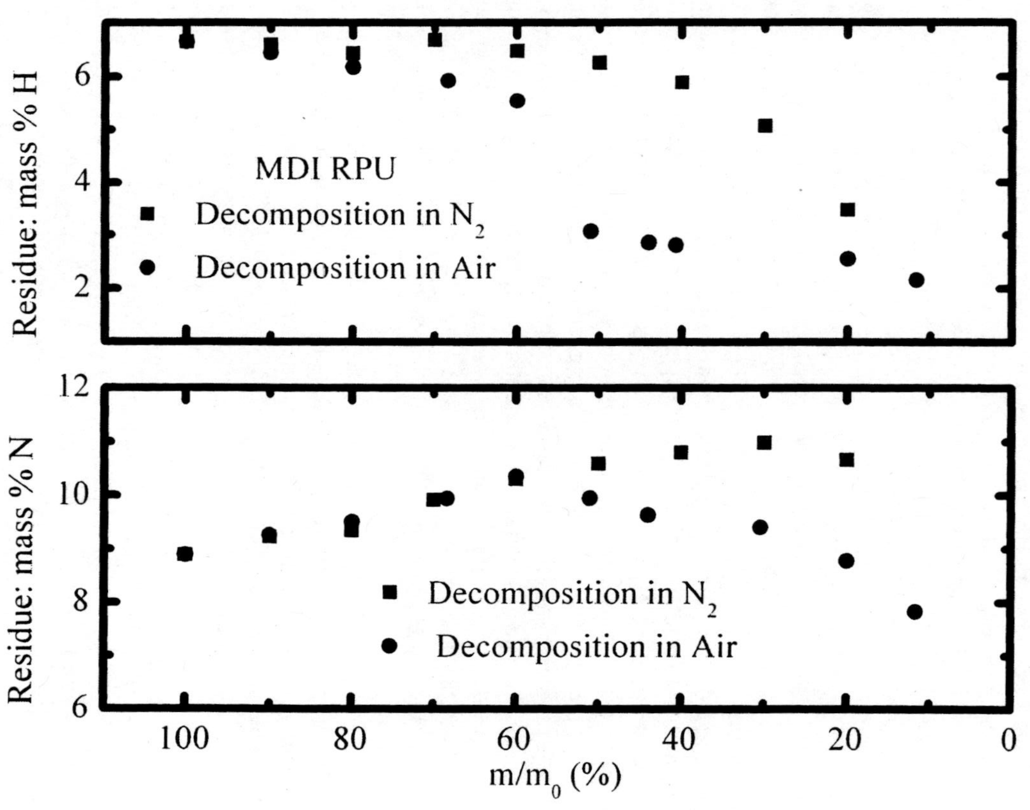


*Figure 6. Results from elemental analysis of MDI RPU residues.*

1308 $cm^{-1}$, probably due to N attached to the aromatic rings in the urethane moiety; (6) the sharp C-O stretching peak at 1223 $cm^{-1}$, due to the C(=O)-O group in the urethane moiety; and (7) the broad C-O-C stretching peak at 1074 $cm^{-1}$, due to the polyether moieties.

In Figure 7, the spectra from partially decomposed samples correspond to values of $m/m_0$ of 80%, 60%, 50%, 40%, and 20%. Relative to the spectrum from the initial material, the spectrum for $m/m_0$=80% showed substantial decreases in the C=O signal at 1726 $cm^{-1}$ and the C-O signals at 1223 $cm^{-1}$ and 1074 $cm^{-1}$. This indicates loss of the C(=O)-O group from the urethane moiety and loss of ether groups from the polyether moieties. Both losses are consistent with the large overlapping $CO_2$-isocyanate peaks and C-O-C stretching peak in the gas-phase spectrum for $m/m_0$=80% in Figure 4. In spectra for $m/m_0$=60%, $m/m_0$=50%, and $m/m_0$=40%, the C=O signal at 1726 $cm^{-1}$ and the C-O signals at 1223 $cm^{-1}$ and 1074 $cm^{-1}$ are very small. The aromatic ring signals at 1595 $cm^{-1}$ and 1520 $cm^{-1}$ are still strong. However, the spectrum for $m/m_0$=20% is significantly different from the other spectra and is dominated by three broad overlapping peaks at 1600 $cm^{-1}$, 1430 $cm^{-1}$, and 1230 $cm^{-1}$.The O-H and N-H stretching peaks and the C-H stretching peaks are no longer distinct. Furthermore, the aromatic ring peak at 1520 $cm^{-1}$ is greatly diminished, which would be the case if some of the aromatic moieties formed polycyclic or heterocyclic moieties (*8*).

In Figure 8, the spectrum for $m/m_0$=80% is similar to the spectrum for $m/m_0$=80% in Figure 7. This is consistent with initial decomposition in both purge gases occurring similarly. The spectrum (Figure 8) for $m/m_0$=60% is significantly different relative to the analogous spectrum in Figure 7; the aromatic ring peak at 1520 $cm^{-1}$ is greatly diminished. The spectrum for $m/m_0$=50% in Figure 8 is very similar to the spectrum for $m/m_0$=20% in Figure 7. This also indicates that the $O_2$ in air accelerated and increased char formation.

## Results: PMDI Based Polyurethane

Results obtained for PMDI RPU were generally similar to those obtained from MDI RPU. Figure 9 shows $m/m_0$ versus temperature from analogous replicate TGA experiments. Figure 10 shows heat flow versus temperature from DSC and SDT experiments using $N_2$, 2% $O_2$ (98% $N_2$), or air for purge gas.

The TGA results in Figure 9 are similar to those in Figure 1 for MDI RPU. However, PMDI RPU decomposition in the presence of $O_2$ began to deviate from decomposition in $N_2$ at larger values of $m/m_0$. In air, the rate of mass loss began to decrease relative to the rate in $N_2$ when $m/m_0$ was about 85% and increasingly deviated from the rate in $N_2$. The char developed less abruptly, was less thermally stable, and was totally consumed by $O_2$ at lower temperatures.

In Figure 10, the DSC and SDT results, obtained using open pans, were similar in form and magnitude to those in Figure 3 for MDI RPU. The onset of

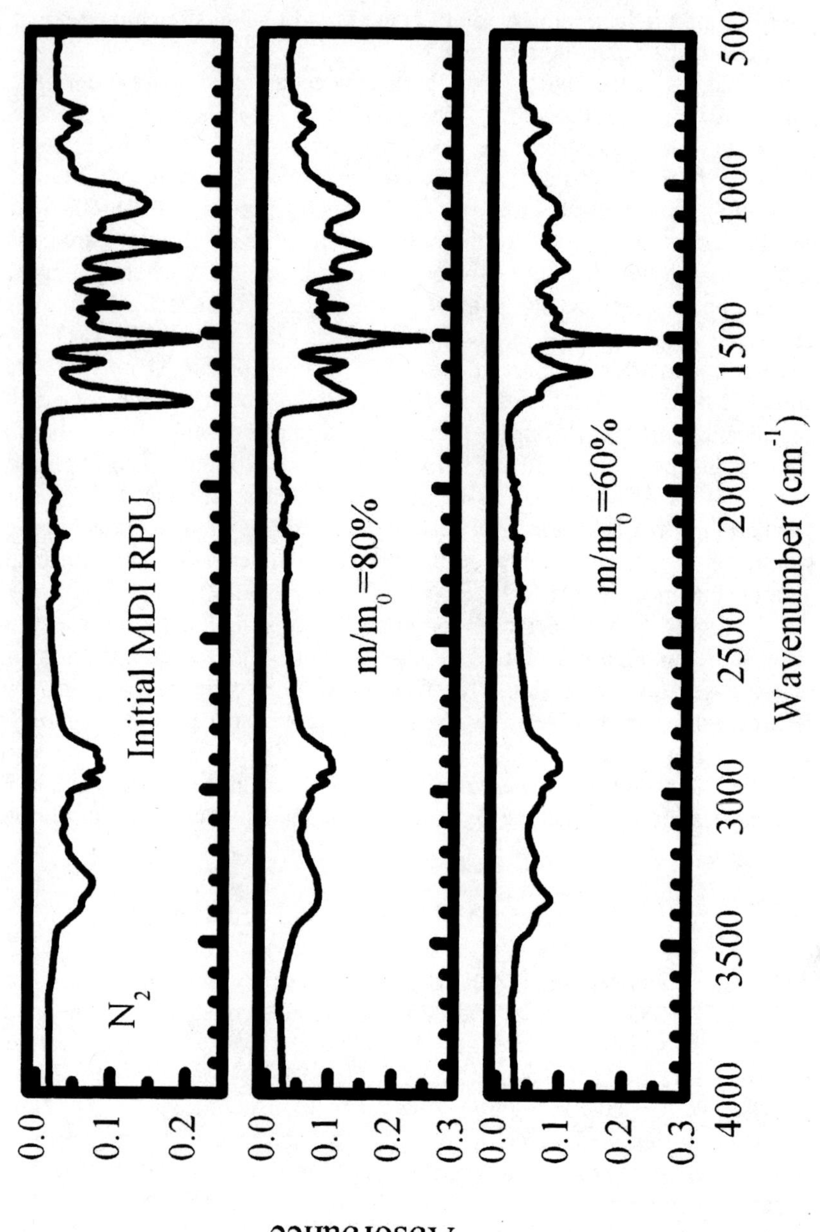
Initial MDI RPU
N2
m/m0=80%
m/m0=60%
Absorbance
0.0
0.1
0.2
0.3
Wavenumber (cm-1)
4000
3500
3000
2500
2000
1500
1000
500

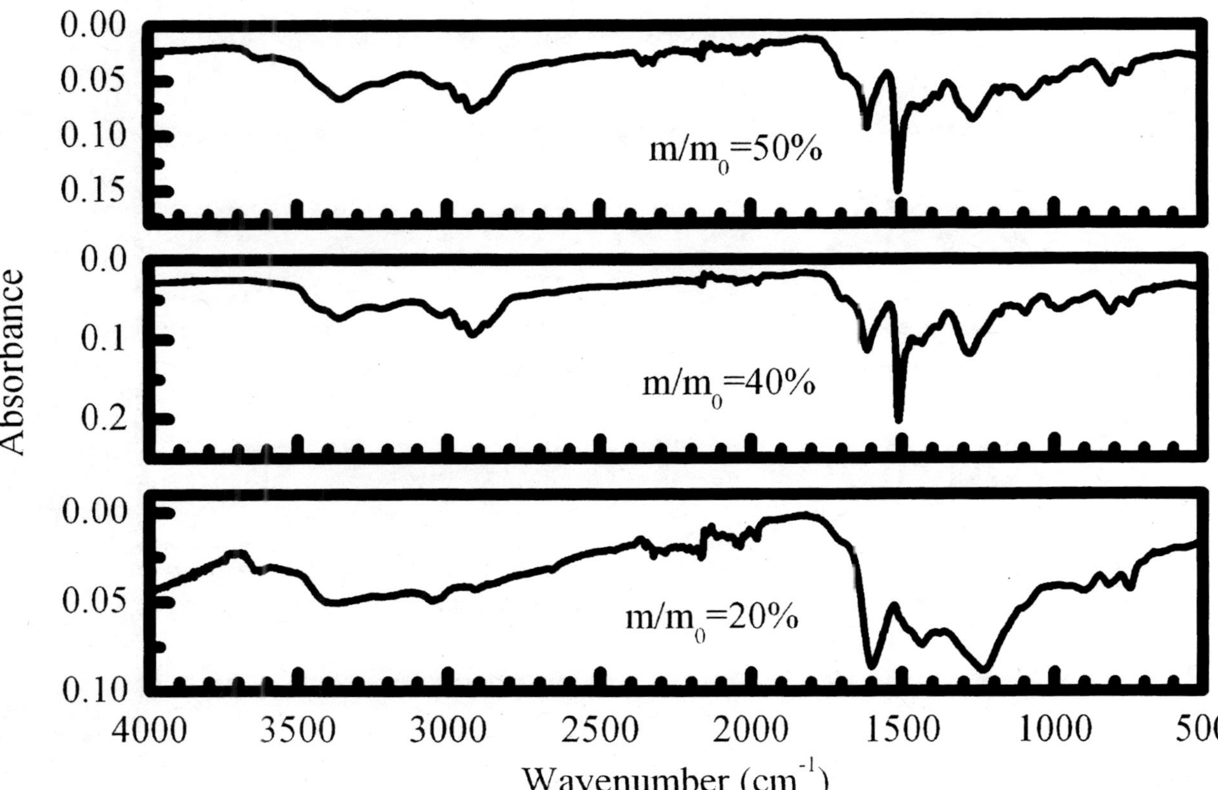


*Figure 7. FTIR spectra from MDI RPU samples partially decomposed in $N_2$.*

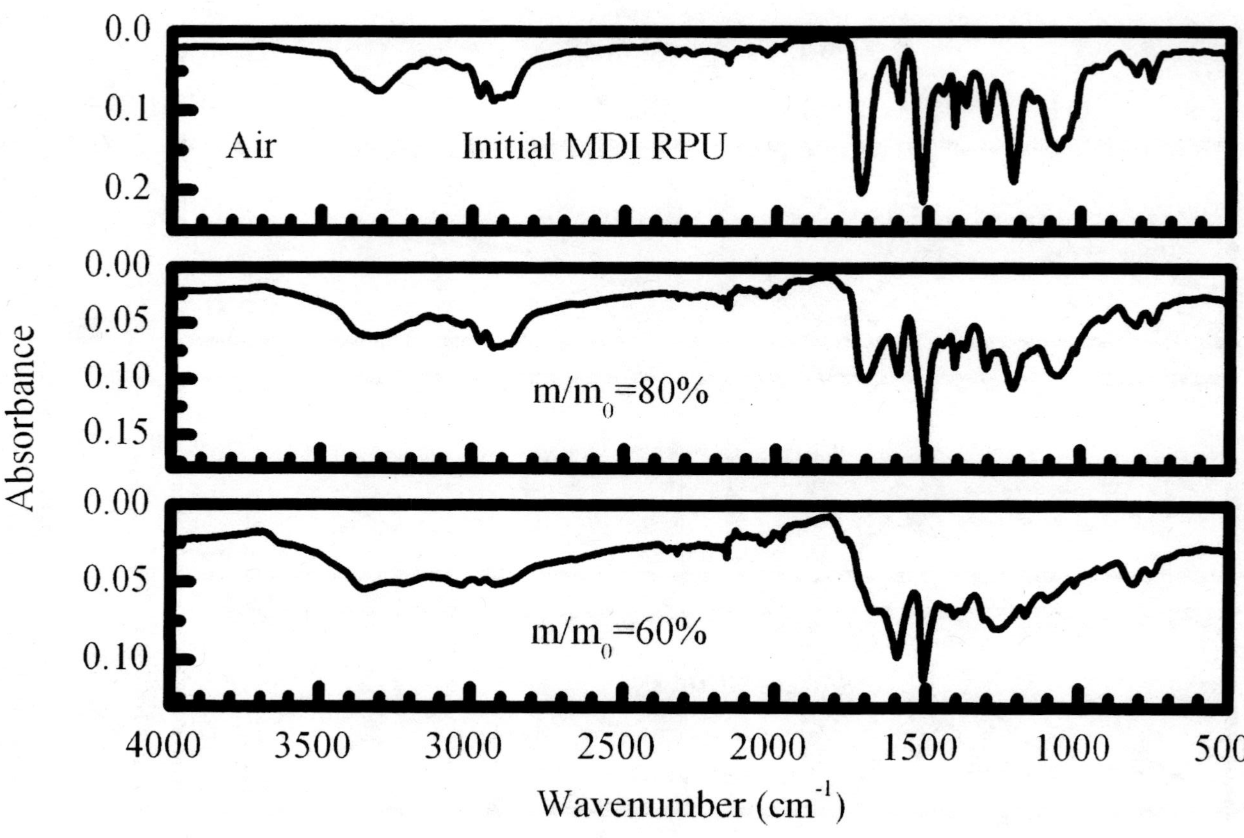
Air
Initial MDI RPU
m/m0=80%
m/m0=60%
Absorbance
0.0
0.1
0.2
0.00
0.05
0.10
0.15
0.00
0.05
0.10
4000
3500
3000
2500
2000
1500
1000
500
Wavenumber (cm-1)

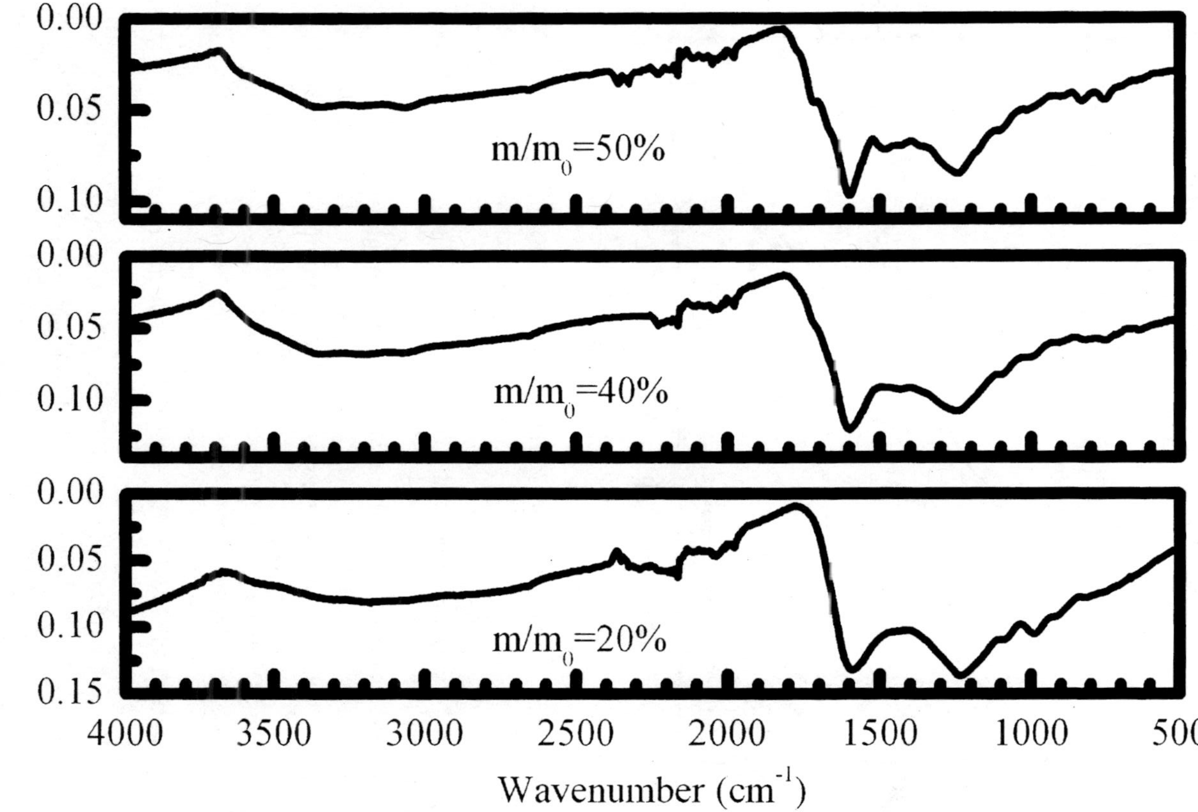


*Figure 8. FTIR spectra from MDI RPU samples partially decomposed in air.*

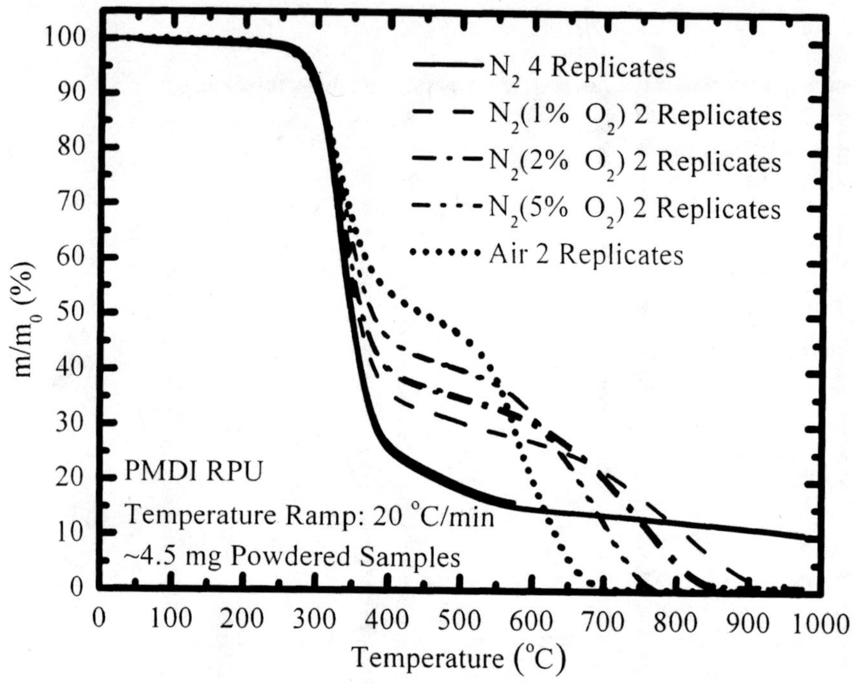


*Figure 9. TGA results from PMDI RPU and several purge gases.*

the exothermic peaks corresponded to the increases in the rate of mass loss observed with the TGA. During reaction of the char to gaseous products, only one exothermic peak occurred, which was broader than the large peak in Figure 3. Also in Figure 10, experiments with both the DSC and SDT were done with $N_2$ as purge gas and with air as purge gas. Results from both the DSC and SDT were essentially identical. The DSC and SDT instruments greatly differ in design, and the SDT is purged at 100 ml/min. It is unlikely that if decomposition products were oxidized in the gas phase, the energy feed back to the sample would be the same for both instruments. Therefore, the exothermic responses observed with both MDI and PMDI RPU's should be due to reaction of $O_2$ with the condensed phase.

The FTIR spectra obtained during TGA experiments with PMDI RPU were similar to those obtained with MDI RPU. Elemental analyses from PMDI RPU residues also were similar to those from MDI RPU residues. However, the H and N concentrations in samples partially decomposed in air began to deviate from concentrations in samples decomposed in $N_2$ at larger values of $m/m_0$, which was consistent with the TGA results. Finally, FTIR spectra from condensed phase PMDI RPU residues were similar to those obtained from MDI RPU residues.

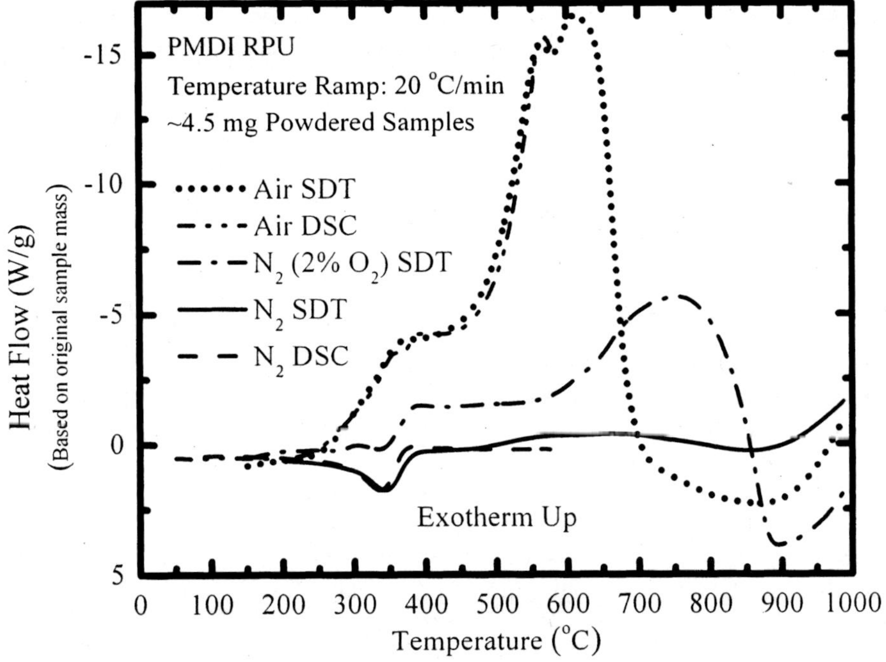


*Figure 10. DSC and SDT results from PMDI RPU and three purge gases.*

## Discussion and Conclusions

The TGA results in Figure 1 (MDI RPU) and Figure 9 (PMDI RPU) showed that the presence of $O_2$ accelerated and increased char formation. The DSC and SDT results in Figure 3 (MDI RPU) and Figure 10 (PMDI RPU) indicated that this occurred by exothermic reaction between $O_2$ and polymer. The extent of char formation and exothermic heat release depended strongly on the ambient $O_2$ concentration. A concentration of only 1% $O_2$ (99% $N_2$) approximately doubled the amount of char that formed relative to the char that formed with $N_2$ purge gas. Ultimate reaction of the char to gaseous products was highly dependent on temperature and oxygen concentration in the purge gas. For decomposition in air, the gas-phase FTIR results (MDI RPU) in Figure 5 indicated species from oxidative decomposition of polyether moieties in the polymer. Those species were not present in Figure 4 ($N_2$ purge).

Results from experiments with MDI RPU and PMDI RPU were generally similar. With MDI RPU, char formation was substantial in TGA experiments with air purge when $m/m_0$=54%, and with $N_2$ purge when $m/m_0$=20% (Figure 1). Elemental analyses (Figure 6) indicated that relative to decomposition in $N_2$, the concentration of H in MDI RPU samples that were partially decomposed in air gradually decreased until an abrupt decrease occurred coincidently with abrupt decreases in the rate of decomposition (mass loss) when $m/m_0$ was between 60% and 50% (Figure 1). The concentration of H for $m/m_0$=50% from decomposition in air is similar to the concentration of H for $m/m_0$=20% from decomposition in $N_2$. Analogously, an abrupt change in condensed-phase FTIR spectra (Figure 7 and Figure 8) also occurred when $m/m_0$ was between 60% and 50%. The condensed-phase FTIR spectrum for $m/m_0$=50% in Figure 8 (decomposition in air) is similar to the spectrum for $m/m_0$=20% in Figure 7 (decomposition in $N_2$).

Although the rate of decomposition (mass loss) decreased when $m/m_0$ was between 60% and 50%, the rate of H loss increased. This would be consistent with increased H abstraction due to formation of $H_2O$ and formation of a greater number of radical moieties caused by reaction with $O_2$. Abstraction of H could cause formation of C=C and C=N moieties consistent with the broad peak at about 1600 $cm^{-1}$ in Figure 8 ($m/m_0$=50%) and the loss of intensity in the C-H stretching peaks. Abstraction of H from the urethane groups could produce tertiary amine moieties (perhaps cross linked), which is consistent with the loss of intensity in the N-H stretching peaks in the FTIR spectra in Figure 8. A decrease in readily extractable H would retard the formation of aniline decomposition products, which proceeds by thermal bond scission and subsequent H abstraction involving the methylene and secondary amine groups associated with the MDI and PMDI moieties in the polymers (*3*).

In closing, the authors gratefully acknowledge the technical support of Ted Borek and Jeanne Barrera (elemental analyses) of Sandia National Laboratories.

## References


1. Hobbs, M.; Erickson, K.; Chu, T.Y. *J. Polym. Degrad. Stabil.* **2000**, *69*, 47.
2. Nicolette, V.; Erickson, K.; Vembe, B.E. In *Proceedings of Interflam 2004*; Interscience Communications Ltd.: London, UK, 2004; pp. 1257-1268.
3. Erickson, K. *J. Therm. Anal. Calor.* **2007**, *89*, 427.
4. Grassie, N.; Zulfiqar, M. *J. Polym. Sci., Polym. Chem. Ed.* **1978**, *16*, 1563.
5. Bajsic, E.; Rek, V. *J. Appl. Polym. Sci.* **2001**, *79*, 864.
6 Erickson, K. In *Proceedings of SAMPE Fall Technical Conference*; Society for Advancement of Material and Process Engineering: Covina, CA, 2007.
7. Pielichowski, K.; Pielichowski, J.; Altenburg, H.; Ballof, H. *Thermochim Acta*, **1996**, *284*, 419.
8. Colthup, N.; Daly, L.; Wiberley, S. *Introduction to Infrared and Raman Spectroscopy*, 3rd ed.; Academic Press: San Diego, CA, 1990; pp. 261-388.

9. Ravve, A *Principles of Polymer Chemistry*, 2nd ed.; Klewer Academic/Plenum Publishers: New York, NY, 2000; pp.598-609.
10. March, J *Advanced Organic Chemistry*, 4th ed.; John Wiley & Sons: New York, NY, 1992; 1174-188.
11. Nicolet Vapor Phase FTIR library; Thermo Electron North America LLC: Madison, WI; library indices 8591, 387, 3873.

# Indexes

# Author Index

# Subject Index

## B

## F

## G



## H

## M

**N**

**O**

## P

**R**

## U

**V**



**W**



**X**



**Z**